Climate Change and Agriculture: Impacts and Adaptations

Climate Change and Agriculture: Impacts and Adaptations

Editor: Edison Harrison

MURPHY & MOORE
www.murphy-moorepublishing.com

www.murphy-moorepublishing.com

MURPHY & MOORE

Cataloging-in-Publication Data

Climate change and agriculture : impacts and adaptations / edited by Edison Harrison.
p. cm.
Includes bibliographical references and index.
ISBN 978-1-63987-753-9
1. Crops and climate. 2. Agricultural ecology. 3. Climatic changes. 4. Climate change mitigation. I. Harrison, Edison.
S600.5 .C55 2023
630.251 5--dc23

Murphy & Moore Publishing
1 Rockefeller Plaza,
New York City,
NY 10020, USA

ISBN 978-1-63987-753-9

Contents

Preface....VII

Chapter 1 **The Nexus of Weather Extremes to Agriculture Production Indexes and the Future Risk**....1
Abdul-Aziz Ibn Musah, Jianguo Du, Thomas Bilaliib Udimal and Mohammed Abubakari Sadick

Chapter 2 **Participatory Sustainability Assessment for Sugarcane Expansion**....25
Heitor Luís Costa Coutinho, Ana Paula Dias Turetta, Joyce Maria Guimarães Monteiro, Selma Simões de Castro and José Paulo Pietrafesa

Chapter 3 **Defining Crop–Climate Departure in West Africa: Improved Understanding of the Timing of Future Changes in Crop Suitability**....40
Temitope S. Egbebiyi, Olivier Crespo and Chris Lennard

Chapter 4 **Effect of Planting Date on Accumulated Temperature and Maize Growth under Mulched Drip Irrigation in a Middle-Latitude Area with Frequent Chilling Injury**....58
Dan Wang, Guangyong Li, Yan Mo, Mingkun Cai and Xinyang Bian

Chapter 5 **On-Farm Evaluation of the Potential use of Greenhouse Gas Mitigation Techniques for Rice Cultivation**....74
Noppol Arunrat, Sukanya Sereenonchai and Nathsuda Pumijumnong

Chapter 6 **Evaluation of the Agronomic Impacts on Yield-Scaled N_2O Emission from Wheat and Maize Fields**....91
Wenling Gao and Xinmin Bian

Chapter 7 **Climate Change-Induced Impacts on Smallholder Farmers**....106
Tafesse Matewos

Chapter 8 **An Economic Assessment of Local Farm Multi-Purpose Surface Water Retention Systems under Future Climate Uncertainty**....123
Pamela Berry, Fuad Yassin, Kenneth Belcher and Karl-Erich Lindenschmidt

Chapter 9 **Geographic Information and Communication Technologies for Supporting Smallholder Agriculture and Climate Resilience**....145
Billy Tusker Haworth, Eloise Biggs, John Duncan, Nathan Wales, Bryan Boruff and Eleanor Bruce

Chapter 10 **The Value of Tactical Adaptation to El Niño–Southern Oscillation for East Australian Wheat**....164
Bangyou Zheng, Scott Chapman and Karine Chenu

Chapter 11 **Resilience for Whom? The Problem Structuring Process of the Resilience Analysis....179**
Hugo Herrera

Chapter 12 **Impact of Climate Change on Twenty-First Century Crop Yields....196**
Lillian Kay Petersen

Chapter 13 **Estimation of the Virtual Water Content of Main Crops on the Korean Peninsula using Multiple Regional Climate Models and Evapotranspiration Methods....213**
Chul-Hee Lim, Seung Hee Kim, Yuyoung Choi, Menas C. Kafatos and Woo-Kyun Lee

Chapter 14 **New Breeding Techniques for Greenhouse Gas (GHG) Mitigation: Plants may Express Nitrous Oxide Reductase....230**
Jordan J. Demone, Shen Wan, Maryam Nourimand, Asbjörn Erik Hansen, Qing-yao Shu and Illimar Altosaar

Permissions

List of Contributors

Index

Preface

Agriculture refers to the practice of cultivating plants and livestock, and it is highly dependent on climate. Climate change and variability are major challenges for agriculture since it can lead to reduced crop yields, decreased nutritional quality of major cereals, and reduced productivity of livestock. The negative impacts of climate change are manifested in the form of increasing temperatures, shifting agro-ecosystem boundaries, weather variability, invasive pests, etc. Climate resilient crops and improved agricultural practices play a crucial role in coping with climate variability. Some major agricultural management strategies include residue incorporation, water harvesting for supplementing irrigation, and water saving technologies such as in-situ and ex-situ moisture conservation. This book explores the impacts and adaptations in agriculture with respect to climate change. It presents researches and studies performed by experts across the globe. The book will serve as a valuable source of reference for graduate and postgraduate students.

The information shared in this book is based on empirical researches made by veterans in this field of study. The elaborative information provided in this book will help the readers further their scope of knowledge leading to advancements in this field.

Finally, I would like to thank my fellow researchers who gave constructive feedback and my family members who supported me at every step of my research.

Editor

1

The Nexus of Weather Extremes to Agriculture Production Indexes and the Future Risk in Ghana

Abdul-Aziz Ibn Musah *, Jianguo Du, Thomas Bilaliib Udimal and Mohammed Abubakari Sadick

School of Management, Jiangsu University, 301 Xuefu Road, Zhenjiang 212013, Jiangsu, China; jgdu2005@163.com (J.D.); tbudimal2007@yahoo.com (T.B.U.); oluzerpublications@yahoo.com (M.A.S.)
* Correspondence: dooziizu@gmail.com

Abstract: The agricultural industry employs a large workforce in Ghana and remains the primary source of food security and income. The consequences of extreme weather in this sector can be catastrophic. A consistent picture of meteorological risk and adaptation patterns can lead to useful information, which can help local farmers make informed decisions to advance their livelihoods. We modelled historical data using extreme value theory and structural equation modelling. Subsequently, we studied extreme weather variability and its relationship to composite indicators of agricultural production and the long-term trend of weather risk. Minimum and maximum annual temperatures have negligible heterogeneity in their trends, while the annual maximum rainfall is homogenous in trend. Severe rainfall affects cereals and cocoa production, resulting in reduced yields. Cereals and cocoa grow well when there is even distribution of rainfall. The return levels for the next 20–100 years are gradually increasing with the long-term prediction of extreme weather. Also, heavy rains affect cereals and cocoa production negatively. All indicators of agriculture had a positive relationship with maximum extreme weather.

Keywords: extreme weather; agriculture production; return level; extreme value theory; weather; risk

---

1. Introduction

Many developing countries particularly those in the tropical regions are sensitive to changing the climate, especially where temperatures are already threatening agricultural production [1–3]. They have restricted access to a human and physical asset that can mitigate its effects [4]. These difficulties are often manifold by the lack of connection to new technologies and established markets [2,4]. Ghana is an example of a country facing these challenges. The irrigated land for agricultural use covers only 1% of farmland, and the majority of the farmers are entirely dependent upon seasonal rainfall [5–7].

This concern about the changing climate is due to its negative impact on the living conditions of humankind. Developing nations, particularly Ghana, is increasingly concerned about the changing climate because they are more vulnerable compared to developed nations. Climate change is a significant issue of risk to sustainable growth in Africa. As such, the efforts of African countries to realise the Millennium Development Goals can be considered as an offer if the adverse effects of climate change are taken seriously by Africa nations. Generally, African states contribute very little to climate change yet they bear the major brunt of it. Also, the Africa continent is more vulnerable to the effects of this changing climate as a result of its excessive reliance on rainfed agriculture, and extreme poverty [8]. The critical long-term effects of climate variation include: change in precipitation leading to reduced agricultural production, reduced food security, deterioration of water security, and reduction of fish stocks due to high temperature and displacement. Also, sea-level upsurge due to climate variation affects coastal areas greatly. The adverse effects of climate change in the form of a reduction in agricultural output ultimately lead to a delay in the development of African countries where a more substantial part of national income comes from agriculture. Also, the agricultural sector functions as a basis of livelihood for most people in Africa [8].

To tackle climate change, Ghana signed the United Nations Framework Convention on Climate Change (UNFCCC) at the Earth Summit in Rio de Janeiro in June 1992, following the adoption of the Convention on 9 May 1992 [9]. In Ghana, three critical physical effects of climate change identified include temperature change, precipitation change, and sea level rise [7]. According to a report [10], there is a shift in the rainfall regime in Ghana towards a longer dry season and vanishing wet season. Despite the signing of the Convention by Ghana, the country continues to face the adverse effects of climate change in the area of health, agricultural, already depletion of coastal areas, and low water levels. For example the country's only hydroelectric dam (which produces 80% of the national electricity supply) due to lower rainfall [11]. The consequence of climate change on the Ghanaian economy is due to the lack of environmental adaptation strategies and the socio-economic costs of adapting those strategies to mitigate the effects of climate change.

Climate change affects the transport system in the areas that are heavily dependent on weather conditions [12,13]. According to Reference [14], climate change adversely affects the critical elements of food production such as soil, water, and biodiversity. As a result, Ghana's economic dependence on areas (as energy, agriculture and forestry) which are particularly susceptible to the changing climate makes it more prone to the adverse effects of weather. In this vein, it is essential to carry out studies on the changing climate and its volatility in Ghana.

Specifically, this article examines the following.

- Examining the trends in extreme maximum rainfall and extreme high/low temperature
- Assessing the variability and weather risk of extreme maximum/minimum
- Analysis of the relationship of extreme weather to agriculture production indexes
    - Effect of exceptionally high rainfall on agriculture production indexes
    - Effect of extremely high temperature on agriculture production indexes
    - Impact of low temperature on agriculture production indexes

Rare weather conditions like severe rainfall, extreme temperature (and heat waves), or strong winds, may have significant effects on sectors such as agriculture and health, which may result in severe risk to human life [15]. Further, risks of extreme heat and drought depend not only on the severity of the event but also on the sensitivity and vulnerability of the exposure system [16].

The existing studies only show regional climate parameters and how the joints of their scales occur. We contend that the environmental parameters if could serve as a tool for eliminating human disasters if their extreme conditions are well understood and managed correctly [17]. Focusing on the regional research, particularly climate system, the influence of climate change and uncertainty in weather conditions could alter and transform societal and institutional behaviours [18,19].

Substantial studies concede extreme value theory as a method that estimates rare event whiles generalised extreme value distribution (GEVD) is capable in determining the probability of events occurrence that fall outside of an observed data range. Given this, GEVD has attracted attention in diverse areas of research such as climatology data analysis [20–23]. Issues relating to Extreme Value Theory gradually implemented in practical covariate approach of non-stationary conditions [15,20,24–28]. An investigation by [29–31] on daily rainfall at various observation sites in West Africa revealed an increasing trend of yearly maximum rainfall. Research has shown variations in extreme rainfall [30]. Thus extreme rain is related to a decline in annual precipitation intensity. In weather forecasting, efforts are made to predict the impact of weather conditions on food security [32]. Such reviews can help planners provide adequate protection and adaptation solutions that contribute to the resilience of the population and the reduction of socio-economic disasters. In the world over, 33% of observed crop production modifiability emanates from a change in climate thereby, a cause of variations of crop yield in Africa [33–35]. The intra-inter yearly rainfall and temperature show considerable effects on crops production and therefore ensures food safety [36].

Similar studies demonstrate that rainfall and temperature adversely affect crop yield. It calls for authorities in Africa to enforce sustainable food security policies [37,38]. In a period of severe soil moisture, flowering development stagnates [39]. Research has shown that drought is inimical to the growth of cocoa. Therefore, there is a causality between rainfall and cocoa yield [40]. Analogously, the sustenance of a bumper harvest is positively related to rainfall distribution than the total amount of rainfall received annually [41]. However, Reference [42] argues on the positive and negative causality of crops production in Ghana.

The yearly rainfall in cocoa growing areas in Ghana is more than 2000 mm. Also, two rainfall seasons are recorded from April to July and September to November, where July to August faces relative dry weather with high humidity condition. There is a dry weather condition between a second month and the eleventh month of the annual calendar [40,43]. Variations in climate pose a threat to the health of animals, and unfavourable heat affects them reproductively [44,45].

The 21st century saw a decline in yields ranging from 2.5% and 10% as temperature rises in some agronomic species [46]. The results of the evaluation of the effect temperature on crop yield at various levels indicate a decrease in yield. For example, the decline in barley production is due to the low temperatures during the vegetative stages and represents about 42% of low yield. The different seasons with low temperatures and high rainfall are unusable conditions for the potato, resulting in reduced yields [47].

Ascertained by [48–50], climate change due to the uncertainty of precipitation has a significant impact on agriculture production. On this account, this study introduces a different dimension into the analysis of weather effects on agriculture by looking at the extremes conditions of temperature and rainfall hence; we aim to fill this gap in the literature.

Given the increasing occurrences of climate change, there is a need for researchers to consider extreme conditions that often occur due to climate variability and its related events. Relying on climate variation in a whole without considering the specifics thus, minimum and maximum extremes have resulted in a situation where policies are formulated but not directed at specific extreme effects. This study looks at weather variability concerning maximum and minimum extreme conditions to enhance the formulation of targeted policies to help curb their impact on agriculture production. Further, we have investigated the relationship between extreme weather events and agriculture production indexes and assessed agricultural risk using extreme value theory (EVT) and structural equation modelling, which are different from previous studies.

2. Materials and Methods

2.1. *Climate Change and Variability in Ghana*

The regional scenarios of seasonal precipitation and temperature changes in 32 regions globally analysed by (IPCC, 2014) show the current variations in climate and the range of variations in 30-year period predicted by GCM, focused in 2025, 2055, and 2085. This background information is critical in explaining the probable effects of climate variation on livestock and crop production.

The IPCC approximate that the past period saw temperatures increased by an average of 0.6 °C. The preceding 25 years, there was no observation of atmospheric temperatures from 1995–2006, 11 out of 12 was the warmest years [51]. Countries are beginning to experience consequences related to global warmings, such as the long-term drought within the Sahel zone in Africa and the expansion of the malaria transmission belt of tropical Africa [52]. Universally, the figure noted for weather-associated natural adversities is fast increasing. From the 1960s, accounts of natural risks have tripled. During 2007, fifteen (14) out of fourteen (15), "emergency appeals" for emergency public-spirited assistance were in the areas of storms, droughts and floods, five times more than in the prior year [53]. Ghana's, climate variation is experiencing increasing unpredictable rainfall and temperatures in all regions [54].

Also, global warming is predicted to show variations in rainfall patterns, acidification, and moisture [55]. In this context, the global effect of climate variation on global life-assistance systems remains uncertain. Some parts experience extreme precipitation resulting in flooding; for example, the Mediterranean areas are experiencing a decline that could result in drought conditions [55]. By some reports [55], the anticipation of global average temperatures will rise between 1.4–5.8 °C by close of the century, as sea levels, increase as melting glaciers melt. Observations recently, however, indicate that many predictions concerning climate change are near the higher limit of the IPCC estimates. Sea levels, for example, have exceeded the IPCC estimates of up to 30 cm [56].

Based on a study by Reference [57], is establish that an estimated 35% of the entire land in Ghana is affected by increasing desertification. The unexpected variability of precipitation patterns is observed for years in Ghana as affirmed by Reference [58]. With the historical data, precipitation was mostly high in the 1960s, but fell to low levels by the end of 1970s and then rose again in 1980s. This fall in precipitation patterns is still prevalent currently, as Reference [59], with 20 years of data, observed this; temperatures are rising throughout Ghana and is precipitation decreasing and becoming gradually unpredictable. The effects of changing climate are anticipated to be severe in Ghana, even though there are rises and fall in both yearly temperatures and precipitation. Conceding to the World Bank's projection, the temperature trend from 2010–2050 shows warming in almost the highest-temperature parts of Ghana, including the North and the Upper Regions.

Nevertheless, the region with the lowest temperature is the Brong Ahafo region. These are base on different climate scenarios [58]. For example, looking at the scenario, it was recognised that the temperatures of the three northern regions would increase by 2.1–2.4 °C by 2050. On the contrary, the predicted increase in Ashanti, West, East, Volta, and Central regions ranges from 1.7–2.0 °C and those of Brong Ahafo 1.3–1.6 °C.

We also reviewed the latest temperature and precipitation forecasts from the Intergovernmental Panel on Climate Change (IPCC) [60] to simulate the impact of climate change on agricultural production in Ghana. These projections are on Phase Five of the Coupled Model Inter-comparison Project (CMIP5), which brings together the results of 39 different global models. We used projections for West Africa until 2035. According to the first scenario, the most optimistic, the temperature should increase by 0.7 degrees and precipitation by 8%. These increases represent the expected minimum increase in temperature and the maximum expected increase in precipitation. The second scenario concerns the median increase in temperature (0.9 degrees) and precipitation (1%). The third scenario, the least optimistic, concerns the maximum expected increase in temperature (1.5 degrees) and the maximum decrease in precipitation (4%). A meta-analysis of crop yield response to climate change, using local average temperature as an indicator of change, concluded that global warming at 2 °C could lead to an increase in wheat, rice, and maize yields, with yields subsequently decreasing with increased warming. The AR4 also showed that crop-level adaptations had a markedly positive effect on all crops, regions, and warming levels [61].

According to Reference [62], Tables 1 and 2 show some of the climate changes in Ghana and the corresponding time periods.

Table 1. The projections of precipitation in Ghana.

Location	Climate Type	Forecast Changes
Accra	Coastal Savanna Zone	From 52% decreases to 44% increases in wet season rainfall by the year 2080.
Kumasi	Deciduous Forest Zone	From 48% decreases to 45% increases in wet season rainfall by the year 2080. Based on their A2 scenario, which generally shows the largest greenhouse gas (GHG) impact, predicts the weakest increase in wet season rainfall, 1.13%.
Tarkwa	Rain Forest Zone	From 45% decreases to 31% increases in wet season rainfall.
Techiman	Forest-Savanna Transition Zone	From 46% decreases to 36% increases in wet season rainfall. The A2 scenario, which generally shows the largest GHG impact, predicts the largest decrease in wet season rainfall, −2.94%.
Tamale	Guinea Savanna Zone	From 36% decreases to 32% increases in wet season rainfall consistent trend toward decreased rainfall.
Walembelle	Northern Guinea Savanna Zone	From 25% decreases to 24% increases in wet season rainfall
Bawku	Sudan Savanna Zone	Range from 28% decreases to 30% increases in wet season rainfall.

Source: Extracted from [8,43].

Table 2. Temperature projections in various climate stations in Ghana.

Location	Climate Type	Temperature Projections	
		Wet Season	Dry Season
Accra	Coastal Savanna Zone	1.68 ± 0.38 °C by 2050 2.54 ± 0.75 °C by 2080	1.74 ± 0.60 °C by 2050 2.71 ± 0.91 °C by 2080
Kumasi	Deciduous Forest Zone	1.71 ± 0.39 °C by 2050 2.60 ± 0.77 °C by 2080	1.81 ± 0.68 °C by 2050 2.83 ± 1.04 °C by 2080.
Tarkwa	Rain Forest Zone	1.69 ± 0.37 °C by 2050 2.56 ± 0.75 °C by 2080	1.76 ± 0.67 °C by 2050 2.76 ± 1.01 °C by 2080.
Techiman	Forest-Savanna Transition Zone	1.77 ± 0.43 °C by 2050 2.71 ± 0.85 °C by 2080	1.95 ± 0.79 °C by 2050 3.05 ± 1.20 °C by 2080.
Tamale	Guinea Savanna Zone	1.84 ± 0.46 °C by 2050 2.83 ± 0.91 °C by 2080	2.05 ± 0.75 °C by 2050 3.18 ± 1.18 °C by 2080.
Walembelle	Northern Guinea Savanna Zone	1.92 ± 0.52 °C by 2050 2.96 ± 0.98 °C by 2080	2.10 ± 0.71 °C by 2050 3.27 ± 1.11 °C by 2080.
Bawku	Sudan Savanna Zone	1.92 ± 0.53 °C by 2050 2.97 ± 0.98 °C by 2080	2.11 ± 0.68 °C by 2050 3.25 ± 1.08 °C by 2080

Source: Extracted from [8,43].

2.2. *Seasonal Changes of Precipitation and Temperature*

The climate of Ghana is tropical, with a dry season in winter and a rainy season during the summer due to an African monsoon. The duration of the rains varies according to the ecological zones. As shown in Figure 1, the rainy season is usually from May to September to the north, from April to October in the centre, and from April to November to the south. However, on the east coast, the rainy season is shorter than the rains from April to June, with no rainfall in July and August, and it picks up slightly in September and October. The south is the coolest part of Ghana, where it has more than

1500 mm (per year), and even more the small west coast, where it reaches 2000 mm (80 inches) per year. The north is the driest in Ghana, where rainfall is about 1000 mm (40 inches) per year and the east coast, including the city of Accra, where it falls below 800 mm (31.5 in).

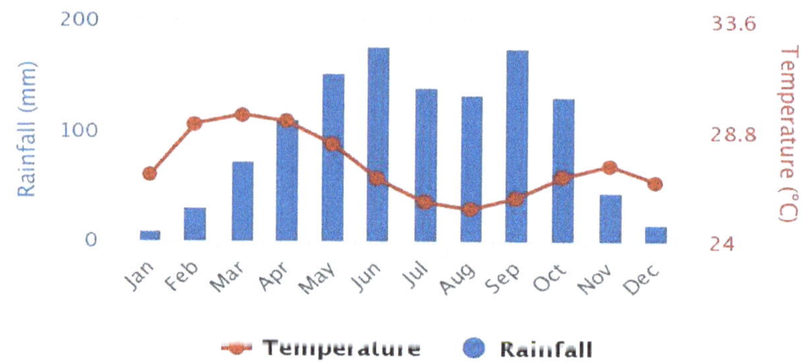

Figure 1. The Monthly trend of temperature and rainfall in Ghana.

2.3. The trend of Climate Change in Ghana

Ghana is located in West Africa, bordered to the north by Burkina Faso, east to Togo, west to Ivory Coast, and south to the Gulf of Guinea. It is located between 4.50 degrees north and 11.50 degrees north and longitude 3.50° west and 1.30° east. The country has an area of 239,460 Km^2 and a surface area of 8520 Km^2 as seen in Figure 2. The country has a population of around 24 million since 2010, with an annual growth rate of about 2.5% [63]. Young people dominate this population. The main exports are cocoa, gold, wood, diamonds, bauxite, manganese, and hydroelectricity. Until recently, the country also began to export crude oil. In 1991/92, the poverty level in Ghana reached 51.7 per cent, and this figure has steadily declined in recent years to 39.5 per cent in 1998/99, 28.5 per cent in 2005/06, and 24.2 per cent in 2012/2013. The country enjoys a high temperature while the average annual temperature is between 24 °C and 30 °C. Despite the average annual temperature, temperatures may be 18 °C and 40 °C in the southern and northern parts of Ghana. Rainfall in Ghana is generally declining from south to north. A more prosperous region in Ghana is the far southwest, with an annual rainfall of about 2000 mm. However, the annual rainfall in northern Ghana is less than 1100 mm. The country has two major systems of rain: the double-twin system and the single maximum regime. For the maximum binary system, the maximum periods are from April to July and from September to November in southern Ghana. While the only maximum system is from May to October in northern Ghana, the prolonged drought lasts from November to May. Over the years, temperatures have risen in all ecological regions of Ghana, while rainfall levels have generally declined and standards have steadily increased [9].

Despite dramatic improvements in technology and crop yields, food production continues to depend heavily on the climate because solar radiation, temperature and rainfall are the critical factors of increase in crop production. The climate is affected by the diseases of plant and the spread of pests, including the supply and demand for irrigation water. For instance, in recent decades, the ongoing drought in the Sahel has caused a continued deterioration in food production [64] in Ghana. The effect of the changing climate on crops was in 1990, where the crop has suffered or decreased. Also, due to drought, climate indicators such as rainfall and average mean temperature are associated with crop change [57]. Table 3 below presented climate change variations experienced.

Figure 2. Location Map of Ghana.

Table 3. Climatic variations experienced in Ghana.

Time Period	Climatic Variations
January–July 1976	Scorching weather conditions
1983–1984	Drought: A yearlong of bushfires
October–December 1989	Scorching weather conditions
1991	Lots of rains throughout the year
1995	About 40 days of intensive rains
2004	Noticeable are frigid winds during March–April (Easter) and November–January was very cold weather
2005	Cold periods resulting in animal deaths
August 2006	One week of intensive rains, and
2007	Lots of rains in August and September.

Source: Extracted from [62].

2.4. *The Generalized Extreme Value Distribution (GEVD)*

The GEVD is part of the family of continuous distribution functions that allows a continuous range of shapes and consists of classes of distribution functions such as Gumbel, Fréchet, and Weibull. Considering the Fisher-Tippett Gnedenko theorem, the GEVD is a limit-form distribution function, which maximises the maxima of the sequence of random variable considered as independent and identical distributed (i.i.d). It, therefore, models the maximum of a finite sequence of random variables. The combined model of maxima is by Equation (1):

$$G_{\gamma,\mu,\sigma} = \exp\left\{-\left(1+\gamma\left(\frac{x-\mu}{\sigma}\right)^{-\frac{1}{\gamma}}\right)\right\} \text{ with, } \gamma \neq 0,\ \sigma > 0 \text{ and } \gamma\left(\frac{x-\mu}{\sigma}\right) > 0 \quad (1)$$

$$\sigma > 0$$

The derivative of Equation (1), give a probability density function in Equation (2) as:

$$g_{\gamma,\mu,\sigma} = \frac{1}{\sigma}\left(1+\gamma\left(\frac{x-\mu}{\sigma}\right)\right)^{-1-\frac{1}{\gamma}} \exp\left\{-\left(1+\gamma\left(\frac{x-\mu}{\sigma}\right)\right)^{-\frac{1}{\gamma}}\right\},\ \gamma \neq 0 \tag{2}$$

where μ and σ are the location and scale parameters, respectively [20].

The GEVD shape parameter γ also termed as the extreme value index. The decay rate of GEVD seen as γ^{-1}. If $\gamma > 0$ for a class of distributions, G fits distributions as; the heavy-tailed Fréchet distribution, Cauchy, Student's t, Pareto class, and mixture other distributions. G fit into the short-tailed Weibel distribution, uniform, and beta distribution if $\gamma < 0$. G fits the right-tailed Gumbel distributions (normal, exponential, gamma, and lognormal) if $\gamma = 0$ [65–67].

2.5. Maximum Likelihood Estimation for GEVD

The assumption that $X_1, \ldots, X_m$ follows an (i.i.d) and also from generalised extreme value distribution with parameter when $\gamma \neq 0$ the log-likelihood function given as:

$$\text{Provided that } 1+\gamma\left(\frac{x_{(i)}-\mu}{\sigma}\right) > 0 \text{ for } i = 1, 2, \ldots, m \tag{3}$$

$$l(\mu,\sigma,\gamma) = -m\ln\sigma - (1+1/\gamma)\sum_{i=1}^{m}\ln\left[1+\gamma\left(\frac{x_{(i)}-\mu}{\sigma}\right)\right] - \sum_{i=1}^{m}\left[1+\gamma\left(\frac{x_{(i)}-\mu}{\sigma}\right)\right]^{-1/\gamma} \tag{4}$$

Parameters combination that deviates from the above conditions (Equation (3)), i.e., in a configuration where at least one of the observed data exceeds the endpoint of the distribution (Equation (4)), the likelihood is zero, and the log-likelihood is equal to $-\infty$. This case $\gamma = 0$ requires separate treatment with GEVD's Gumbel restriction leading to logarithmic log-likelihood as in Equation (5);

$$l(\mu,\sigma) = -m\ln\sigma - \sum_{i=1}^{m}\left(\frac{x_{(i)}-\mu}{\sigma}\right) - \sum_{i=1}^{m}\exp\left\{-\left(\frac{x_{(i)}-\mu}{\sigma}\right)\right\} \tag{5}$$

Equations (2) and (3) are differentiated and maximised concerning the parameter vector (μ,σ,γ), Solving for (μ,σ,γ), results to the maximum likelihood estimates for the whole GEVD model [20,28,68,69]. Maximum likelihood estimation offers the advantage of estimation of the three parameters together and applicable to the series of maxima per block [70].

Model Checking for GEVD

The model fit of GEVD measure after estimating the parameters by utilising residual plots function as defined by Equation (6),

$$res = \begin{cases} \left(1+\frac{\gamma}{\sigma}(x-\mu)\right)^{-1/\gamma} & if\ \gamma = 0 \\ \exp\left[-\exp\left(-\frac{x-\mu}{\sigma}\right)\right] & if\ \gamma \neq 0 \end{cases} \tag{6}$$

Ascertain by Reference [20] conversion of data to unit exponential distributed residuals is on the null assumption that GEVD fits the data.

2.6. Return Period or Level Estimates

The frequency of extreme quantiles incidence estimated with a fixed value of return level. The return level is the mean number of events taking place within a unit period, e.g., one year [71]. Return levels are essential for prediction purposes and estimated from stationary models. The expected return time is the number of time (years) one is expected to wait on average before the observation of another extreme event of at least the same intensity. If a threshold exceedance of a given probability of an observed extreme incidence in any given time (year) is p, then the mean return period T is such that $T = 1/p$.

2.7. Test for Stationarity and Seasonality

The stationarity of the data conducted by the augmented Dickey-Fuller (ADF) stationarity test on the assumption that there is no trend [72]. The quality of convergence of the weather extremes is access using the Kolmogorov–Smirnov (K-S) and Anderson–Darling goodness-of-fit tests. The K-S test, relying on the empirical study of the cumulative distribution function, is used to determine whether the sample is from the hypothesised continuous distribution. The K-S approach is less sensitive for normal distribution [72]. The Anderson-Darling test, an enhancement of the K-S test, compares the fit to the expected cumulative distribution function of the observed cumulative distribution function. This test gives more substantial weight to the tail of the distribution than the K-S test [72].

The assumption is that the data is from a population which is independent identically distribution (i.i.d). The alternative hypothesis is a two-tail test on the assumption that the data follow a monotonic trend. Thus, the following test statistics by Mann-Kendall determine by Equation (7):

$$S = \sum_{k}^{n-1} \sum_{j=k+1}^{n} sgn(x_j - x_k) \tag{7}$$

with sgn the signum function.

3. Methodology

This paper analyses past Composite Indexes in Agriculture ranges from 1961–2016: crops production, cocoa production, livestock production, cereal production, and food production in Ghana. The data also consider records of maximum rainfall, maximum temperature, and minimum temperature value as weather indicators from January 1965 till July 2016. We sourced the data from the Ghana Meteorological Agency for climate data, and agriculture production indexes from the Food and Agriculture Organization also in Ghana. Rainfall and temperature are assumed to be the primary determinants of weather in Ghana as seen in Figure 1. The first task was to check for stationarity of the weather variables using Augmented Dickey-Fuller (ADF) unit root test and then the Mann-Kendall Trend Test of seasonality. It was necessary to apply methods that explicitly allow for testing non-stationarity in the distribution parameters of climate variables [20].

Next step was to model from the dataset of the weather indicators employing the Block Maximum Method for the weather extremes under Generalized Extreme Value Distribution (GEVD). There were two approaches to the modelling of Block minima data for the minimum temperature. Either the GEVD for minima fitted to this data or the data negated and the GEVD for maxima fitted [20]. The latter approach was adopted since the Extremes Toolkit does not include a routine to estimate the GEVD for minima directly. The block maxima method is a parametric approach to Extreme Value Theory. It entails fitting the GEVD to a specific group of maximum values chosen in a given sample of data. It focuses on the statistical behaviour of the largest or smallest value in a sequence of independent random variables. Assume that the sequence is grouped into blocks of size N (with a reasonably large number) and that only the maximum score M_i $(i = 1, 2, 3, \ldots, n)$ of each block extracted. Each M_i $(i = 1, 2, 3, \ldots, n)$ of the weather indicators is then used to estimate the relationship between the composite indexes of agriculture production.

The mean return period defines the amount of time (e.g., years) that is expected to pass on average before a new extreme with the same or increased intensity. Given the likelihood that events past a certain threshold will follow an extreme of a particular security at any given time (year) is defined as p, then the mean return period T can be calculated as $T = 1/p$.

Food production index includes food crops that are considered edible, and that contain nutrients with the exclusion of coffee and tea because they have no nutritive value although edible (FAO). Figure 3 shows the primary crop food calendar.

Finally, we investigated the relationship between extreme weather events and agriculture production using SEM software to evaluate the potential impacts of weather extremes on Agriculture production. We used SEM regression for the paths equation modelling analysis with the partial least squares (SEM) estimation technique [73]. SEM is a modelling approach with a flexible procedure, which can handle data with missing values, strongly correlated variables, and small samples. SEM-regression works with both continuous and discrete observed variables as indicators. The SEM estimates loading and path parameters between variables and maximises the variance explained for the outcome variables [73].

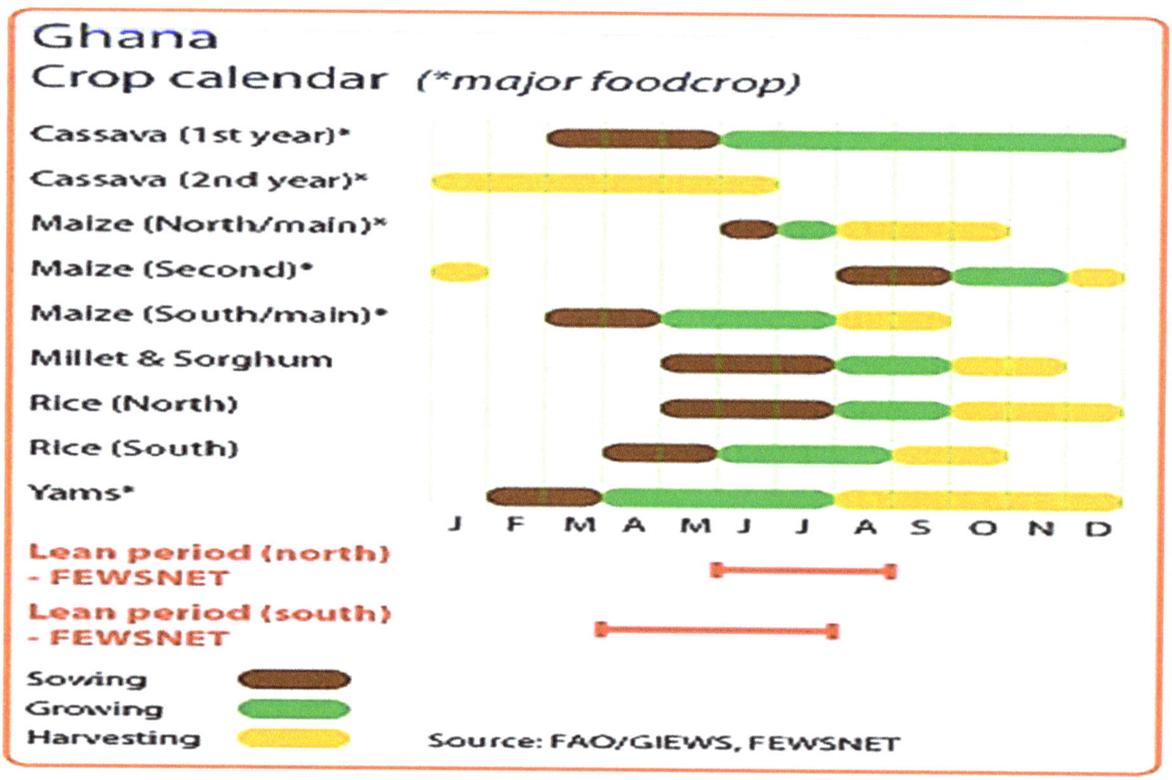

Figure 3. Major food crops calendar in Ghana.

4. Results and Discussion

4.1. Stationarity Test for the Weather Indicators

The ADF test is captured in Table 4 indicating the significance of the *p*-value statistics. The premise of non-stationary at 1%, 5%, and 10% rejected, and therefore we conclude the stationarity of the weather indicators.

It is reported by scholars that, Mann-Kendall Trend Test of stationarity is reliable and efficient. In line with this, analysing environmental data demands the exposure of movements of events on separate points [74]. Based on this, the test outcome illustrates high or low trends in weather conditions of a particular jurisdiction.

Table 4. Stationarity and Seasonality test.

Augmented Dickey-Fuller Stationarity Test					
Test Variable	**Test's Critical Values**			**Test Statistics**	***p*-Value**
	1%	**5%**	**10%**		
Annual maxi. Rainfall	−3.958	−3.410	−3.127	−16.350	0.0000
Annual maxi. Temperature	−10.007	−3.431	−2.862	−2.567	0.0000
Annual mini. Temperature	−12.482	−3.431	−2.862	−2.567	0.0000

Seasonal Mann-Kendall Trend Test						
Series	**Statistics**		***p*-value**		**tau**	**Slope 95% CI**
	z (trend)	**z (Het)**	**p (trend)**	**p (Het)**		
Maxi. Rainfall	0.434	22.376	0.664	0.0216	0.0019	0.0044 [−0.0194,0.0308]
Maxi. Temperature	21.842	4.779	<0.001	0.9410	0.1320	0.0318 [0.0286,0.0346]
Mini. Temperature	25.123	23.894	<0.001	0.1320	0.1520	0.0231 [0.0212,0.0250]

95% confidence interval in parenthesis.

In Table 4, the estimated annual trend is 0.0044 mm/year, a yearly increase in the maximum annual rainfall. The p-value based on the Kendall seasonal trend test is $p = 0.6640$, which shows no importance. The 95% confidence interval on both sides for the trend (−0.014,0.0307), the chi-square test for heterogeneity (Het) gave a p-value of 0.0216. Therefore, there is a difference in the level of a trend in the different seasons of the maximum annual rainfall. As shown in Table 4, the estimated annual trend is 0.0318 degrees Celsius (°C)/year, which is a yearly increase in the yearly maximum temperature. The p-value corresponds to the Kendall seasonal test for the $p < 0.001$ trends, indicating that it is statistically significant. The 5% level of significance on both sides for the trend is (0.0286, 0.0346). The chi-square heterogeneity test (Het) provides a p-value of 0.9410, so there is no evidence for different sets of stresses at different times of the maximum annual temperature. The estimated annual trend is 0.0231 degrees Celsius (°C)/year, a yearly increase in the maximum annual temperature. The p-value of the Kendall seasonal trend test, $p < 0.001$, indicating that it is statistically significant. The 5% level of significance on both sides for the trend is (0.0212,0.0250). The chi-square test for heterogeneity (Het) gives a p-value of 0.1318, i.e., no indication of the different trend in different seasons of the minimum annual temperature.

4.2. GEVD Model for Extreme Maximum Rainfall

In Table 5, the estimated return periods of maximum rainfall likely to occur over the next 5, 10, 20, 50 or even 100 years fitted by GEVD. The estimated results are (μ, σ, γ) (149.03,23.98,0.0024), with standard errors (3.758, 2.718, 0.1002). The approximate 95% confidence intervals for the parameters are thus (141.67, 156.39) for μ, (18.65, 29.31) for σ, and (−0.193,0.198) for γ.

Table 5. Generalised extreme value estimates of maximum rainfall.

GEV	**Maximum Rainfall**		
	Location	**Scale**	**Shape**
Estimates	$\mu = 149.03$	$\sigma = 23.98$	$\gamma = 0.0024$
Std error	3.758	2.718	0.1002
95% CI (normal app)	(141.67,156.39)	(18.65,29.31)	(−0.193,0.198)
Estimated Return Levels	**95% Lower**	**Estimate**	**95% Upper**
5-year return level	173.14	185.06	196.98
10-year return level	186.59	203.13	219.68
20-year return level	196.99	220.50	244.03
50-year return level	206.49	243.04	279.57
100-year return level	210.72	259.95	309.05

The validity and reliability of the extrapolation of GEVD fit is assessed base on the observed data. Four graphical analyses assist with model checking [20,75]. Figure 4 shows diagnostic plots assessing the accuracy of the GEVD model fitted. Neither the quantile plot nor the density plot has any reason to doubt the validity of the fitted model: each drawn set of points is almost linear. The return level plots asymptotically converge to a determinate value due to the positive estimates, with the curve approaches a straight line. The sample variable under consideration provides an adequate representation graphically of the empirical estimates. Finally, the corresponding density estimate appears to be consistent with the density curve. As a result, all four diagnostic diagrams support the GEVD model as in Figure 4 (Top-left: empirical plot; Top-right: empirical quartile plot; Bottom-left: density plot; Bottom-right: return level plot).

The determination of the limiting distribution by maximising the GEV negative log-likelihood for annual maximum rainfall leads to the following function in Equation (8):

$$G(z) = \exp\left\{ -\left[1 + 0.00243\left(\frac{z - 23.98}{149.03}\right)\right]^{\frac{-1}{0.00243}} \right\} \tag{8}$$

Equation (8) gave estimates of return levels for 5, 10, 20, 50, and 100-years and their 5% significant level as shown in Table 3. Thus, based on the data from 1965 to 2016, once in 50 years we should expect to see an extreme annual maximum rainfall hit between 206.5 and 279.6 mm. The upper bound of the model prediction for the 50-years Return level of 279.6, but 510 mm extreme rainfall recorded in 1968. Of course, this is undoubtedly extreme beyond regular extreme events, which is not expected based on the model's predictions. Results from Table 2, indicates that extreme maximum rainfall is steadily increasing significantly over the 100 years.

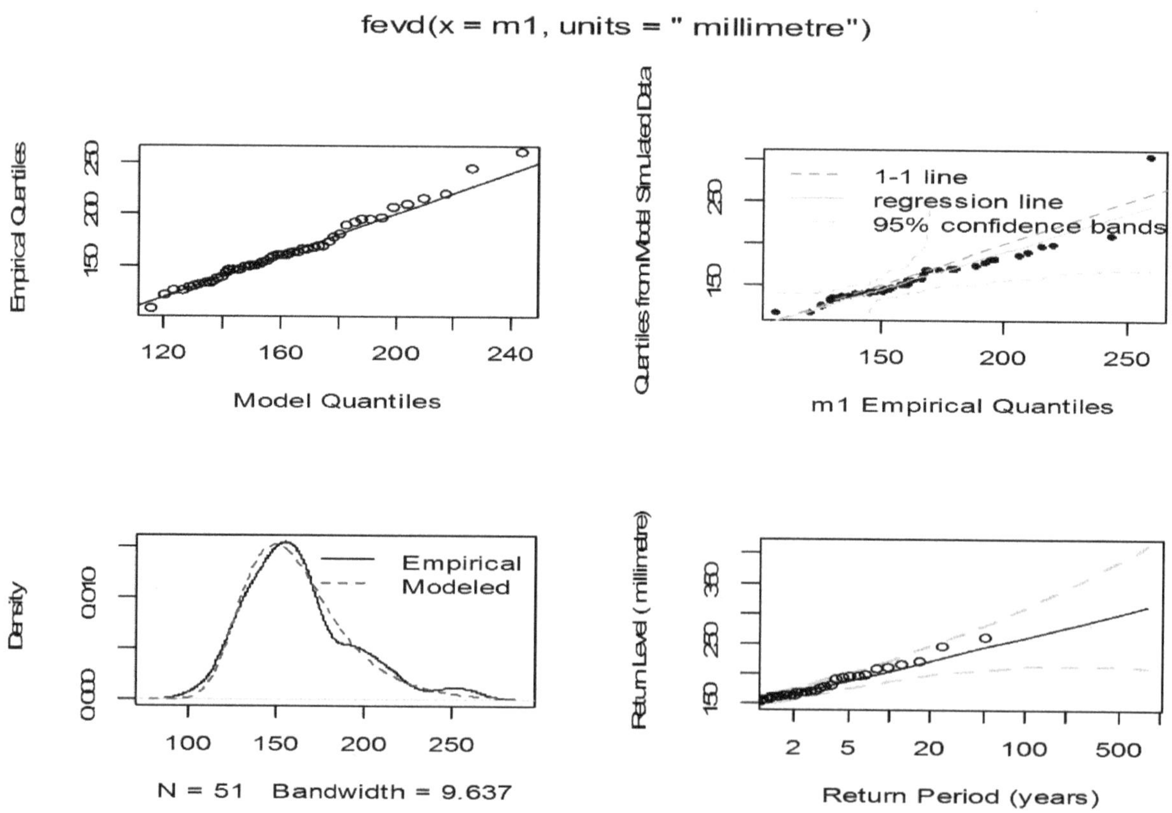

Figure 4. Diagnostic annual maximum rainfall plots.

4.3. GEVD Model for Extreme Maximum Temperature

As shown in Table 6, the estimated return level of maximum temperature likely to occur over the next 5, 10, 20, 50, or even 100 years by fitting these data to the GEVD. The maximum rainfall data yield estimates for (μ, σ, γ) of (41.933,0.892, 0.203), with standard errors (0.137,0.105,0.079). The approximate 95% confidence intervals for the parameters are thus (41.66,42.20) for μ, (0.686,1.098) for σ, and (0.0463,0.359) for γ.

Table 6. GEVD estimates of maximum temperature.

GEVD	Maximum Temperature		
	Location	**Scale**	**Shape**
Estimates	$\mu = 42.08$	$\sigma = 0.826$	$\gamma = -0.292$
Std error	0.128	0.0912	0.0942
95% CI (normal app)	(41.664,42.202)	(0.686,1.098)	(0.046,0.359)
Estimated Return Levels	**95% lower**	**Estimate**	**95% upper**
5-year return level	42.82	43.08	43.35
10-year return level	43.16	43.44	43.72
20-year return level	43.39	43.72	44.04
50-year return level	43.59	44.00	44.41
100-year return level	43.67	44.17	44.67

Analytic plots used in estimating the accuracy of the GEVD model fitted to the annual maximum temperature data shown in Figure 5 (Top-left: empirical plot; Top-right: empirical quartile plot; Bottom-left: density plot; Bottom-right: return level plot). All four diagnostic schemes provide support for fitting the GEVD to the maximum annual temperature.

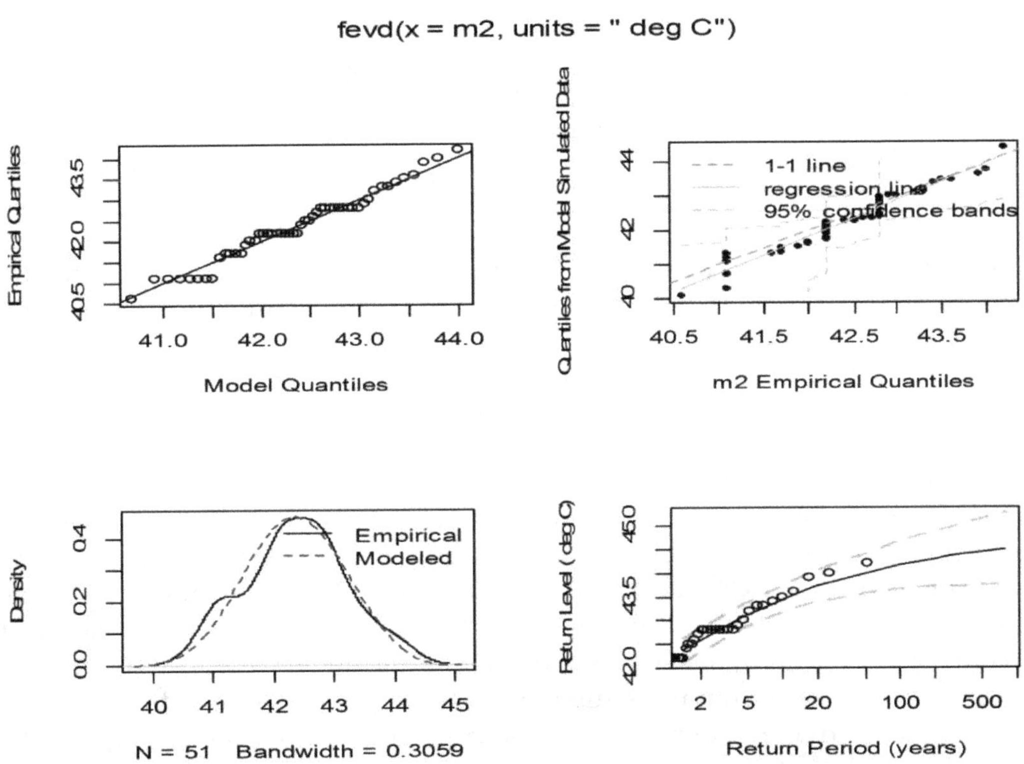

Figure 5. Diagnostic annual maximum temperature plots.

The determination of the limiting distribution by maximising the GEV negative log-likelihood for annual maximum temperature leads to the following function, Equation (2): (μ, σ, γ) of (42.081, 0.826, −0.292)

$$G(z) = \exp\left\{-\left[1 + 0.826\left(\frac{z - 0.892}{42.08}\right)\right]^{\frac{-1}{-0292}}\right\} \quad (9)$$

From Equation (9), estimates of return periods for 5, 10, 20, 50, and 100-years and their confidence intervals at 95% as shown in Table 6. Thus, based on the data from 1965 to 2016, once in 100 years we should expect to see an extreme annual maximum temperature hit between 43.6 °C and 44.4 °C maximum temperature. The upper bound of the model prediction for the 100-years return is 44.4 °C, but 65 °C extreme annual temperature recorded in 1989. Of course, this is also undoubtedly extreme beyond regular extreme events, which is not expected based on the model's predictions. It is revealed by Table 6, that extreme maximum temperature consistently increasing marginally over the 100 years.

4.4. GEVD Model for Extreme Minimum Temperature

In Table 7 below, the estimated return periods of minimum rainfall likely to occur over the next 5, 10, 20, 50 or even 100 years fitted to the GEVD. The maximum rainfall variable yields estimates for (μ, σ, γ) of (6.408, 5.261, −0.632), with standard errors (0.817, 0.758, 0.148) respectively. Approximate 95% confidence intervals for the parameters are thus (4.806, 8.011) for μ, (3.774, 6.747) for σ, and (−0.922, −0.342) for γ.

Table 7. GEV estimates of Minimum Temperature.

GEV	Minimum Temperature		
	Location	**Scale**	**Shape**
Estimates	$\mu = 6.408$	$\sigma = 5.261$	$\gamma = -0.632$
Std error	0.817	0.758	0.148
95% CI(normal app)	(4.806,8.011)	(3.774,6.747)	(−0.922,−0.342)
Estimated Return Levels	**95% lower**	**Estimate**	**95% upper**
5-year return level	10.355	11.506	12.657
10-year return level	11.882	12.723	13.564
20-year return level	12.761	13.456	14.151
50-year return level	13.233	14.022	14.812
100-year return level	13.333	14.274	15.216

Equation (10) is the determination of the limiting distribution by maximising the GEV negative log-likelihood for annual minimum temperature leads to the following function:

$$G(z) = \exp\left\{-\left[1 - 0.632\left(\frac{z - 5.261}{6.408}\right)\right]^{\frac{-1}{(-0.632)}}\right\} \quad (10)$$

Supposing the relative stability of the GEVD process producing estimates for annual minimum temperature in degree Celsius (°C), the model estimates that the 5-year return level is 11.5 °C with 95% confidence interval (10.4, 12.7). For ten years it is a 12.7 °C extreme minimum temperature with 95% confidence interval (11.9, 13.6), and for 50 years it is 14.0 °C extreme minimum temperature with 95% confidence interval (13.2, 14.8). Thus, based on the data from 1968 to 2016, once in 100 years we should expect to see an extreme annual minimum temperature hit between 13.3 °C and 15.2 °C. For the period under annual extreme minimum temperature, there was no extreme beyond normal extreme events. In Table 7, the extreme minimum temperature is consistently increasing over the 100 years' duration. In Figure 6 (Top-left: empirical plot; Top-right: empirical quartile plot; Bottom-left: density plot; Bottom-right: return level plot), all four diagnostic schemes provide support for fitting the GEVD to the minimum annual temperature.

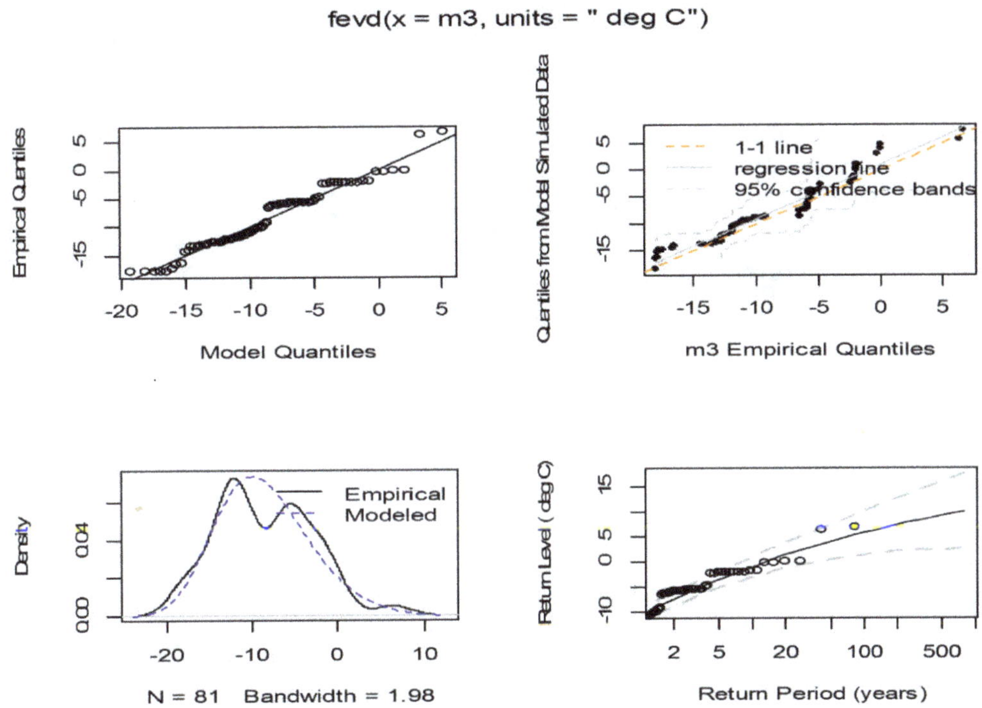

Figure 6. Diagnostic annual minimum temperature plots.

4.5. *Return Level*

Given 50-year return level for each of the indicators of extreme weather (for the year 2076), the return levels of extreme maximum rainfall in Ghana is higher than 150 mm reaching a warning line of extremely torrential rain, as defined by the Meteorological Service of Ghana. Similarly, the 50 years return level for maximum temperature exceeds 40 °C reaching a warning line of unusual temperature as defined by the Meteorological Service of Ghana. Also, the 50 years return level for extreme minimum temperature is lower than 20 °C reaching a warning line of frigid cold, as defined by the Meteorological Service of Ghana.

4.6. *Structural Equation Modeling (SEM)-Regression Analysis*

The term "structural equation modelling" (SEM) conveys two significant phases of the process: (a) causal effects under the research epitomised by a lot of structural equations (i.e., regression), and (b) these structural relationships can be presented to enable more specific concepts of theory studying. The assumed model (Figure 7) can then be statistically tested in a simultaneous analysis of the entire variables system to determine its compatibility with the data. If the suitability is appropriate, the model argues for the acceptance of assumed interactions between the variables; if inappropriate, the likelihood of such relationships fails to accept [76]. We chose PLS-SEM in present work for the following reasons: It is suitable for studies of theory construction [77,78]. It is appropriate to assess the sophisticated models of the cause-effect interaction [79,80]. The PLS-SEM assume a non-boundary approach, with fewer restrictions regarding sample size and data distribution [77].

SEM-regression estimation procedure was used to examine the hypothesised relationships as shown in Figure 4 between weather indicators and agriculture production. The results of SEM analysis showed a significant correlation between extreme weather and Agriculture production.

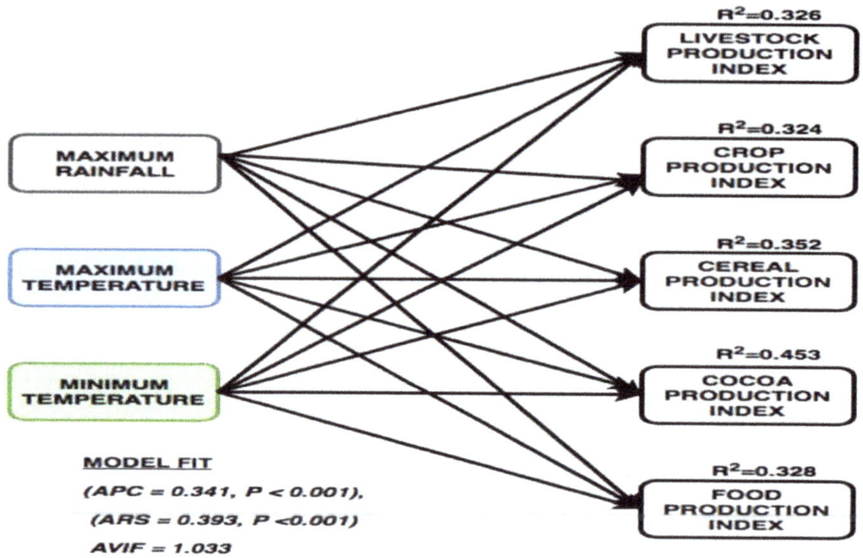

Figure 7. The Conceptual frame of the relationship of extreme weather on agriculture production indexes.

4.6.1. The relationship between Maximum Rainfall and Composite Agriculture Indexes

The analysis as showed in Table 8 is that, Livestock production index ($\beta = -0.1840$, $p = 0.144$), crop production ($\beta = -0.189$, $p < 0.133$), Cereal production ($\beta = -0.266$, $p < 0.031$), Cocoa ($\beta = -0.461$, $p < 0.001$), and food production index ($\beta = -0.190$, $p < 0.131$). Each is influenced by the effect of extreme maximum rainfall negatively on all composite agriculture indexes with no significant effect on crop production, food production, and livestock indexes. There has been a significant effect on cereal production and cocoa production indexes.

Table 8. Standardised Regression Weights and significance of correlations.

Predictor	Outcome	Path Coefficient	*p*-Values
Maximum Rainfall	Livestock Production Index	−0.184	0.144
	Crop production index	−0.189	0.133
	Cereal Production index	−0.266 *	0.031
	Cocoa production	−0.461 ***	<0.001
	Food Production Index	−0.190	0.131
Maximum Temperature	Livestock Production Index	0.305 *	0.015
	Crop production index	0.263 *	0.037
	Cereal Production index	0.276 *	0.025
	Cocoa production	0.424 *	0.023
	Food Production Index	0.268 *	0.033
Minimum Temperature	Livestock Production Index	0.457 ***	<0.001
	Crop production index	0.482 ***	<0.001
	Cereal Production index	0.415 ***	<0.001
	Cocoa production	−0.211 *	0.038
	Food Production Index	0.484 ***	<0.001

Significance of coefficient: *** $p < 0.001$ and * $p < 0.050$.

The results as shown in Table 8, shows each index is influenced by the effect of extreme maximum rainfall negatively, with no significant impact on crop production, food production, and livestock indexes. There has been a considerable effect on cereal production and cocoa production indexes. Maximum extreme rainfall hurts the performance of cereals.

Consequently, a unit increase in maximum extreme rainfall leads to a decrease in cereal production by 0.266 units. Maximum extreme rain leads to filtration of essential nutrients necessary for grain growth. Under such condition, any nutrient whether organic or inorganic leached beyond the reach of the roots, will result in reduced yields.

For cereals to bear maximum yields, rainfall, especially during tasseling for maize, is needed in moderation, inter-sparse with sunlight for maximum yields. Torrential rains do not favour most crops production and most especially cereals. Several studies have shown the importance of rainfall variability in crop production in various spatial scales [33,38].

Excessive rain has an adverse impact on agriculture. These effects run via different mechanisms. Heavy rains and floods have resulted in crop damage and the creation of poor conditions for harvesting, storage and transport of agricultural products. It is not astonishing that maximum rainfall has a negative association with all the variables under consideration, but only cereal and cocoa production indexes are statistically significant. Rainfall affects more variations in cocoa yields from year to year than with any other climatic factor. Trees are prone to a soil water shortage. The rain should be abundant and well distributed throughout the year. The annual precipitation between 1500 mm to 2000 mm is generally preferred. Droughts with rainfall below 100 mm per month should not exceed three months. The flooding of farmland leads to the leaching of nutrients needed for the growth of cocoa trees. If the phenomenon occurs over a period, this often leads to the death of cocoa trees or poor yields are observed [81]. It affects the flowering of cocoa trees and leads to flower aborting in some instances.

4.6.2. The Relationship between Maximum Temperature and Composite Agriculture Indexes

As shown in Table 8, Livestock production index (β = 0.305, p = 0.015), crop production (β = −0.263, p = 0.037), Cereal production (β = 0.276, p = 0.025), Cocoa (β = 0.424, p = 0.023), and food production index (β = 0.268, p < 0.033). Each is influenced significantly by the effect of extreme maximum temperature positively on all agriculture production indexes.

As shown in Table 8, each outcome is influenced significantly by the effect of extreme maximum temperature positively on all agriculture production indexes. The result indicates that a unit change in the maximum temperature will result in about 0.305 change in livestock production index. The nature of Ghana's livestock production immune it from the effects of extreme temperature conditions. Most animals are subject to a free or semi-intensive management system where animals are about to move freely.

Also, most cattle raised in Ghana are more adaptable to the state of the coast. As a result, maximum temperatures in Ghana does not affect them negatively since most of the animal rearing areas are almost in the coastal savannah region where the temperatures are not as high as the actual Sahel regions.

Breeding animals are sensitive to climate change and are severely affected by heat stress with an adverse effect on reproductive function [44,82]. According to Reference [83], high temperature and radiant heat load affect the reproductive rhythm through the hypothalamohypophyseal-ovarian axis. The primary factor in regulating ovarian activity is GnRH of thalamus and gonadotropin, i.e., FSH and LH of the anterior pituitary wall.

Research by [84,85] showed that the LH pulse amplitude and frequency of heat stressed cattle decreased. However, this is not the case in Ghana as shown in the results. Extreme temperatures that result in detrimental conditions not recorded in Ghana. High extreme temperatures hurt the crop production index, cereals production index, cocoa production, and food production index.

The maize pollen viability declines at temperatures above 35 °C [86–88]. Temperature increases in the 21st century may lead to yield losses of between 2.5% and 10% in some agronomic species [46]. Other assessments of crop yield due to temperature have produced different outcomes. Studies conducted by [89,90] showed estimates of yield between 3.8% and 5% decreases According to [90], crop growth for maize, soybeans and cotton will increase gradually with temperatures ranging from 29 °C to 32 °C and then sharply decrease as temperatures rise above this limit. It is however not surprising that maximum temperature in Ghana does not have adverse effects on yields. Maximum temperatures in Ghana is from 29 °C to 32 °C recorded in a dry season where no cultivation is taking place.

The period for production is the rainy season where temperatures hardly get close to 29 °C to 32 °C. Cocoa especially requires much heat, but direct sunshine damages it. As a result, some level of protection is necessary, especially when trees are young. Cocoa trees respond well to moderately high temperatures with a maximum yearly mean of 30 °C to 32 °C [91]. It is however not surprising that maximum temperature associate positively with cocoa production in Ghana where the maximum temperature falls within the acceptable range for cocoa.

4.6.3. The Relationship between Minimum Temperature and Composite Agriculture Indexes

As shown in Table 8, livestock ($\beta = 0.457, p < 0.001$), crop ($\beta = 0.482, p < 0.001$), Cereal ($\beta = 0.415, p < 0.001$), Cocoa ($\beta = -0.211, p = 0.038$), and food ($\beta = 0.439, p < 0.001$). Each is influenced significantly by the effect of extreme minimum temperature adversely on cocoa and positively on food, livestock, cereal, and crop production indexes.

As shown in Table 8, each outcome is influenced significantly by the effect of extreme minimum temperature adversely on cocoa production index and positively on (food, livestock, cereal, and crop) production index. Except for the cocoa sector, which is associated negatively with minimum extreme temperature the remaining areas are associated positively with low temperature. Average monthly temperatures below 23 °C are considered to suppress flowering.

The range in the average monthly temperature of the mainstream of cocoa-growing regions is found to be from 15 °C to 32 °C and considered to be the optimum for cocoa growth. The absolute minimum for any reasonable period is taken to be 10 °C, below which frost injury is likely [82]. Temperatures below the absolute minimum have a devastating impact on cocoa yields, as the results show.

Low arable yields caused by unfavourable weather conditions during certain stages of the growing season. The effects of unfavourable weather situations have shown reduced arable yields in recent decades. During the vegetative stage, low temperatures cause a reduction in barley yields. Low temperatures account for about 42% of the decrease in yield. Estimates show low temperatures in April, high rainfall in May and a heat wave in July followed by a cold and rainy August created unfavourable growth conditions for potatoes resulting in a decrease in yields [47].

Low yields of corn associated with a combination of low amounts of irradiation during the growing season (64% of low yields) and cold and wet spring (79% of low yields) cause delayed planting and slow biomass growth. Delayed frost has often worsened this situation (36% of low returns). Also, low yields contributed to the stress of drought and heat in flowering (21 per cent of low yields) and the recording of water during harvesting (29% of low yields) [47]. The type of low temperatures that often result in yield reduction is not the type often recorded in Ghana. Shallow temperatures experienced during the growing seasons in Ghana, hence its positive association with all the parameters except cocoa.

Regression estimates showed in Figure 7, extreme weather could explain almost 35.2% of the variance seen in cereal production ($R^2 = 0.352$), 45.3% of the variance seen in cocoa production ($R^2 = 0.453$), 32.6% of the variance seen in livestock production ($R^2 = 0.326$), 32.4% of the variance seen in crop production ($R^2 = 0.324$), and 32.9% of the variance seen in food production index ($R^2 = 0.328$). The whole model demonstrated an acceptable fit to the data for (APC = 0.341, $p < 0.001$), (ARS = 0.393, $p < 0.001$) and AVIF = 1.033

4.6.4. Paths Equations

As seen below, Equations (11)–(15) are the path equations for prediction of the agriculture production indexes and weather extremes.

Let X_1 = Extreme Maximum Rainfall, X_2 = Extreme Maximum Temperature

X_3 = Extreme Minimum Temperature

Thus, obtained are the following regression models for indexes prediction

$$Livestock\ production\ index = -0.184\ MaxRain + 0.305\ MaxTemp + 0.457\ MinTemp \quad (11)$$

$$Crop\ production\ index = -0.189\ MaxRain + 0.206\ MaxTemp + 0.482\ MinTemp \quad (12)$$

$$Cereal\ production\ index = -0.266\ MaxRain + 0.276\ MaxTemp + 0.455\ MinTemp \quad (13)$$

$$Cocoa\ production\ index = -0.461\ MaxRain + 0.257\ MaxTemp - 0.211 MinTemp \quad (14)$$

$$Food\ production\ index = -0.190\ MaxRain + 0.268\ MaxTemp + 0.484\ MinTemp \quad (15)$$

5. Conclusions

In this present study, we created and examined a model that could contribute to understanding the linkage, and predictability of severe weather and agriculture production in Ghana. The model and structure outlined, tested the nature of extreme maximum rainfall, extreme maximum temperature, extreme minimum temperature and the relationship that exist on agriculture production.

The annual maximum rainfall showed a decreasing trend. However, the yearly maximum temperature and minimum temperature exhibited a significant increase. As observed, there appears no significant trend heterogeneity for each month of the yearly minimum and maximum temperatures, while the annual maximum rainfall shows homogeneity for precipitation in each month. The results show that Extreme Value Theory (EVT) is a reliable tool for climate extreme scenarios construction, where maximum likelihood method supported the evaluation of distribution parameters for weather extreme. Generalised extreme value model is found to be the most suitable model with fulfilling all statistical selection criteria. The return level for the model is constructed to predict the weather extremes for a long run in future. There is generally an increase in weather extreme as it consistently increasing from time to time for the next 100 years.

Evidence from results indicated extreme maximum rainfall adversely affects cereal and cocoa production. Cereals and cocoa thrive well when the rainfall is well distributed and not concentrated in some months and leaving other months virtually without rains.

Maximum extreme temperatures contribute positively to all the indicators under consideration. Minimum extreme temperatures also except cocoa production have a positive impact on the remaining parameters. In the case of cocoa minimum extreme temperatures result in black pod diseases which causes yield reduction. The effect of the temperature and rainfall that is maximum or minimum on food production index depends on their impact on other cereals, livestock and crop productions. Where their respective measures are positive, it results in a positive outcome for food production index. To help improve the food production index of the country there is the need to consider investing in other production sectors. Based on the results the following recommendations are proposed for consideration by policymakers.

The planting time for cereals should be considered going forward, to avoid the detrimental effects of maximum extreme rainfalls. By so doing the yields of cereals will not be affected since they will avoid the period of torrential rains, which affects yields. The diversification of cereals production will help guide against the effect of maximum extreme rainfall on the cereals sector. Some cereals can withstand the impact of maximum extreme rains; diversification into those areas will help reduce the impact of maximum extreme rains if not eliminated.

Minimum extreme temperatures are reported to have detrimental effects on cereals. We recommend the developing of resistant varieties that can withstand the minimum extreme temperatures, which are negatively affecting cereals production. In the case of developing a resistant variety for cocoa, it will help deal with the situation. Since cocoa it is a perennial crop, it will be impossible to use planting period to help deal with the effects of maximum extreme rainfalls. Developing a resistant cocoa variety that will be able to withstand both extreme conditions will be a key in mitigating extreme effects on cocoa yields.

Other research focuses on the more complex problem of catastrophic agricultural risk. To some degree, the catastrophic agricultural risk is the result of extreme weather events. However, the catastrophic agricultural risk is not the same as extreme weather risk. Factors such as environment, agricultural investment, and farmer management should be of interest. To this extent, the distribution of potential damages and losses after a particular type of extreme weather condition should be of interest.

Improving the resilience of Ghanaian agriculture sector is essential. To help do this, farmers and stakeholders in the food production chain should consider the options for adaptation. Adaptation is highly context-specific; this is important for crop, region and climatic zone to use specific adaptation strategies to help minimise the effect of weather extremes on agriculture. The ability of the agricultural sector to deal with climate events will assume a downward trend as the globe warms, and is likely to exceed or fall at specific temperatures and rainfalls. Therefore, farmers need to get used to measures for effective, sustainable, and resilient crop and animal production. Thereby enhancing farmers understanding of growing seasons, improved crop rotation systems, adaptive water management techniques, and higher quality weather forecasts.

For further study, researchers can make a long-term prediction for another weather parameter which indirectly affects the agriculture sector like, production industry on which human life is dependent.

Author Contributions: J.D., conceptualised and initiated the study, M.A.S. and T.B.U. conducted a literature review on the study, whiles A.-A.I.M. carried out analysis on extreme value distribution and SEM-regression analysis. All other authors coordinated and provided comments on this manuscript.

Acknowledgments: This work was supported in part by the National Science Foundation of China under grants 71671080, 71471076, 71704066, 71501084 and 71701090. This work was also supported by the Special Funds of the National Social Science Fund of China [18VSJ038] and by Jiangsu Provincial Natural Science Foundation of China (Grant No. BK20170542).

References

1. Da Cunha, D.A.; Coelho, A.B.; Féres, J.G. Irrigation as an adaptive strategy to climate change: An economic perspective on Brazilian agriculture. *Environ. Dev. Econ.* **2015**, *20*, 57–79. [CrossRef]
2. Kurukulasuriya, P.; Mendelsohn, R.; Hassan, R.; Benhin, J.; Deressa, T.; Diop, M.; Eid, H.M.; Fosu, K.Y.; Gbetibouo, G.; Jain, S. Will African agriculture survive climate change? *World Bank Econ. Rev.* **2006**, *20*, 367–388. [CrossRef]
3. Mendelsohn, R.; Dinar, A.; Williams, L. The distributional impact of climate change on rich and poor countries. *Environ. Dev. Econ.* **2006**, *11*, 159–178. [CrossRef]
4. Di Falco, S. Adaptation to climate change in Sub-Saharan agriculture: Assessing the evidence and rethinking the drivers. *Eur. Rev. Agric. Econ.* **2014**, *41*, 405–430. [CrossRef]
5. Ministry of Food and Agriculture. *Medium Term Agriculture Sector Investment Plan (METASIP)—2011–2015*; Ministry of Food and Agriculture: Accra, Ghana, 2010.
6. MOFA. *Agriculture in Ghana: Facts and Figures 2010*; MOFA: Accra, Ghana, 2011.
7. World Bank Group. *Economics of Adaptation to Climate Change-Ghana*; World Bank Group: Washington, DC, USA, 2010.
8. Asante, F.A.; Amuakwa-Mensah, F. Climate change and variability in Ghana: Stocktaking. *Climate* **2014**, *3*, 78–99. [CrossRef]
9. EPA (Environmental Protection Agency). *Ghana's Second National Communication to the UNFCCC*; Environmental Protection Agency and Ministry of Environment, Science and Technology: Accra, Ghana, 2011.
10. Owusu, K.; Waylen, P.; Qiu, Y. Changing rainfall inputs in the Volta basin: Implications for water sharing in Ghana. *GeoJournal* **2008**, *71*, 201–210. [CrossRef]

11. Bernstein, L.; Bosch, P.; Canziani, O.; Chen, Z.; Christ, R.; Davidson, O.; Hare, W.; Huq, S.; Karoly, D.; Kattsov, V.; et al. *Fourth Assessment Report*; Longer Report; IPCC: Geneva, Switzerland, 2007.
12. Al-Hassan, R.; Poulton, C. Agriculture and Social Protection in Ghana. 2009. Available online: http://opendocs.ids.ac.uk/opendocs/handle/123456789/2340 (accessed on 22 October 2018).
13. Senaratne, A.; Scarborough, H. Coping with climatic variability by rain-fed farmers in dry Zone, Sri Lanka: Towards understanding adaptation to climate change. In *AARES 2011: Australian Agricultural & Resource Economics Society 55th Annual Conference Handbook*; AARES: Reston, VA, USA, 2011; pp. 1–22.
14. FAO. The State of Food Insecurity in the World 2008: High Food Prices and Food Security–Threats and Opportunities 2008. Available online: http://www.fao.org/docrep/011/i0291e/i0291e00.htm (accessed on 22 October 2018).
15. Garcia-Aristizabal, A.; Bucchignani, E.; Palazzi, E.; D'Onofrio, D.; Gasparini, P.; Marzocchi, W. Analysis of non-stationary climate-related extreme events considering climate change scenarios: An application for multi-hazard assessment in the Dar es Salaam region, Tanzania. *Nat. Hazards* **2015**, *75*, 289–320. [CrossRef]
16. Jin, Z.; Zhuang, Q.; Wang, J.; Archontoulis, S.V.; Zobel, Z.; Kotamarthi, V.R. The combined and separate impacts of climate extremes on the current and future US rainfed maize and soybean production under elevated CO_2. *Glob. Chang. Biol.* **2017**, *23*, 2687–2704. [CrossRef] [PubMed]
17. Berezuk, A.G.; da Silva, C.A.; Lamoso, L.P.; Schneider, H. Climate and Production: The Case of the Administrative Region of Grande Dourados, Mato Grosso do Sul, Brazil. *Climate* **2017**, *5*, 49. [CrossRef]
18. Adger, W.N.; Arnell, N.W.; Tompkins, E.L. Successful adaptation to climate change across scales. *Glob. Environ. Chang.* **2005**, *15*, 77–86. [CrossRef]
19. Withanachchi, S.S.; Köpke, S.; Withanachchi, C.R.; Pathiranage, R.; Ploeger, A. Water resource management in dry zonal paddy cultivation in Mahaweli River Basin, Sri Lanka: An analysis of spatial and temporal climate change impacts and traditional knowledge. *Climate* **2014**, *2*, 329–354. [CrossRef]
20. Coles, S. *An Introduction to Statistical Modeling of Extreme Values*; Springer: Berlin, Germany, 2001; ISBN 1-85233-459-2.
21. Field, C.B.; Barros, V.; Stocker, T.F.; Qin, D.; Dokken, D.J.; Ebi, K.L.; Mastrandrea, M.D.; Mach, K.J.; Plattner, G.K.; Allen, S.K. A special report of working groups I and II of the intergovernmental panel on climate change. In *Managing the Risks of Extreme Events and Disasters to Advance Climate Change Adaptation*; IPCC: Geneva, Switzerland, 2012.
22. Mannshardt, E.; Craigmile, P.F.; Tingley, M.P. Statistical modeling of extreme value behavior in North American tree-ring density series. *Clim. Chang.* **2013**, *117*, 843–858. [CrossRef]
23. Rahimpour, V.; Zeng, Y.; Mannaerts, C.M.; Su, Z. (Bob) Attributing seasonal variation of daily extreme precipitation events across The Netherlands. *Weather Clim. Extrem.* **2016**, *14*, 56–66. [CrossRef]
24. Thiombiano, A.N.; El Adlouni, S.; St-hilaire, A.; Ouarda, T.B.M.J.; El-jabi, N. Nonstationary frequency analysis of extreme daily precipitation amounts in Southeastern Canada using a peaks-over-threshold approach. *Theor. Appl. Climatol.* **2016**. [CrossRef]
25. Minkah, R. An application of extreme value theory to the management of a hydroelectric dam. *Springerplus* **2016**, *5*, 96. [CrossRef] [PubMed]
26. Ouarda, T.; El-Adlouni, S. Bayesian nonstationary frequency analysis of hydrological variables. *JAWRA J. Am.* **2011**, *47*, 496–505.
27. Seidou, O.; Ramsay, A.; Nistor, I. Climate change impacts on extreme floods II: Improving flood future peaks simulation using non-stationary frequency analysis. *Nat. Hazards* **2012**, *60*, 715–726. [CrossRef]
28. Katz, R.W. Statistics of extremes in climate change. *Clim. Chang.* **2010**, *100*, 71–76. [CrossRef]
29. Lekina, A.; Chebana, F.; Ouarda, T.B.M.J. Weighted estimate of extreme quantile: An application to the estimation of high flood return periods. *Stoch. Environ. Res. Risk Assess.* **2014**, *28*, 147–165. [CrossRef]
30. Panthou, G.; Vischel, T.; Lebel, T. Recent trends in the regime of extreme rainfall in the Central Sahel. *Int. J. Climatol.* **2014**, *34*, 3998–4006. [CrossRef]
31. Zahiri, E.P.; Bamba, I.; Famien, A.M.; Koffi, A.K.; Ochou, A.D. Mesoscale extreme rainfall events in West Africa: The cases of Niamey (Niger) and the Upper Ouémé Valley (Benin). *Weather Clim. Extrem.* **2016**, *13*, 15–34. [CrossRef]
32. Yabi, I.; Afouda, F. Extreme rainfall years in Benin (West Africa). *Quat. Int.* **2012**, *262*, 39–43. [CrossRef]
33. Ray, D.K.; Gerber, J.S.; MacDonald, G.K.; West, P.C. Climate variation explains a third of global crop yield variability. *Nat. Commun.* **2015**, *6*, 5989. [CrossRef] [PubMed]

34. Jones, P.G.; Thornton, P.K. The potential impacts of climate change on maize production in Africa and Latin America in 2055. *Glob. Environ. Chang.* **2003**, *13*, 51–59. [CrossRef]
35. Pachauri, R.K.; Allen, M.R.; Barros, V.R.; Broome, J.; Cramer, W.; Christ, R.; Church, J.A.; Clarke, L.; Dahe, Q.; Dasgupta, P. *Climate Change 2014: Synthesis Report. Contribution of Working Groups I, II and III to the Fifth Assessment Report of the Intergovernmental Panel on Climate Change*; IPCC: Geneva, Switzerland, 2014; ISBN 9291691437.
36. Warner, K.; Afifi, T. Where the rain falls: Evidence from 8 countries on how vulnerable households use migration to manage the risk of rainfall variability and food insecurity. *Clim. Dev.* **2014**, *6*, 1–17. [CrossRef]
37. Adamgbe, E.M.; Ujoh, F. Effect of variability in rainfall characteristics on maize yield in Gboko, Nigeria. *J. Environ. Prot. (Irvine, Calif.)* **2013**, *4*, 881. [CrossRef]
38. Ogunrayi, O.A.; Akinseye, F.M.; Goldberg, V.; Bernhofer, C. Descriptive analysis of rainfall and temperature trends over Akure, Nigeria. *J. Geogr. Reg. Plan.* **2016**, *9*, 195–202.
39. Nair, K.P.P. *The Agronomy and Economy of Important Tree Crops of the Developing World*; Elsevier: Amsterdam, The Netherlands, 2010; ISBN 0123846781.
40. Anim-Kwapong, G.J.; Frimpong, E.B. Vulnerability and Adaptation Assessment Under the Netherlands Climate Change Studies Assistance Programme Phase 2 (NCCSAP2). *Cocoa Res. Inst. Ghana* **2005**, *2*, 1–30.
41. Rao, G.P. *Climate Change Adaptation Strategies in Agriculture and Allied Sectors*; Scientific Publishers: Valencia, CA, USA, 2011; ISBN 9386347474.
42. Ali, F.M. Effects of rainfall on yield of cocoa in Ghana. *Exp. Agric.* **1969**, *5*, 209–213. [CrossRef]
43. Stanturf, J.A.; Warren, M.L.; Charnley, S.; Polasky, S.C.; Goodrick, S.L.; Armah, F.; Nyako, Y.A. *Ghana Climate Change Vulnerability and Adaptation Assessment*; United States Agency for International Development: Washington, DC, USA, 2011.
44. Amundson, J.L.; Mader, T.L.; Rasby, R.J.; Hu, Q.S. Environmental effects on pregnancy rate in beef cattle 1. *J. Anim. Sci.* **2006**, *84*, 3415–3420. [CrossRef] [PubMed]
45. Sprott, L.R.; Selk, P.G.E.; Adams, D.C. Factors affecting decisions on when to calve beef females. *Prof. Anim. Sci.* **2001**, *17*, 238–246. [CrossRef]
46. Hatfield, J.L.; Boote, K.J.; Kimball, B.A.; Ziska, L.H.; Izaurralde, R.C.; Ort, D.; Thomson, A.M.; Wolfe, D. Climate impacts on agriculture: Implications for crop production. *Agron. J.* **2011**, *103*, 351–370. [CrossRef]
47. Gobin, A. Weather related risks in Belgian arable agriculture. *Agric. Syst.* **2018**, *159*, 225–236. [CrossRef]
48. Piao, S.; Ciais, P.; Huang, Y.; Shen, Z.; Peng, S.; Li, J.; Zhou, L.; Liu, H.; Ma, Y.; Ding, Y. The impacts of climate change on water resources and agriculture in China. *Nature* **2010**, *467*, 43. [CrossRef] [PubMed]
49. Adger, W.N.; Huq, S.; Brown, K.; Conway, D.; Hulme, M. Adaptation to climate change in the developing world. *Prog. Dev. Stud.* **2003**, *3*, 179–195. [CrossRef]
50. Schlenker, W.; Lobell, D.B. Robust negative impacts of climate change on African agriculture. *Environ. Res. Lett.* **2010**, *5*, 14010. [CrossRef]
51. United Nations Framework Convention on Climate Change (UNFCCC). Climate Change: Impacts, Vulnerabilities and Adaptation in Developing Countries 2007. Available online: https://www.preventionweb.net/publications/view/2759 (accessed on 22 October 2018).
52. Wealth, C. *Economics for a Crowded Planet*; Allen Lane: London, UK, 2008.
53. Organization, W.H. Protecting the health of vulnerable people from the umanitarian consequences of climate change and climate related disasters. In Proceedings of the 6th session of the Ad Hoc Working Group on Long-Term Cooperative Action under the Convention (AWG-LCA 6), Bonn, Germany, 1–12 June 2009; pp. 1–12.
54. NCCPF (National Climate Change Policy Framework); Policy, C.; Goals, M.D.; African, W. Government of Ghana Ghana Goes for Green Growth Discussion Document—Summary Climate Change in Ghana. 2015. Available online: https://cdkn.org/wp-content/uploads/2011/04/NCCPF-Summary-FINAL.pdf (accessed on 22 October 2018).
55. Intergovernmental Panel on Climate Change. IPCC Climate Change 2007: The physical science basis. *Agenda* **2007**, *6*, 333.
56. Kearney, M. Hot rocks and much-too-hot rocks: Seasonal patterns of retreat-site selection by a nocturnal ectotherm. *J. Therm. Biol.* **2002**, *27*, 205–218. [CrossRef]
57. Agyemang-Bonsu, W.K.; Minia, Z.; Dontwi, J.; Dontwi, I.K.; Buabeng, S.N.; Baffoe-Bonnie, B.; Frimpong, E.B. *Ghana Climate Change Impacts, Vulnerability and Adaptation Assessments*; Environmental Protection Agency: Accra, Ghana, 2008.

58. World Bank Group. *The Costs to Developing Countries of Adapting to Climate Change—New Methods and Estimates*; World Bank Group: Washington, DC, USA, 2010; p. 20433.
59. Cameron, C. Climate Change Financing and Aid Effectiveness: Ghana Case Study. 2011. Available online: http://www.eldis.org/document/A61721 (accessed on 22 October 2018).
60. Christensen, J.H.; Kanikicharla, K.K.; Marshall, G.; Turner, J. Climate Phenomena and Their Relevance for Future Regional Climate Change. 2013. Available online: https://www.ipcc.ch/pdf/assessment-report/ar5/wg1/WG1AR5_Chapter14_FINAL.pdf (accessed on 22 October 2018).
61. Challinor, A.J.; Watson, J.; Lobell, D.B.; Howden, S.M.; Smith, D.R.; Chhetri, N. A meta-analysis of crop yield under climate change and adaptation. *Nat. Clim. Chang.* **2014**, *4*, 287. [CrossRef]
62. Codjoe, S.N.A.; Owusu, G. Climate change/variability and food systems: Evidence from the Afram Plains, Ghana. *Reg. Environ. Chang.* **2011**, *11*, 753–765. [CrossRef]
63. Ghana Statistical Service. Ghana Multiple Indicator Cluster Survey with an Enhanced Malaria Module and Biomarker. 2011; Final Report. Available online: https://www.popline.org/node/577148 (accessed on 22 October 2018).
64. Rosenzweig, C.; Iglesius, A.; Epstein, P.R.; Chivian, E. Climate Change and Extreme Weather Events—Implications for Food Production, Plant Diseases, and Pests. 2001. Available online: https://digitalcommons.unl.edu/cgi/viewcontent.cgi?article=1023&context=nasapub (accessed on 22 October 2018).
65. Huang, C.; Lin, J.-G. Modified maximum spacings method for generalized extreme value distribution and applications in real data analysis. *Metrika* **2014**, *77*, 867–894. [CrossRef]
66. Karmakar, M.; Shukla, G.K. Managing Extreme Risk in Some Major Stock Markets: An Extreme Value Approach. *Int. Rev. Econ. Financ.* **2015**, *35*, 1–25. [CrossRef]
67. Switzer, L.N.; Wang, J.; Lee, S. Extreme risk and small investor behavior in developed markets. *J. Asset Manag.* **2017**, *18*, 457–475. [CrossRef]
68. Davison, A.C.; Padoan, S.A.; Ribatet, M. Statistical Modeling of Spatial Extremes. *Stat. Sci.* **2012**, *27*, 161–186. [CrossRef]
69. Bader, B. Automated, Efficient, and Practical Extreme Value Analysis with Environmental Applications. 2016. Available online: https://arxiv.org/pdf/1611.08261.pdf (accessed on 22 October 2018).
70. Chinhamu, K.; Huang, C.K.; Huang, C.S.; Hammujuddy, J. Empirical analyses of extreme value models for the South African mining index. *S. Afr. J. Econ.* **2015**, *83*, 41–55. [CrossRef]
71. Yee, T.W. *Vector Generalized Linear and Additive Models: With an Implementation in R*; Springer: Berlin, Germany, 2015; ISBN 9781493928187.
72. Statistics, J. Distribution Fitting 2. Pearson-Fisher, Kolmogorov-Smirnov, Anderson- Darling, Wilks-Shapiro, Cramer-von-Misses and Jarque-Bera Statistics. *Bull. UASVM Hortic.* **2009**, *66*, 691–697.
73. Hair, J.F.; Sarstedt, M.; Ringle, C.M.; Mena, J.A. An assessment of the use of partial least squares structural equation modeling in marketing research. *J. Acad. Mark. Sci.* **2012**, *40*, 414–433. [CrossRef]
74. Pohlert, T. Non-parametric trend tests and change-point detection. *CC BY-ND* **2016**, *4*. Available online: http://cran.stat.upd.edu.ph/web/packages/trend/vignettes/trend.pdf (accessed on 29 October 2018).
75. Castillo, E.; Hadi, A.; Balakrishnan, N.; Sarabia, J. *Extreme Value and Related Models with Applications in Engineering and Science*; Wiley: New York, NY, USA, 2005.
76. Byrne, B.M. *Structural Equation Modeling with AMOS: Basic Concepts, Applications, and Programming*; Routledge: London, UK, 2016; ISBN 131763313X.
77. Vinzi, V.E.; Trinchera, L.; Amato, S. PLS path modeling: From foundations to recent developments and open issues for model assessment and improvement. In *Handbook of Partial Least Squares*; Springer: Berlin, Germany, 2010; pp. 47–82.
78. Hair, J.F.; Ringle, C.M.; Sarstedt, M. Partial least squares: The better approach to structural equation modeling? *Long Range Plan.* **2012**, *45*, 312–319. [CrossRef]
79. Richter, N.F.; Sinkovics, R.R.; Ringle, C.M.; Schlaegel, C. A critical look at the use of SEM in international business research. *Int. Mark. Rev.* **2016**, *33*, 376–404. [CrossRef]
80. Lowry, P.B.; Gaskin, J. Partial least squares (PLS) structural equation modeling (SEM) for building and testing behavioral causal theory: When to choose it and how to use it. *IEEE Trans. Prof. Commun.* **2014**, *57*, 123–146. [CrossRef]
81. Field, C.B.; Barros, V.; Stocker, T.F.; Qin, D.; Dokken, D.J.; Ebi, K.L.; Mastrandrea, M.D.; Mach, K.J.; Plattner, G.-K.; Allen, S.K.; et al. *Managing the Risks of Extreme Events and Disasters to Advance Climate*

Change Adaptation. A Special Report of Working Groups I and II of the Intergovernmental Panel on Climate Change; Cambridge University Press: Cambridge, UK; New York, NY, USA, 2012.

82. Montagnon, C.; Leroy, T.; Eskes, A. Amélioration variétale de# Coffea canephora#. 2: Les programmes de sélection et leurs résultats. *Plant. Rech. Dév.* **1998**, *5*, 89–98.
83. Madan, M.L.; Prakash, B.S. Reproductive endocrinology and biotechnology applications among buffaloes. *Soc. Reprod. Fertil. Suppl.* **2007**, *64*, 261–281. [CrossRef] [PubMed]
84. Gilad, E.; Meidan, R.; Berman, A.; Graber, Y.; Wolfenson, D. Effect of heat stress on tonic and GnRH-induced gonadotrophin secretion in relation to concentration of oestradiol in plasma of cyclic cows. *J. Reprod. Fertil.* **1993**, *99*, 315–321. [CrossRef] [PubMed]
85. Wise, M.E.; Armstrong, D.V.; Huber, J.T.; Hunter, R.; Wiersma, F. Hormonal Alterations in the Lactating Dairy Cow in Response to Thermal Stress1. *J. Dairy Sci.* **1988**, *71*, 2480–2485. [CrossRef]
86. Dupuis, I.; Dumas, C. Influence of temperature stress on in vitro fertilization and heat shock protein synthesis in maize (Zea mays L.) reproductive tissues. *Plant Physiol.* **1990**, *94*, 665–670. [CrossRef] [PubMed]
87. Herrero, M.P.; Johnson, R.R. High Temperature Stress and Pollen Viability of Maize 1. *Crop Sci.* **1980**, *20*, 796–800. [CrossRef]
88. Schoper, J.B.; Lambert, R.J.; Vasilas, B.L.; Westgate, M.E. Plant factors controlling seed set in maize: The influence of silk, pollen, and ear-leaf water status and tassel heat treatment at pollination. *Plant Physiol.* **1987**, *83*, 121–125. [CrossRef] [PubMed]
89. Rowhani, P.; Lobell, D.B.; Linderman, M.; Ramankutty, N. Climate variability and crop production in Tanzania. *Agric. For. Meteorol.* **2011**, *151*, 449–460. [CrossRef]
90. Schlenker, W.; Roberts, M.J. Nonlinear temperature effects indicate severe damages to U.S. crop yields under climate change. *Proc. Natl. Acad. Sci. USA* **2009**, *106*, 15594–15598. [CrossRef] [PubMed]
91. Growing Cocoa—The International Cocoa Organization. Available online: https://www.icco.org/about-cocoa/growing-cocoa.html (accessed on 5 August 2018).

2

Participatory Sustainability Assessment for Sugarcane Expansion in Goiás, Brazil

Heitor Luís Costa Coutinho [1,†], Ana Paula Dias Turetta [1,*], Joyce Maria Guimarães Monteiro [1], Selma Simões de Castro [2] and José Paulo Pietrafesa [3]

1. Embrapa Solos, Rua Jardim Botânico 1024, Rio de Janeiro-RJ 22460-000, Brazil; heitor.coutinho@embrapa.br (H.L.C.C.); joyce.monteiro@embrapa.br(J.M.G.M.)
2. Laboratory of Geomorphology, Pedology and Physical Geography—LABOGEF, IESA, UFG, Campus Samambaia, Goiânia, GO 74.690-900, Brazil; selma@ufg.br
3. Education Faculty of UFG—Program of Pos Graduation in Education (PPGE). Rua 235, s/n-Setor Universitário, Goiânia, GO 74.605-050, Brazil; jppietrafesa@gmail.com

* Correspondence: ana.turetta@embrapa.br

† Deceased.

Abstract: The sugarcane expansion in Brazil from 1990 to 2015 increased crop area by 135.1%, which represents more than 10 million hectares. Brazilian ethanol production hit a record high in 2015, reaching 30 billion liters, up 6% compared to 2014. In 2009, the Sugarcane Agroecology—ZAE-CANA—was launched to be a guideline to sustainable sugarcane production in Brazil. However, although it aims at sustainable production, it only considered natural aspects of the country, such as soil and climate. It is still necessary to develop instruments for studies on sustainability in all pillars. The aim of this study is to present the results regarding the application of the FoPIA (Framework for Participatory Impact Assessment) methodology in the Southwestern Goiás Planning Region (SGPR). FoPIA is a participatory methodology designed to assess the impacts of land use policies in regional sustainability, and the results showed the capacity of FoPIA to assess the impacts of land use change of the sugarcane expansion in that area. The major advantage of FoPIA is its participatory method feature, as it is possible to join stakeholders to debate and define sustainability guidelines.

Keywords: sugar cane expansion; sustainability assessment; FoPIA methodology

---

1. Introduction

Bio-ethanol from Brazil is an attractive type of biofuel because of its low price and relatively large greenhouse gas emissions reduction potential [1,2].

In the late 1970s, Brazil's National Bioethanol Program (PROALCOOL) ordered the mixture of anhydrous bioethanol (BE) in gasoline (blends up to 25%) and encouraged automakers to produce engines running on pure hydrated ethanol (100%) [3]. The Brazilian adoption of mandatory regulations to determine the amount of BE to be mixed with gasoline was essential to the success of the program [4]. The goal was to reduce oil imports which consumed one half of the total hard currency from exports. Although it was a decision made by the federal government during a military regime, it was well accepted by civil society, the agricultural sector, and car manufacturers [5].

Taking advantage of all the learning and experience of that period, the Brazilian government undertook some responsibilities against the international scenario related to climate change. In Brazil's Nationally Determined Contribution (NDC), submitted during COP21 in December 2015 and ratified in September 2016, the country agreed to reduce greenhouse gas (GHG) emissions by 37% by 2025 and 43% by 2030, with the 2005 emissions as a reference. To do so, the government agreed to increase biofuel (biodiesel and ethanol) participation by 18% in the energy matrix by 2030 [6]. Other commitments were related to actions to reduce GHG emissions by some 37% by 2020 [7]. The sugarcane expansion in Brazil increased 135.1% in crop area from 1990 to 2015, which represents more than 10 million hectares [8].

In 2009, with Decree 6961 [9], the Sugarcane Agroecology Zoning—ZAE-CANA [10] was created to guide the sustainable sugarcane expansion in Brazil. ZAE-CANA's main goals are to provide technical subsidies to policy makers to direct sugarcane expansion into legally recommended areas and sustainable production in Brazil. To achieve these goals, the study followed the guidelines that will allow the expansion of production: indication of areas with agricultural potential for sugarcane harvesting without environmental restrictions; exclusion of areas with original vegetation and indication of areas currently under anthropic use; exclusion of areas for cultivation in the Amazon, Pantanal biomes and the Upper Paraguay basin; reduction of direct competition with food production areas; indication of areas with agricultural potential (soil and climate) for the cultivation it means with slopes below 12%, by mechanical harvesting.

However, although it aims at sustainable production, the zoning only considered natural aspects of the country, such as soil, climate and relief. Also, the governance continues through contracts, to guarantee the productive supply and effectiveness of the productive chain, thus enhancing uncertainty regarding the sustainability of the Brazilian biodiesel production program [11].

Since the 1990s, environmental studies identified a wide range of reflections on sustainability and agricultural production systems. These reflections converged to the idea that economic growth, environmental preservation and social equity should be considered together to achieve a satisfactory development level [12–15].

Also at that time, the concept of sustainable development (SD), also in the agro-energy sector, was widespread, despite shortcomings in making SD operational. Therefore, policymakers are increasingly demanding comprehensive and reliable analyses of policy impacts on the economic, social and environmental dimensions of SD [16,17].

The use of criteria on sustainability allowed the assessment of the impacts caused by development processes, both in urban and rural areas. This process has contributed to regional assessments such as the implementation of public policies aimed at developing measures to mitigate social and environmental liabilities [18,19]. For instance, the European Union and some of its countries have specific directives to access the sustainability of biofuels such as the Renewable Energy Directive 2009/28/EC [20] and the Fuel Quality Directive 2009/30/EC [21], which established sustainability criteria to meet EU targets and to be eligible for financial support.

The development of instruments for studies on sustainability criteria and indicators, as well as those on the impacts on land use, is quite recent and shows gaps that are still under analysis. However, they are important instruments to understand changes that take place in social, environmental or economic phenomena. They can drive a particular need or even resources indicating trends that are undetectable in the processes. Therefore, it is necessary to develop instruments to help land managers to assess the social, economic and environmental impacts caused by land use-related public and corporate policies. These instruments may be quantitative, based on indicators' response models and functions built according to scientific knowledge and census databases; as well, they may be qualitative instruments based on technical knowledge integrated to that of the local stakeholders.

Particularly, the FoPIA (Framework for Participatory Impact Assessment) methodology has been useful to prepare for the participatory assessment of significant changes in land use and in the possibility of sustainability. The FoPIA is designed to enable assessments of policy impacts that are sensitive to national, regional and local sustainability priorities by harnessing the knowledge and expertise of national, regional and local stakeholders who play a central role in the analytical process. The analysis of specific sustainability problems gives rise to realistic national and regional policy and land use change scenarios [17].

The FoPIA was originally developed for application in the European Union to conduct stakeholder-based impact assessments of alternative land use policies, for example, to assess the policy options for biodiversity conservation in Malta [17]. This approach has been adapted for the assessment of land use policies in developing countries, with experiences in Indonesia, Tunisia and China [22–24].

Considering the assessment of impacts on the sustainability of sugarcane expansion policies in the southwestern region of Goiás state, Brazil, a participatory consultation was held to promote a structured interdisciplinary discussion about the sugarcane expansion in the region, to select public policy instruments for the construction of sugarcane expansion scenarios, as well as to define land use functions and indicators to be used in FoPIA. Hence, the aim of this study is to present the results regarding the application of this methodology.

2. Material and Methods

2.1. Study Area

The Southwestern Goiás Planning Region (SGPR) was chosen as the focus area because of its prominent expansion of sugar cane. In 2012, the sugarcane planted area in the SGPR was 286,512 ha and, in 2015, it was nearly twice that area (412,466 ha) [8]. The region spreads across an area of 61,498,463 km^2 and 26 municipalities (Figure 1).

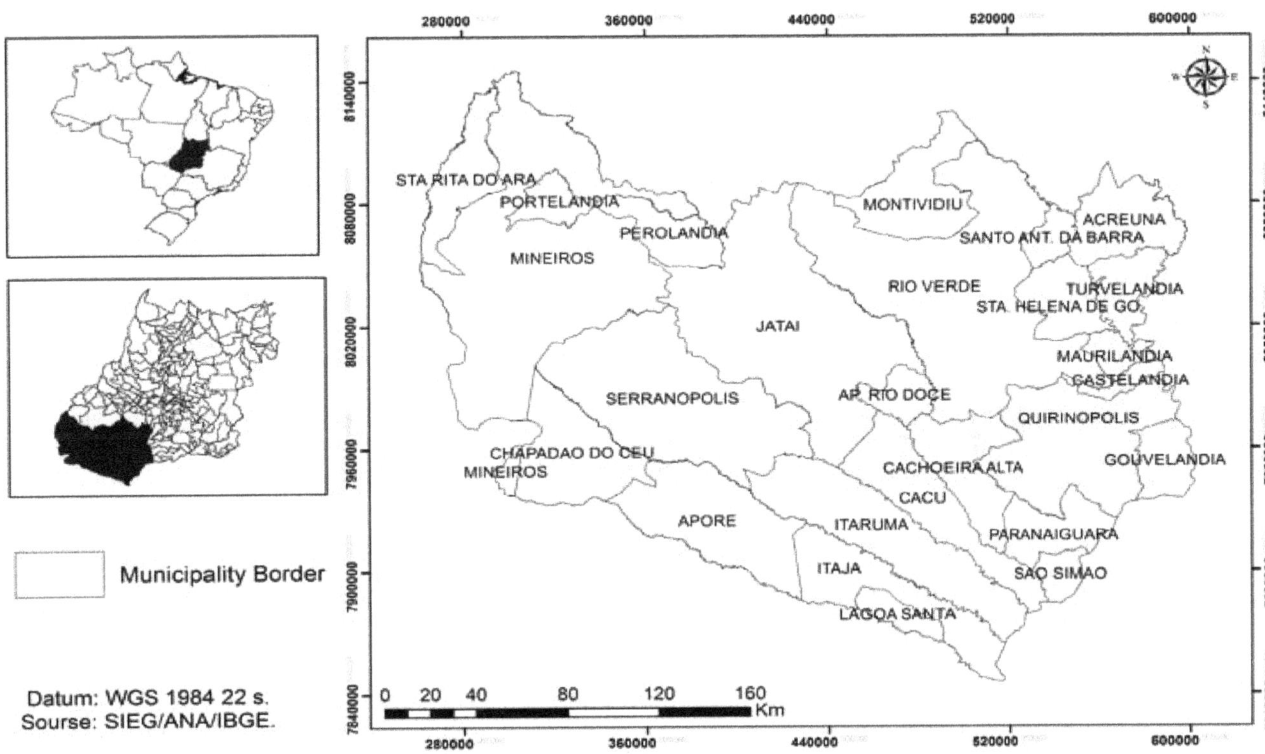

Figure 1. The Southwestern Goiás Planning Region and its municipalities. Source: SIEG—GO. Prepared by Trindade [25].

The sugarcane expansion in the large southern Goiás mesoregion is considered recent, as it started in 2004, mainly due to the advance of sugarcane agribusiness. The expansion of agribusiness in Goiás is characterized by high competition for land, favoring the leasing of large plots to harvest sugarcane for the sugar industry. This shows land concentration, mainly in parts of Southwestern Goiás involved in soybean and sugar cane production, and the exclusion of crops like rice, beans and, more recently, corn [26].

The SGPR has regional economic importance in the state, as well as consolidated logistics. However, it needs to be recovered and expanded to support the sugarcane expansion. At first

(2004–2008), much of the sugarcane expansion replaced soybean plantations, which, in turn, have shifted to pasture areas.

2.2. Methodology—FoPIA Background

The FoPIA methodology was developed as part of the EU-SENSOR project ("Sustainability Impact Assessment: Tools for Environmental, Social and Economic Effects of Multifunctional Land Use in European Regions") and applied to assess the impact of changing land use on different socioenvironmental situations of the European continent [17]. Subsequently, the FoPIA was adapted and used in the participatory impact assessment of different decision-making contexts and of environmental problems associated with land use and management in different Asian and African countries [22–24].

The FoPIA is a participatory methodology designed to assess the impacts of land use policies on regional sustainability. Its conceptual model, in which instrument users and public policy makers indicate policy scenarios to be assessed, considers variables such as driving forces, pressure, states, impacts, and responses (DPSIR). Each scenario generates different economic, fiscal or legislative conditions that, in turn, become driving forces of changes in land use. The pressures are changes in land use and management resulting from the implementation of policies. The pressures act on the states, as well as on the social, economic and environmental features of the regions subjected to changes, and they are represented by indicators. The impacts on sustainability are assessed through changes in the values of indicators as a response to the pressures, and through its relation with the sustainability limits or goals set to the region. Then, the decisions (responses) concerning the mitigation of or the adaptation to the impacts are then taken by the instrument users. However, due to limitations during the project, it was not possible to apply the entire methodology (Figure 2).

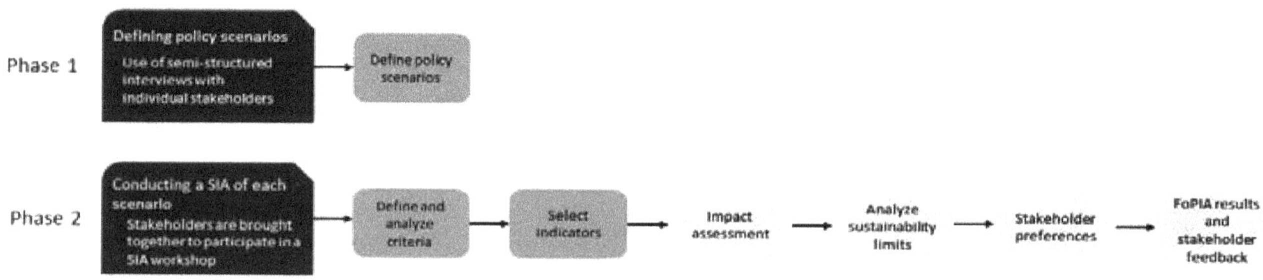

Figure 2. Framework for Participatory Impact Assessment (FoPIA) methodology steps [17]. Light gray boxes indicated what was not applied in this study.

In this study the FoPIA methodology comprises three stages: (1) The development of policy implementation scenarios and the consequent changes in land use, as well as the preliminary assessment of sustainability issues in the case study; (2) The definition of the Land Use Functions [27] suitable to the case study on a regional scale, by structuring the sustainability issue in the social, economic and environmental dimensions similarly balanced; (3) The definition of indicators for each Land Use Function, their responses to each scenario presented, followed by the integrated analysis of the results.

2.3. The Methodological Construction of FoPIA for Sugarcane Expansion in SGPR

In order to apply the FoPIA methodology to assess the impacts of land use change due to sugarcane expansion in the southwestern region of Goiás, a workshop for the participatory assessment of impacts on the sustainability of sugarcane expansion policies in Southwestern Goiás, was held at the Institute of Social and Environmental Studies (IESA—Instituto de Estudos Sócio-Ambientais) of the Federal University of Goiás (UFG), in Goiania, Goiás - Brazil, on 12 December 2012. The goals of this consultation were to promote a structured interdisciplinary discussion on the sugarcane expansion in SGPR, to select public policy instruments for the construction of sugarcane occupation scenarios, and to define land use functions and indicators to be used in the FoPIA. It condensed phase 1 and two steps of phase 2 of the FoPIA methodology (Figure 2). These steps concern topics to be discussed by experts, who have experience and knowledge in the study subject, to promote the technical base for the furthers steps. The goal was the establishment of the sugarcane expansion drivers and their potential indicators. The further steps—unfortunately, not considered in this paper—would include other stakeholders' consultation, such as government representatives, farmers and practitioners.

The workshop was a consultation activity to regional experts to subsidize the construction of policy scenarios, as well as the selection of Land Use Functions and impact indicators, by taking into consideration the case study of the expansion of sugarcane in the SGPR.

The workshop for the participatory assessment of impacts on the sustainability of sugarcane expansion policies in Southwestern Goiás included 32 experts. The number of participants the research team considered ideal to the required dynamic activities of the FoPIA workshop. The criteria to define the experts was their performance in research projects at the SGPR or their experience in studies on the impacts generated by land use changes of sugarcane expansion in the region, favoring researchers, professors, Master's and PhD students experts in geography, agronomy, ecology, climatology, soil and rural sociology.

The workshop structure was based on guiding lectures, followed by group works and plenary discussions. The workshop was divided into three study sessions: (1) Public policy scenarios; (2) land use functions; and (3) impact indicators. These steps followed the FoPIA methodology [17] (Figure 2). A previous material contend technical information to support discussions during the workshop sessions was prepared and distributed to the participants. Thus, they could work on this material and upgrade it with their knowledge. The following sections describe each session and their outcome are described in the results.

2.3.1. Session I: Public Policy Scenarios

Many factors are driving the increase in sugar cane production to meet the increase in ethanol demand in the national and international markets, such as government incentives (laws, decrees, public plans, programs, and policies) and foreign investment and market pressure (demand × supply). Based on past trends and on the experts' opinions it is possible to draw up reference scenarios. These scenarios represent developments without interference—in the absence of policy changes—they are the counterfactual, the background against which the impact of a policy can be evaluated. These scenarios are needed to know what the situation in the target year would be if policies did not change [28].

In the workshop one of the objectives was to identify the driving forces (key trends) that expand sugarcane in the SGPR, and select which would be considered in the baseline scenario. Also, the main public policies that promote sugarcane expansion sustainability were identified and selected to elaborate public policy scenarios.

Thus, six work groups were randomly formed for this session in order to discuss and summarize their findings around four guiding questions. These questions intended to exchange knowledge on sustainability concepts and how this knowledge applies to the case study and to the sugarcane expansion in the SGPR. These questions also aimed to point out the most appropriate public policies to the assessment, namely:

1. What are the main issues related to sugarcane expansion sustainability?
2. What are the main factors driving the sugarcane expansion in SGPR?
3. What are the main public policies promoting this process?
4. Consensually define which public policy would have greater influence on the sugarcane crop expansion and spatial distribution in the SGPR. Present two options for the implementation of this policy to build scenarios.

After the discussion, each group wrote down their answers in cards, which were presented at a plenary sitting and posted on a board. The session ended with a debate including all participants.

2.3.2. Session II: Land Use Functions

The definition and use of Land Use Functions (LUFs) contribute to the aggregation and prioritization of indicators according to different social, economic and environmental functions performed by land use [29], that is, the LUFs summarize the relationship between the sustainability dimensions and the indicators to be built.

The SENSOR project considered nine LUFs, three for each sustainability dimension, namely: (a) the social dimension: Labor supply; human health and recreation; cultural (landscape identity, scenic beauty, cultural heritage); (b) the economic dimension: Industrial activities and construction; rural production and mining; transport; and (c) the environmental dimension: Supply and conservation of abiotic resources; support, provision and conservation of biotic resources; maintenance of ecosystem processes (Table 1). However, the LUFs are flexible and allow changes to better meet the goals of each case study. Therefore, the original LUFs were presented and discussed by the audience. It was reviewed in terms of relevance and suitability considering the investigated object of study—the sugarcane expansion in SGPR.

Table 1. Land Use Functions (LUF) defined by the SENSOR Project team based on the LUFs suggested to the European Union [26].

Sustainability Dimension	LUF
SOCIAL	Labor supply Quality of life Human health and recreation
ECONOMIC	Industrial activities and construction Rural production Infrastructure
ENVIRONMENTAL	Conservation of abiotic resources Conservation of biotic resources Maintenance of ecosystem processes

The working groups were reorganized into two groups for knowledge and experience representation in each sustainability dimension (social, economic and environmental). A panel was set up and the groups presented their results in plenary by posting cards containing land use functions defined by them, along with their supporting justifications.

2.3.3. Session III: Impact Indicators

The experts participating in the workshop used the following criteria to select the sustainability indicators, built according to the recommendations of the Organization for Economic Cooperation and Development [30,31].

- Relevance to the formulation of policies
- Simplicity, conciseness, and ease of interpretation
- Analytical robustness
- Measurability
- Operability
- Availability (spatial and temporal).

The same work groups formed in the previous session applied the above-mentioned criteria to point out the most appropriate indicators to represent each LUF. Each group was asked to select three indicators per LUF. The results of each group were presented in plenary and the indicators were grouped—eliminating redundancies—and systematized for further analysis.

3. Results and Discussion

The activities carried out in the three sessions produced a summary of results obtained from the experts' work. Subsequently, a set of aspects that should compose the policy scenarios for participatory evaluation was presented by taking into consideration the feasibility of implementing the FoPIA methodology to assess social, environmental and economic impacts on the sugarcane expansion process in the SGPR.

3.1. Session I: Public Policy Scenarios

To build reference scenarios, it was necessary to draw up guiding questions based on the local reality considering the development of the biofuel industry in the region. The responses given by the work groups were summarized for each question, as follows.

- What are the main issues related to the sustainability of sugarcane expansion?

There was strong transversality between the responses and comments of the work groups, and this shows the importance of the sustainability issues in their three dimensions (social, economic and environmental).

Regarding the social aspect, the factor "concentration of lands controlled by large companies (sugar and alcohol plants) and landowners for cattle breeding and soybean crops, at the expense of small/medium farmers and family farmers" was pointed out by nearly all groups. One group reported that the process started through land leases (for soybean, and later, sugarcane), as the price was attractive, especially for small producers. During the successive sugar plantation renewals, for example, the lease price fell due to the reduction in the sucrose content linked to the lease price. As the situation continued, the plants made a proposal to purchase the lands and thus consolidate the concentration of rural properties in the studied region. There was strong impact on the land ownership structure. It mainly affected small farmers, who lost their identity as rural producers and it significantly changed the local agricultural profile.

This reality has given rise to several issues related to the growth of the area cultivated with sugarcane monocultures in the SGPR: The transition from manual to mechanized harvesting systems without burning led to sugarcane cutters losing their jobs. These workers had no training and thus were not employed by the sugar and alcohol industry. These sugarcane cutters, and the small farmers who sold their lands, migrated to cities with insufficient infrastructure to absorb the new population. Another critical factor was the inadequate working conditions for the sugarcane cutters who remained in the non-mechanized plantations. As many of these workers are required to cut a very large amount of sugarcane per day, the physical strain often leads to exhaustion and occupational diseases [32]. There is also labor supply seasonality, a characteristic of the conventional sugarcane production system.

As for the economic aspect, the issue of increasing land prices was highlighted, since the sale of small properties caused strong impacts. Such impacts also occur in the urban real estate sector, as the demand for housing also increases due to the arrival of new workers trained to work in plants and the rural exodus of farmers who lease or sell their properties.

Tax evasion was reported as relevant to the local economy, as most of the income generated from the sugar and sugarcane production does not stay in the municipality. It is transferred to the plants' centers of origin installed in the SGPR, which are mostly located in the Brazilian southeastern and northeastern regions. Income concentration, characterized by the low equity in the distribution of the economic benefits generated by sugarcane production, was identified as an important sustainability issue, despite the large tax collection increase in the state and municipalities.

It was reported that the local food production has been strongly impacted by sugarcane expansion due to reduced family production or land use change of these families' lands for sugarcane production.

The study groups presented some environmental aspects. They highlighted soil compaction, worsened by the standardization of mechanized crop management techniques that replaced the extensive pastures and the annual crops of family farmers previously managed through manual techniques. Mechanical harvesting includes heavy machinery and its successive use in the fields, which leads to soil detachment contributing to its compaction. In addition to soil quality loss, water infiltration and retention capacity is reduced. This leads to increased rainwater runoff, as well as increased runoff of the water used in irrigation systems. The outcome is soil erosion and transportation of nutrients and pesticides to the beds of streams, creeks and rivers [33].

The change in sugarcane field drainage dynamics increases the risk of groundwater and aquifers contamination with pesticides and industrial wastes such as vinasse and heavy metals, dumped into the soil through fertirrigation. This is worse in soils presenting sandier texture, and the experts showed concern regarding this topic. The strong water footprint resulting from high evapotranspiration—typical of the sugarcane culture—is another relevant issue to the overall environmental impacts, since it threatens water availability to humans, fauna and flora [34].

In addition to these facts, the experts mentioned that burning-based sugarcane harvest systems are still used in the SGPR, and this leads to severe air quality issues.

The issue of biodiversity loss, which becomes more evident when livestock areas are converted into sugarcane fields, was also mentioned. Although these areas are already deforested, livestock coexists with small forest fragments and scattered trees, as they provide shade and shelter for farm animals, as well as landing to several bird species, especially in livestock systems with low technological input. These systems are quite evident in large areas that have not been converted into sugarcane crops or grain monocultures.

- What are the main factors driving the sugarcane expansion in Southwestern Goiás?

The agribusiness model in the SGPR and in the state of Goiás is similar to that used in the São Paulo countryside—the largest national sugarcane producer—facilitating crop expansion. The high domestic and international demand for energy sources as alternatives to fossil fuels was considered an important sugarcane expansion factor. This expansion was supported by public policies and international agreements that favored the transformation of ethanol into an agricultural commodity. The other expansion factors mentioned were the technological development of the sugarcane and bioethanol productive chain and land availability.

Infrastructure and logistics such as agricultural flow paths and the ethanol pipeline installation plan, were considered the main sugarcane expansion factors in the SGPR. In addition, the agroecological suitability (soil, climate, topography) of most of the territory for sugarcane cultivation was considered average. This scenario qualified owners and entrepreneurs to receive government economic incentives for such purpose Manzatto et al. [10] (Figure 3).

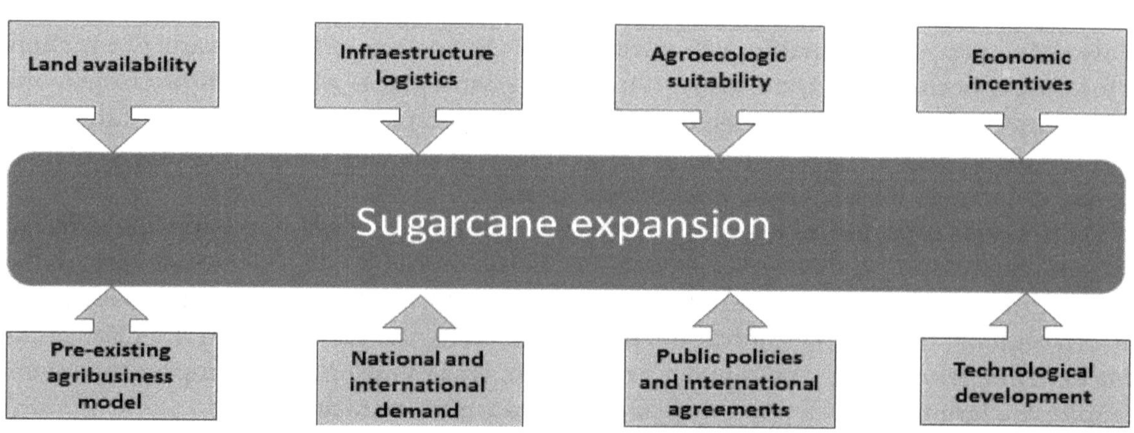

Figure 3. Sugarcane crop expansion drivers in the Southwestern Goiás Planning Region (SGPR).

- What are the main public policies promoting this process?

Some information on macro and micro state intervention policies were identified to reflect on intervention policies in the sugarcane industry and to simultaneously develop mechanisms to assess the industry impacts. The participants noticed that a set of policies should be highlighted in the SGPR: the Kyoto Protocol (at the global level); the National Agro-Energy Plan and the ZAE Cana (at the federal level); the Goiás Industrial Development Program ("Programa Produzir") and Grants and Taxes on services (ISS) at the state level.

The Kyoto Protocol (KP) is an international agreement linked to the United Nations Framework Convention on Climate Change (UNFCCC), which commits developed countries by setting internationally emission reduction targets. Brazil does not have a mandatory GHG emission reduction target; however, it has participated in the KP through the Clean Development Mechanism (CDM) projects. Clean energy generation projects, biofuels and other renewable sources could generate carbon credits to be traded on the carbon market. In this context, the National Plan for Climate Change (NPCC) aimed, among other things, "to foster the sustainable increase in the share of biofuels in the national transport matrix and work to structure an international market of sustainable biofuels".

One of the goals of the National Agro-Energy Plan [35] was to create conditions to internalize and regionalize the development based on agro-energy expansion and on the value added to the supply chain. The guidelines for this expansion were provided by the Agro-Ecological Sugarcane Zoning—ZAE-CANA [10].

Goiás state programs, such as *Programa Produzir* [36], which focuses on to the implementation and expansion of industries, also have a strong influence on the sugarcane expansion and reduce the due value-added tax (VAT) installments.

Finally, municipal policies were mentioned, including conveniences such as land grants and taxes on services (ISS) to implement new industries in Goiás. In addition, policies that have restricted sugarcane advancement, such as those found in the Rio Verde, one of the municipalities of the SGPR, which limited the sugarcane occupation to 10% of the municipal territory to protect areas planted with soybean, as well as the local food industry supply.

- Policy proposals that could be implemented.

The last item discussed in the sessions to create scenarios concerned proposals for public policy instruments to be implemented. These instruments would result in the effective transition of the production chains and allow the creation of sustainable logic-related actions.

Among public policy instruments, it was suggested that the ZAE-CANA should be further detailed to condition the government incentive contributions under the National Agro-Energy Plan, as this mechanism is considered critical to the process of sustainability. The agroecological zonings of the main crops in Goiás (scale 1:100,000) would enable a specific regulatory benchmark for Goiás and for the SGPR through the definition of sugarcane production priority areas. According to Manzatto et al. [10], many municipalities in the SGPR have large areas with medium/high suitability for sugarcane production, for example, Rio Verde (72%) and Quirinópolis (73%).

It was suggested that the environmental planning should be considered when formulating tax incentive policies.

Also, the development of an environmental education policy to encourage better understanding of the use and impacts related to soil, water resources and biodiversity was suggested. The ecological VAT could be used to reward municipalities that encourage land management changes. Accordingly, the development of a certification policy with economic incentives to certified systems was suggested. Figure 4 summarizes the main proposals built in public policies discussions.

Figure 4. Strategies and public policy themes suggested for the sustainable development of sugarcane expansion in the SGPR.

3.2. Session II: Land Use Functions

The work groups presented their views on LUFs initially adopted by Sensor project (Table 1) and suggested some modifications. A plenary consensus was obtained and a set of three LUFs was proposed for each sustainability dimension, as well as the potential indicators to achieve each LUF (Table 2).

Regarding the social dimension, the experts showed concern about the quality of the manual labor found in the SGPR agricultural production systems. These activities were often considered inhospitable, especially in the case of crops using sugarcane burning before the harvest. The proposal for the inclusion of the health and quality of life aspects in order to compose LUFs to assess this issue was accepted unanimously. The inclusion of the local socio-cultural development LUF was also suggested as LUFs hold recreational, educational, religious, scientific and cultural land use functions. The SGPR is notable for its natural beauty and its rich, exuberant landscape, mainly due to the heterogeneity of its elements.

Regarding the economic dimension, rural production along with the local consumption expansion issue were considered important to comprise another LUF as a way to reflect on the local impacts of the sugarcane expansion, and about the difficulty to produce and consume other food products. The income evasion was pointed out in Session I. If only the agricultural production is assessed, a positive result might hide a local negative impact, such as the reduced consumption of locally produced agricultural products. A similar situation can be found in the Mid-Goiás Planning Region, where sugarcane crops expanded much between 2000 and 2011, increasing from 23,000 to 123,000 hectares, which pushed up either the land value or the food supply in the region [37].

The region hosts one of the largest protected areas in the Cerrado biome, the Parque Nacional das Emas, whose integrity is threatened by the land use in its surroundings. Plateaus, cliffs and gorges, forest fragments, and water sources that drain their waters to three major watersheds (Paraguay, Araguaia and Paraná rivers), as well as cultural and archaeological sites, provide the local population and visitors with the scenic beauty and the cultural wealth of the region. Monocultures devoid of the tree element—which some cattle ranches still have in their pastures—are dominated by heavy machinery and, in some cases, use burning as an agricultural practice. Thus, they represent a high risk to the integrity of rural landscapes in the SGPR.

As for the environmental dimension, by taking into consideration the importance given to water resources in both productive activities and in the supply for human consumption, as well as the high water demand in the sugarcane culture, the insertion of the conservation of abiotic resources LUF was proposed. It was considered comprehensive and included issues such as water resources and erosion and loss of soil and nutrients.

Table 2. Indicators to assess sugarcane expansion sustainability in the SGPR.

Sustainability	LUF	Indicators
SOCIAL (a)	Labor quality and supply	Hiring, firing and balance Average income Employment rate by sector Use of local labor
	Health care and quality of life	Access to basic sanitation Hospital facilities and beds/100,000 inhabitants Student attendance per school year Crime rate
	Local socio-cultural development	Number of public leisure facilities GINI index Cultural groups Number of Municipal Councils Number of high school graduates
ECONOMIC (b)	Industrial activities and construction	Urbanization rate Industrial diversification Industrial gross domestic product (GDP) Availability of public transportation Electricity consumption
	Rural production and local consumption	Agricultural GDP Agricultural diversification Area occupied by crops Consumption of local agricultural products
	Infrastructure	Electrical power generation Electric transmission network Road network diversity Electricity cogeneration
ENVIRONMENTAL (c)	Conservation of abiotic resources	Consumption of pesticides and fertilizers Use/ Agroecology Zoning (ZAE) discrepancy Percentage of preserved PPA Total use/sugarcane expansion rate Soil loss Burned area/harvested area
	Conservation of biotic resources	Percentage of preserved permanent preservation area (PPA) Total use/sugarcane expansion rate Burned points/year Deforestation rate—pasture clearing Number of fragments (measures of associated landscape) Pesticide consumption
	Maintenance of ecosystem processes	Percentage of preserved PPA Carbon stock and sequestration Water body sedimentation rate Percentage of contiguous production area (landscape/permeability matrix) Fragmentation level of the remaining forests

3.3. *Session III: Impact Indicators*

The work groups in the Session III were the same as session II and, according to criteria such as relevance, simplicity, robustness, measurability, operability and spatial and temporal availability, they chose the indicators that would be most appropriate to represent each LUF defined in the previous session. Each work group proposed three indicators for each LUF, according to the sustainability dimension they represented (Table 2).

Fifty-three indicators were pointed out by the six work groups. The analysis and removal of redundancies and inconsistencies resulted in 43 proposed indicators: 13 social, 13 economic and 17 environmental indicators (Table 2). The experts understood that the indicators must reflect the impacts of land use changes on the municipality or region, based on a broader scale than that observed exclusively in production areas such as plants or rural properties.

Regarding the social dimension, most experts agreed to include indicators capable to evaluate the quality of issues such as job opportunities, education, public security and leisure for the people.

Similarly, not only did the suggested economic indicators measure the industrial and agricultural production values, but they also assessed the diversification of these products. They measured power generation and assessed its distribution. They also measured the availability of public transportation, as well as the capacity of the road network to meet the demand for transport. Thus, the experts selected variables to identify the relationship between social and economic factors in order to expand the potential to improve the local population's quality of life.

The environmental issue is reflected in the indicators representing the relevant impacts to sugarcane monocultures, namely the contamination of soils and water resources by pesticides and excessive nutrients (especially nitrogen) [38]; the biodiversity loss due to the sugarcane monoculture and the way it is managed by agricultural companies and landowners [39,40]; the air quality, which is strongly compromised by the smoke from post-harvest burning [41]; and the soil compaction and its subsequent incapacity to retain sediments, nutrients and water [42].

4. Conclusions

The sugarcane expansion scenarios in the SGPR indicate the dynamics of the expansion and point out some weaknesses and potentialities of ethanol production in this region. This information can be analyzed in the light of the influence of the driving forces that operate in the local industrial and agro-industrial sectors, which may favor or restrict the cultivation of sugarcane in the region. The study collected information that allows us to conclude that the edaphoclimatic conditions, the availability of areas for cultivation, and governmental policies at the federal and state levels are among the main attraction factors for the implantation, expansion and revitalization of the sugar and alcohol industry in the SGPR.

The ZAE-CANA was the main instrument of public policies selected for the construction of sugarcane expansion scenarios in SGPR. Other public policies and drivers should be considered in scenarios for the expansion of sugarcane. However, due to the socio-environmental diversity of the region, the same set of public policy instruments can result in very different social, economic and environmental impacts. The participatory methodological approach provided the basis for evaluating the sustainability impacts caused by the expansion process of sugarcane cultivation and the implementation of ethanol agribusiness in the SGPR.

Based on technical knowledge integrated with the stakeholders of the local society, it was possible to indicate the LUF and the indicators that should be considered in the FoPIA in the SGPR sugarcane expansion scenarios. The LUF defined in this study were similar to those originally proposed, demonstrating that the LUF set by the experts address the main sustainability issues of the sugarcane expansion in SGPR. Participants, however, placed a special emphasis on the health care issue of quality of life as a new LUF proposal. They also stressed the local socio-cultural development, which would include recreational, educational, religious, scientific and cultural land use as the region shelters landscapes with remarkable scenic beauty. The prioritization stage of the indicators was not carried out due to the low availability of time during the workshop so that the participating experts could have access to more information and reflect on the issue. However, they showed a clear need to select a minimum set of indicators in a participatory way associated with each LUF to evaluate and monitor the sustainability of scenarios for expansion of sugarcane cultivation geared to the industrial production of sugar and ethanol in SGPR.

Finally, the results indicate the potential of the FoPIA methodology as a tool to assess sustainability in a participatory way, bringing together stakeholders to discuss and promote guidelines to achieve sustainability. However, as the methodology was not fully applied, it was not possible to access stakeholders' preferences and feedback.

FoPIA has proven to be a powerful tool—although it is complex and demands a great amount of energy input. We highly recommend this tool; however, we must stress that studies that use it must require thorough previous planning, especially concerning the workshops and stakeholders mobilization.

Acknowledgments: We gratefully acknowledge the time and efforts of the experts who joined the workshop. We also thank the Sixth EU's Framework Programme for the financial support and the anonymous reviewers for their critical and constructive comments.

Author Contributions: Heitor Coutinho designed the FoPIA application for the sugarcane expansion in SGPR; Ana Turetta wrote the manuscript and compiled the figures and tables; Joyce Monteiro wrote the manuscript mainly the public policies section and conclusions; Selma Castro and José Pietrafesa contributed with the workshop organization, map elaboration and text suggestions. All authors contributed to the discussion.

References

1. Buckeridge, M.S.; de Souza, A.P.; Arundale, R.A.; Teixeira, K.J.; De Lucia, E. Ethanol from sugarcane in Brazil: a 'midway' strategy for increasing ethanol production while maximizing environmental benefits. *GCB Bioenergy* **2012**, *4*, 119–126. [CrossRef]
2. Martinelli, L.A.; Filoso, S. Expansion of Sugarcane Ethanol Production in Brazil: Environmental and Social Challenges. Available online: http://onlinelibrary.wiley.com/doi/10.1890/07-1813.1/epdf (accessed on 14 August 2017).
3. Walter, A.; Galdos, M.V.; Scarpare, F.V.; Scarpare, F.V.; Leal, M.R.L.V.; Seabra, J.E.A.; da Cunha, M.P.; Picoli, M.A.; de Oliveira, C.O.F. Brazilian sugarcane ethanol: Developments so far and challenges for the future: Brazilian sugarcane ethanol. *Wiley Interdiscip. Rev. Energy Environ.* **2014**, *3*, 70–92. [CrossRef]
4. Marin, R.M. Understanding sugarcane production, biofuels, and market volatility in Brazil—A research perspective. *Outlook Agric.* **2016**, *45*, 75–77. [CrossRef]
5. Goldemberg, J. Ethanol for a sustainable energy future. *Science* **2007**, *315*, 808–810. [CrossRef] [PubMed]
6. Brasil. Pretendida Contribuição Nacionalmente Determinada. Available online: http://www.itamaraty.gov.br/images/ed_desenvsust/brasil-indc-portugues.pdf (accessed on 19 June 2017).
7. Brasil. Política Nacional Sobre Mudança do Clima—PNMC e dá Outras Providências. Available online: http://www.planalto.gov.br/ccivil_03/_ato2007-2010/2009/lei/l12187.htm (accessed on 19 June 2017).
8. IBGE/SIDRA Pesquisa Agrícola Municipal. Available online: http://www.sidra.ibge.gov.br/bda/tabela/listabl.asp?z=t&o=11&i=P&c=1181 (accessed on 5 November 2016).
9. Brasil. Aprova o Zoneamento Agroecológico da Cana-de-açúcar e Determina ao Conselho Monetário Nacional o Estabelecimento de Normas para as Operações de Financiamento ao setor Sucroalcooleiro, nos termos do Zoneamento. Available online: http://www.planalto.gov.br/ccivil_03/_Ato2007-2010/2009/Decreto/D6961.htm (accessed on 19 June 2017).
10. Manzatto, C.V.; Assad, E.D.; Mansilla Bacca, J.F.; Zaroni, M.J.; Pereira, S.E.M. Zoneamento Agroecológico da Cana-de-Açúcar—Expandir a Produção, Preservar a Vida, Garantir o Future. Available online: https://ainfo.cnptia.embrapa.br/digital/bitstream/CNPS-2010/14408/1/ZonCana.pdf (accessed on 14 August 2017).
11. Rathmann, R.; Silveira, S.J.C.; Santos, O.I.B. Governança e configuração da cadeia produtiva do biodiesel no Rio Grande do Sul. *Rev. Ext. Rural* **2008**, *15*, 69–101.
12. Hardi, P.; Zdan, T.J. *Assessing Sustainable Development: Principles in Practice*; International Institute for Sustainable Development: Winnipeg, MB, Canada, 1997.
13. Vezzoli, C.; Manzini, E. *O Desenvolvimento de Produtos Sustentáveis: Os Requisitos Ambientais dos Produtos Industriais*; EDUSP: São Paulo, Brazil, 2002.
14. Lawn, P. *Sustainable Development Indicators in Ecological Economics*; Edward Elgar: Cheltenham, London, UK, 2006.
15. Sachs, I. Barricadas de ontem, campos de futuro. *Estud. Av.* **2010**, *24*, 25–38. [CrossRef]
16. Helming, K.; Diehl, K.; Bach, H.; Dilly, O.; König, B.; Kuhlman, T.; Pérez-Soba, M.; Sieber, S.; Tabbush, S.; Tscherning, K. Ex ante impact assessment of policies affecting land use, Part A: Analytical framework. *Ecol. Soc.* **2011**, *16*, 29. [CrossRef]
17. Morris, J.B.; Tassone, V.; de Groot, R.; Camilleri, M.; Moncada, S. A framework for participatory impact assessment: Involving stakeholders in European policy making, a case study of land use change in Malta. *Ecol. Soc.* **2011**, *16*, 12. [CrossRef]
18. Van Bellen, H.M. Indicadores de Sustentabilidade: Uma Análise Comparativa. Ph.D. Thesis, Programa de Pós-Graduação em Engenharia de Produção, Universidade Federal de Santa Catarina, Florianópolis, Brazil, 2002.

19. Veiga, J.E. *Indicadores de Sustentabilidade*; Estudos Avançados: São Paulo, Brazil, 2010.
20. EU—European Union. Directive 2009/28/EC—Renewable Energy Directive. Available online: http://eur-lex.europa.eu/legal-content/EN/ALL/?uri=CELEX:32009L0028 (accessed on 6 April 2017).
21. EU—European Union. Directive 2009/30/EC Amending Directive 98/70/EC on Fuel. Available online: http://ec.europa.eu/environment/air/transport/pdf/art7a.pdf (accessed on 6 April 2017).
22. König, H.J.; Schuler, J.; Suarma, U.; Mcneill, D.; Imbernon, J.; Damayanti, F.; Dalimunthe, S.A.; Uthes, S.; Sartohadi, J.; Helming, K. Assessing the impact of land use policy on urban-rural sustainability using the FoPIA approach in Yogyakarta, Indonesia. *Sustainability* **2010**, *2*, 1991–2009. [CrossRef]
23. König, H.; Sghaier, M.; Schuler, J.; Abdeladhim, M.; Helming, K.; Tonneau, J.P.; Ounalli, N.; Imbernon, J.M.; Wiggering, H. Participatory Impact assessment of soil and water conservation scenarios in Oum Zessar Watershed, Tunisia. *Environ. Manag.* **2012**, *50*, 153–165. [CrossRef] [PubMed]
24. König, H.J.; Zhen, L.; Helming, K.; Uthes, S.; Yang, L.; Cao, X.; Wiggering, H. Assessing the impact of the sloping land conversion programme on rural sustainability in Guyuan, Western China. *Land Degrad. Dev.* **2014**, *25*, 385–396. [CrossRef]
25. Trindade, S.P. Aptidão Agrícola, Mudanças de usos dos Solos, Conflitos e Impactos Diretos e Indiretos da Expansão da Cana de Açúcar na Região Sudoeste Goiano. Tese Doutorado Ufg/Programa de Pós Graduação em Ciências Ambientais. Available online: http://repositorio.bc.ufg.br/tede/handle/tede/5047 (accessed on 29 August 2017).
26. Lima, D.L. Estrutura e Expansão da Agroindústria Canavieira no Sudoeste Goiano: Impactos No Uso Do Solo e na Estrutura Fundiária a Partir de 1990. Ph.D. Thesis, IE/UNICAMP, Campinas, Brazil, 2010.
27. Pérez-Soba, M.; Petit, S.; Jones, L.; Bertrand, N.; Briquel, V.; Omodei-Zorini, L.; Contini, C.; Helming, K.; Farrington, J.H.; Mossello, M.T. Land use functions: a multifunctionality approach to assess the impact of land use changes on land use sustainability. In *Sustainability Impact Assessment of Land Use Changes*; Helming, K., Pérez-Soba, M., Tabbush, P., Eds.; Springer: Berlin, Germany, 2008; cap. 19; pp. 375–404.
28. Kuhlman, T. Scenarios: Driving forces and policies. In *Sustainability Impact Assessment of Land Use Changes*; Helming, K., Pérez-Soba, M., Tabbush, P., Eds.; Springer: Berlin, Germany, 2008; cap. 19; pp. 131–157.
29. Paracchini, M.L.; Pacini, C.; Jones, M.L.M.; Pérez-Soba, M. An aggregation framework to link indicators associated with multifunctional land use to the stakeholder evaluation of policy options. *Ecol. Indic.* **2011**, *11*, 71–80. [CrossRef]
30. OECD—Organisation for Economic Co-Operation and Development. *Core Set of Indicators for Environmental Performance Reviews*; OECD: Paris, France, 1993. Available online: http://enrin.grida.no/htmls/armenia/soe2000/eng/oecdind.pdf (accessed on 23 June 2017).
31. OECD—Organisation for Economic Co-Operation and Development. *Environmental Indicators: Development, Measurement and Use*; OECD: Paris, France, 2003. Available online: https://www.oecd.org/env/indicators-modelling-outlooks/24993546.pdf (accessed on 23 June 2017).
32. Ficarelli, T.R.d.A.; Ribeiro, H. Queimadas nos canaviais e perspectivas dos cortadores de cana-de-açúcar em macatuba, São Paulo. *Saúde Soc.* **2010**, *19*, 48–63.
33. Vieira, J.N.d.S. A Agroenergia e os Novos Desafios para a Política Agrícola no Brasil. In *O Futuro da Indústria: Biodiesel*; Série Política Industrial, Tecnológica e de Comércio Exterior; Ministério do Desenvolvimento, Indústria e Comércio Exterior-MDIC/Instituto Euvaldo Lodi-IEL/Núcleo Central: Brasília, Brazil, 2006; pp. 37–48.
34. Rosillo-Calle, F.; Cortez, L.A.B. Towards Proalcool II: A review of the Brazilian Bioethanol Programme. *Biomass Bioenergy* **1998**, *14*, 115–124. [CrossRef]
35. Brasília, D.F. Plano Nacional de Agroenergia 2006–2011/Ministério da Agricultura. In *Pecuária e Abastecimento, Secretaria de Produção e Agroenergia*, 2nd ed.; Embrapa Informação Tecnológica: Brasília, Brazil, 2006.
36. Goiás. Aprova o Regulamento do Programa de Desenvolvimento Industrial de Goiás—PRODUZIR (Decreto No 5.265). Available online: http://www.gabinetecivil.goias.gov.br/decretos/numerados/2000/decreto_5265.htm (accessed on 19 June 2017).
37. Pietrafesa, J.P.; Sauer, S.; Santos, A.E.A. Políticas e recursos públicos na expansão dos agrocombustíveis em Goiás: ocupação de novos espaços em áreas de Cerrado. In *Transformação do Cerrado: Progresso, Consumo e Natureza*; Pietrafesa, J.P., e Silva, S.D., Eds.; Editora da PUC Goiás: Goiânia, Brazil, 2011; pp. 93–121.

38. Silva, M.A.M.; Martins, R.C. *Produção de Etanol e Impactos Sobre os Recursos Hídricos*; IBASE/BNDES: Hong Kong, China; Brasília, Brazil, 2008; pp. 50–63.
39. Carvalho, S.P.; Marin, J.O.B. Agricultura familiar e agroindústria canavieira: Impasses sociais. *Rev. Econ. Sociol. Rural* **2011**, *49*, 3. [CrossRef]
40. Pasqualetto, A.; Zito, R.K. *Impactos Ambientais da Monocultura da Cana-de-Açúcar*; UFG: Goiânia, Brazil, 2000.
41. Gonçalves, D.B. Sob as cinzas dos canaviais: o perigoso impasse das queimadas no estado de São Paulo. *Inf. Econ.* **2005**, *35*, 32–44.
42. Severiano, E.C.; Oliveira, G.C.; Dias Júnior, M.S.; Costa, K.A.P.; Castro, M.B.; Magalhães, E.N. Potencial de descompactação de um Argissolo promovido pelo capim-tifton 85. *Rev. Bras. Eng. Agríc. Ambient.* **2010**, *14*, 39–45. [CrossRef]

3

Defining Crop–Climate Departure in West Africa: Improved Understanding of the Timing of Future Changes in Crop Suitability

Temitope S. Egbebiyi *, Olivier Crespo and Chris Lennard

Climate System Analysis Group (CSAG), Department of Environmental and Geographical Science, University of Cape Town, Private Bag X3, Rondebosch, Cape Town 7701, South Africa

* Correspondence: EGBTEM001@myuct.ac.za. or temitope@csag.uct.ac.za

Abstract: The future climate is projected to change rapidly with potentially severe consequences for global food security. This study aims to improve the understanding of future changes in the suitability of crop growth conditions. It proposes a definition of crop realization, of the climate departure from recent historical variability, or crop–climate departure. Four statistically downscaled and bias-corrected Global Climate Models (GCMs): CCCMA, CNRM5, NOAA-GFDL, and MIROC5 performed simulations for the period 1960–2100 under the Representative Concentration Pathway RCP8.5 scenario to compute 20 year moving averages at 5-year increments. These were used to drive a crop suitability model, Ecocrop, for eight different crops across the three Food and Agriculture Organizations (FAO) AgroEcological Zones (AEZs) of West Africa (Guinea, Sahel, and Savanna). Simulations using historical climate data found that all crops except maize had a suitability index value (SIV) ≥0.50 outside the Sahel region, equivalent to conditions being suitable or strongly suitable. Simulations of future climate reveal that warming is projected to constrain crop growth suitability for cassava and pineapple in the Guinea zone. A potential for the northward expansion of maize is projected by the end of the century, suggesting a future opportunity for its growth in the southern Sahel zone. Crop growth conditions for mango and pearl millet remain suitable across all three AEZs. In general, crops in the Savanna AEZ are the most sensitive to the projected changes in climate. The changes in the crop–climate relationship suggests a future constraint in crop suitability, which could be detrimental to future food security in West Africa. Further studies to explore associated short- and long-term adaptation options are recommended.

Keywords: climate-departure; crop–climate departure; crop suitability; Ecocrop; food security; West Africa

---

1. Introduction

The livelihood and economies of most Sub Saharan African (SSA) countries are driven by rainfed agriculture [1–3]. About 96% of agricultural lands in SSA are rainfed [1,4]. Agriculture employs over 65% of the active labour force of the region, the majority of whom are practicing subsistence rainfed farming [5]. The agricultural sector is also responsible for 75% of SSA domestic trade [6,7]. It adds significantly to the economy of the region by contributing up to 15–20% to the Gross Domestic Product (GDP) [1,4,8,9]. In 2000, about 80% of the cereals consumed in SSA were domestically produced locally [4]. However, West Africa has been identified as one of the most vulnerable regions of the world owing to its low adaptive capacity and a fast-growing population, with many citizens whom are faced with malnourishment [1,10–12]. An adverse change in the climate over West Africa, both spatially and temporally, coupled with inadequate institutional and economic capacity to cope or adapt to its impact could become a determinant threat to agricultural production. Thus, food security and socio-economic activities across the region may be affected [13–17].

Climate strongly affects rainfed agriculture with direct consequences on food security [18–20]. This has resulted in different studies focusing on the response of crops and agriculture to the impact of increased greenhouse gas emissions across different regions of SSA owing to malnutrition and the need to improve food security [1,11,20–22]. Extreme changes in climate are projected to increase [23], translating into increased occurrence of both droughts and floods, which already account for 70% of economic losses through soil erosion and drought in West Africa [24,25]. The fifth Intergovernmental Panel on Climate Change (IPCC) assessment reported a projected warming across the different seasons over SSA to be larger than the global annual mean temperature increase [23]. The projected warming (1.5–4 °C by 2100) is likely to affect the agricultural sector by a reduction of up to 50% in crop yield and 90% in revenue across the region by the end of the century [7,23]. However, this may be further aggravated in regions like West Africa where the climate is warming faster and may lead to a radical departure from the regions' historical variability [26].

The definition of departure varies across disciplines. Broadly, a departure refers to a deviation or variation from a norm, standard rule, or behaviour. It can also mean starting out on a new course of action. In climate science, climate departure can be defined as a shift in the climate pattern of a region outside the range of historical variability and may be described in terms of mean local temperature exceeding historical highs [27–29]. Mora et al. [29] described climate departure as the year in which the average temperature of the coldest year after 2005 was warmer than the historic hottest year at a given location. Here we have defined climate departure as a deviation from the historical mean and/or variance of the local climate of an area or region induced by global warming [30].

The projected global warming level and the timing over the continent may intensify the impact of climate change on crop suitability. Severe temperature fluctuations and other extreme weather conditions such as droughts and floods may also threaten crop suitability thresholds. They vary spatially, resulting in potential yield declines where crop growth conditions are currently suitable and possible yield increases in other areas [31,32]. Challinor et al. [33], for instance, projected a future decline in crop yield of up to 5% for every degree of warming above the historical level in Africa.

Given the current state of climate departure research and the direct impact of climate on crop production systems (particularly rainfed), we are interested in the climate change induced crop realizations when climate departs from historical variability, which we term crop–climate departure. This study explores and proposes the information value of a comprehensive definition of crop–climate departure as "a departure from historical crop suitability threshold, whether in terms of variability, mean or both, due to warming of the climate over a location both in space and time resulting from climate change whether of radical climatic nature or not". This is in the context of recent climate historical variability and future climate projections using three West African weather stations, within three Food and Agriculture Organizations (FAO) Agro-Ecological Zones (AEZs). Mora et al. [29] suggests that West Africa will experience a climate departure with a mean temperature about two decades (2029) earlier than the global mean temperature (2047). Thus, we use the region as our proof of concept and to examine any likely large-scale crop suitability consequent changes the region may already be experiencing. Section 2 describes the data and methods used. Results from the study are outlined in Section 3. The discussion of the results and concluding remarks and recommendations for future are in Sections 4 and 5, respectively.

2. Data and Methods

2.1. Study Area

The demonstration area for this work is West Africa (Figure 1), which has rainfed agriculture as its mainstay economy. It is located at latitude 4–20° N and 16° W–20° E. The region comprises of 15 countries namely Benin, Burkina Faso, Gambia, Ghana, Guinea Bissau, Guinea, Ivory Coast, Liberia, Mali, Mauritania, Niger, Nigeria, Senegal, Sierra Leone, and Togo. It is divided into three FAO AEZs: Guinea (4–8° N), Savanna (8–12° N), and the Sahel (12–20° N) [25,34]. The temperature increases to the north of the region, while precipitation increases towards the south [25,34,35]. The Sahel zone is the warmest and driest, while the Guinea zone is the coolest and wettest of the three AEZs in West Africa. The climate of the region is mainly controlled by the West African Monsoon (WAM) which accounts for about 70% of the annual rainfall [20,34]. The WAM is an important and dynamic characteristic of the West African climate during the summer period [36]. It is produced from the reversal of the land and ocean differential heating and dictates the seasonal pattern of rainfall over West Africa between latitudes 9° and 20° N. The WAM is characterized by winds that blow south–westerly during the warmer months (June–September) and north-easterly during the cooler months (January–March) of the year [25,36]. It is the major system that influences the onset, variability, and pattern of rainfall over West Africa [3,37]. This affects rainfall producing systems with an impact on rainfed agriculture, which influences crop growth suitability and consequently food production in the region.

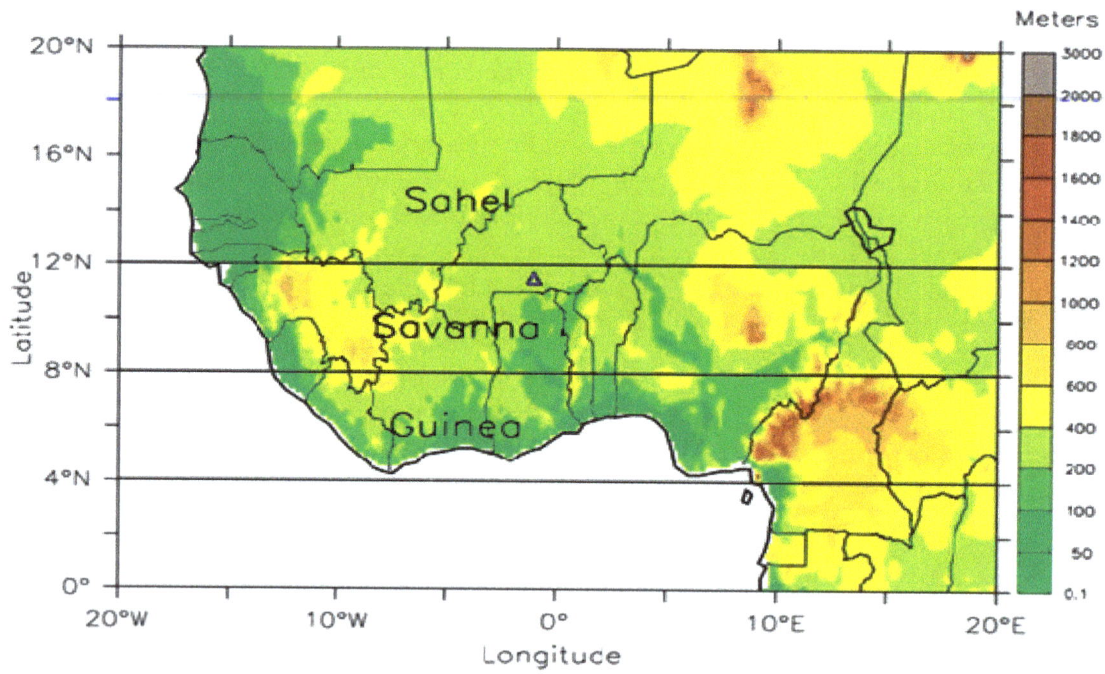

Figure 1. West African topography and the three Food and Agriculture Organizations (FAO)-Agro-ecological zones (AEZs): the Guinea, Savanna, and Sahel zones [25,34].

Different crops are grown in various parts of West Africa. Some of the major crops grown in the region are cassava, groundnut, millet, maize, sorghum, yam, plantain, cocoa, rice, and cowpea [20,38–40]. Millet and sorghum accounted for 64% of cereal production within the region in the year 2000, making them among the more important staple crops in West Africa [20,41]. Cassava is also an important staple food crop in terms of production in West Africa owing to its high resilience to drought [20,40,42]. This also applies to yam production, which accounts for about 91% of the world's production [20,41,43]. Maize provides about 20% of the calorie intake in West Africa and is adjudged the most important staple food overall in SSA [20,41]. Other crops such as cocoa and plantain, to mention a few, contribute significantly to the economy of the region.

2.2. Data

This study used three dataset types: observational weather station data, climate modelled data (statistically downscaled at the weather station level), and crop suitability data. The observed weather station data validated the mean monthly temperature and total monthly rainfall across the three AEZs. The crop suitability time series were simulated based on output from the Ecocrop suitability model [43] and the modelled climate data.

2.2.1. Climatic Variables

Temperatures (minimum and mean) and rainfall are important climate variables used in determining the impacts of climate change at subcontinental to global scales [44,45]. These two climate variables also have a significant effect on crop yield [46]. While rainfall affects the crop production in relation to its photosynthesis activities and leaf area, temperature affects the length of the crop growing season [47,48]. For this study, we used mean monthly minimum temperature (t-min) and mean monthly temperature (t-mean) and total monthly precipitation (prec.) of weather station data from Tabou, Ivory Coast; Sokode, Togo; Magaria, Niger. These weather stations each lie in three AEZs, Guinea, Savanna, and Sahel, respectively over West Africa. For the study, we used four statistically downscaled and bias corrected Global Climate Models (GCMs) in our analysis (CCCMA, CNRM5, GFDL, and MIROC) under a high-end climate change emission scenario (no adaptation), RCP8.5 (See Table 1 below for a description of the model). The GCMs were statistically downscaled using the Conditional Interpolation method as described in Hewitson and Crane [49]. The Conditional Interpolation downscaling method calculates the local phase relationships (PMI) for each weather station and each synoptic state combination. The bias relationship (BSI) between the weather station and its surroundings is then calculated. The method estimates the spatial extent of precipitation accurately and derives spatially referenced values representative of the area average. Overall, the interpolation conditioned by the synoptic state appears to better estimate realistic gridded values appropriate for use with model simulation output. For the temperature variable, the conditional interpolation employs the information content of the source data coupled with additional assumptions that may be physically justified (such as lapse-rate effects). The climate data were sourced from the Climate Information Portal (CIP) of the Climate System Analysis Group (CSAG), University of Cape Town (http://www.csag.uct.ac.za/climate-services/cip/). Data from this portal are at the station scale and weather stations in each AEZ are representative of that area.

Table 1. List of statistically downscaled and bias-corrected GCMs used in the study.

Modelling Institution	Institute ID	Model Name	Resolution
Canadian centre for climate modelling and analysis	CCCMA	CanESM2	2.8° × 2.8°
Centre National de Recherches Meteorolo-Giques/Centre Europeen de Recherche et Formation Avanceesencalcul scientifique	CNRMCERFACS	CNRM-CM5	1.4° × 1.4°
National Oceanic and Atmospheric Administration Geophysical Fluid Dynamic Laboratory	NOAAGDFL	GFDL_ESM2M	2.5° × 2.0°
Japan agency for Marine-Earth Science and Technology	MIROC	MIROC5	1.4° × 1.4°

2.2.2. Crop Thresholds to Suitability

The results of field experiments apply globally. A database of crop thresholds that translate into climate suitability has been collected and used in many locations to describe the suitability range of many plant and crop species using prec., t-min, t-mean, and the length of the growing season [43,50]. The climate threshold hosted by FAO dataset was obtained from the "dismo" package of the cran R software [51] (https://cran.r-project.org/web/packages/dismo/index.html). It was used in computing the climate suitability of each crop evaluated. It is acknowledged that thresholds will vary depending on finer resolution of the species (e.g., different varieties) or location (e.g., different soil, different rain distribution). However, the concept of crop suitability and the general validation of the thresholds makes this a useful tool to assess the impact of climate change and the emergence of novel regional climates on crop suitability over large areas examining the concept of crop–climate departure. The Ecocrop suitability model assessed four broad crop types and eight crops in total: cereals (pearl millet and maize); horticultural crops (tomato and pineapple); root and tuber crops (plantain and cassava) and fruit crops (mango and orange), using Ecocrop. The crop thresholds are listed in Table 2.

Table 2. Crop growth thresholds for eight crops as generated by the Ecocrop model.

Crop Name	Growing Duration (Days)	Temperature (°C)				Rainfall (mm)			
		Tmin	Topmin	Topmax	Tmax	Rmin	Ropmin	Ropmax	Rmax
Pearl millet	60–120	12	25	35	40	200	400	900	1700
Maize	65–365	10	18	33	47	400	600	1200	1800
Cassava	180–365	10	20	29	35	500	1000	1500	5000
Plantain	365	16	23	28	38	1000	1300	2000	5000
Pineapple	330–365	10	21	30	36	550	800	2500	3500
Tomato	70–150	7	20	27	35	400	600	1300	1800
Orange	180–365	13	20	38	38	450	1200	2000	2700
Mango	150–365	8	24	30	48	300	600	1500	2600

Where Tmin, Topmin, Topmax, and Tmax represents monthly minimum temperature, minimum optimum temperature, maximum optimum temperature, and maximum temperature, respectively; Rmin, Ropmin, Ropmax, and Rmax represents total monthly minimum rainfall, minimum optimum total monthly rainfall, maximum optimum total monthly rainfall, and maximum total monthly rainfall, respectively; optimum values represent the most suitable period for crop planting.

2.2.3. Model Description

The Ecocrop model is a crop suitability model. It uses a crop growth suitability threshold dataset hosted by the FAO [43]. It is an empirical model originally developed by Hijmans et al. [43] and based on the FAO-Ecocrop database [50] (Figure 2). The computation of optimal, suboptimal, and non-optimal conditions based on these datasets allows for the simulation of the suitability of crops in response to 12-month climate via t-min, t-mean, and prec. [43] (Figure 2). The Ecocrop model evaluates the relative suitability of crops in response to a range of climates including rainfall, temperature, and the growing season for optimal crop growth. A suitability index is generated as follows: 0 < 0.25 (not suitable), 0.25 < 0.5 (marginally suitable), 0.5 < 0.75 (suitable), and 0.75 < (highly suitable) [50,52]. The default Ecocrop parameters were assumed. Although those thresholds may vary with different geographical and/or climatic conditions, previous studies report a close correlation between the Ecocrop model and the climate change impact projections from other crop models [16,27,33,50]. A paucity of data over regions of interest like SSA limits the validation of these processes [53]. Nevertheless, the method contributes to the demand for regional scale assessment of crop response to future climate projections. The 12 coloured lines observed for the Ecocrop climate suitability simulations in Figure 2a,b below represents 12 months. Each describes the most suitable conditions for the crop under consideration in any given month. A highly seasonal crop (e.g., maize) has suitable growth conditions for a limited number of months. The conditions for non-seasonal crops are suitable throughout the year. Crops with a growing cycle longer than a year (e.g., pineapple and plantain) are represented by a single 12-month period.

2.3. Method

Four GCMs for the period 1960–2100 under the RCP8.5 scenario computed a 20-year moving average at 5 year time increments. It generated one mean 12-month value per 20-year window period for t-min, tmean, and prec. The mean 12-month climate value informed the crop suitability model, Ecocrop, for each GCM based on the methodologies described in Ramirez-Villegas, Jarvis, and Läderbach [50] for eight crops across the three AEZs of West Africa. The Ecocrop model simulated crop suitability indices characterized the crop–climate relationship and the impact global warming has on this relationship for each AEZ both spatially and temporally for each climate window. The suitability index scores were calculated for a range of climate variables for the period 1980–2000 using observed weather station data. This was used as a baseline to evaluate the downscaled GCM results spanning 1960–2100 at the three West African weather stations. It assessed the crop growth suitability in the zone for past climate conditions in reference to the published literature. Present day climate data was used as the preference for this zone owing to the constraint and paucity of weather station data and this data being the best available data to overlap with the Ecocrop model for the zone in the given study period.

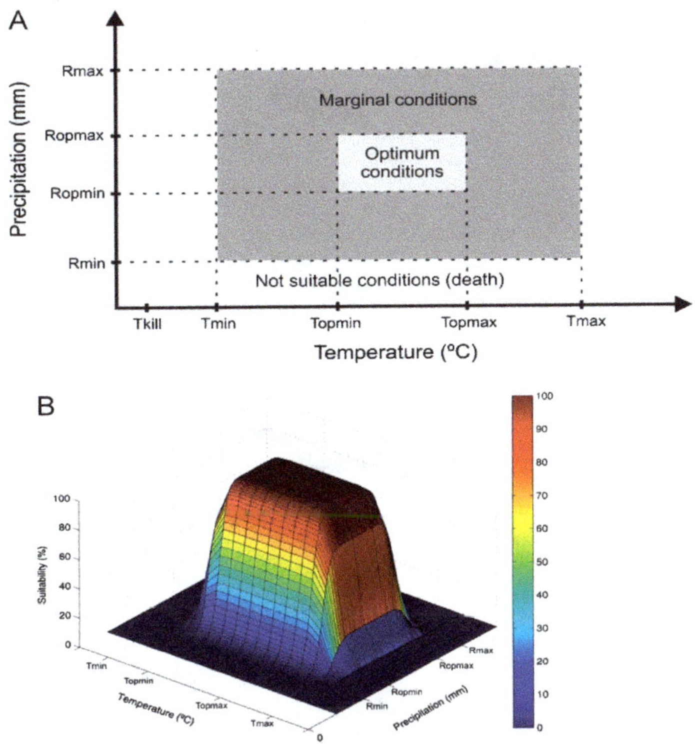

Figure 2. (**a**) two dimensional and (**b**) three dimensional diagram describing climate thresholds and its translation into crop suitability (Adapted from [50]).

3. Results

3.1. Evaluation of the GCMs in Simulating Rainfall and Temperature over West Africa

The downscaled climate data was firstly validated with the observed weather station data. Where missing records occurred, the corresponding month in the model's data were removed before computing the relationship between the datasets. Each GCM was correlated with the observed data for prec. and t-mean over the three weather stations, despite some discrepancy in precipitation over the Guinea zone (Figures 3 and 4). For temperature, the four models were correlated ($r \geq 0.6$) with the observed t-mean across the three AEZs of West Africa with the highest correlation ($r = 0.9$) over Magaria. The models were also correlated ($r \geq 0.6$) with the observed prec. in the Savanna and Sahel AEZs. A moderate correlation ($r \geq 0.3$) with observed weather station data was evident in the Guinea AEZ. This weak correlation may be due to the low resolution of the GCMs in capturing the total monthly rainfall in the Guinea zone. The validated GCM data was then input into the Ecocrop model.

3.2. GCMs Representation of AEZs, Seasons, and Suitability over West Africa AEZs

A correlation exists between the Ecocrop suitability model simulated with climate inputs from four GCMs, CCMA, CNRM, GFDL, and MIROC (hereafter Eco-GCMs), although with minor variations in amplitude and time. However, it is worth stating that the variation in simulated suitability by the four GCMs may be attributed to the inter-annual variability of the GCMs or the GCMs parametrization scheme. Nevertheless, Eco-GCMs simulated crop suitability is similar across the three AEZs over West Africa for the eight crops considered in the study. For example, cassava shows a similar suitability pattern across the AEZs (Figure 5). It is unsuitable (Eco-CCMA and CNRM) to marginally suitable (Eco-GFDL and MIROC) for cassava crop growth in the Sahel AEZ. In the Savanna AEZ, it is currently highly suitable for cassava, but this is predicted to decline in the future to become marginally unsuitable. The Guinea AEZ suitability for cassava does not change. The variability in crop growth suitability curves may be attributed to the variation in yield and production of cassava across the region due to the impact of climate change, corroborating previous studies [8,38].

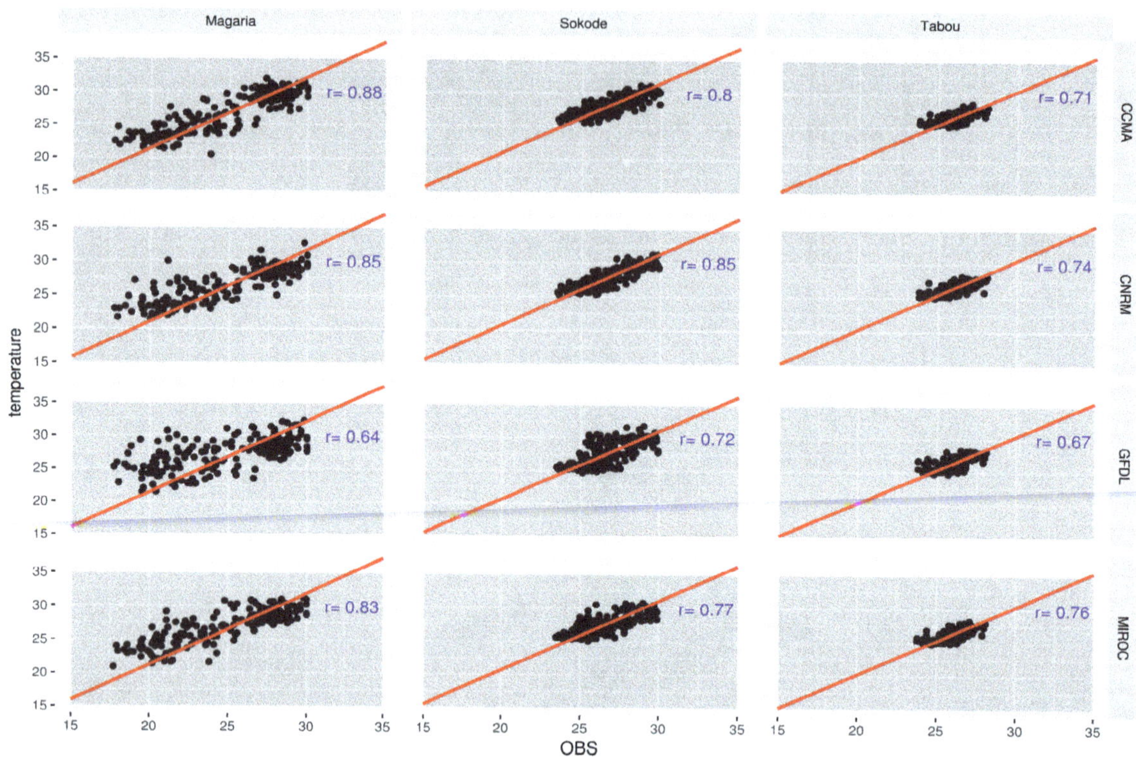

Figure 3. Mean monthly temperature (°C) as depicted by station observations and statistically downscaled CMIP5 GCMs (CCMA, CNRM5, GFDL, and MIROC) across the three agro ecological zones, Tabou, Sokode, and Magaria, of West Africa for the period 1980–2000. The top right corner r-values in each panel represent the correlations between the simulated and observed mean monthly temperature.

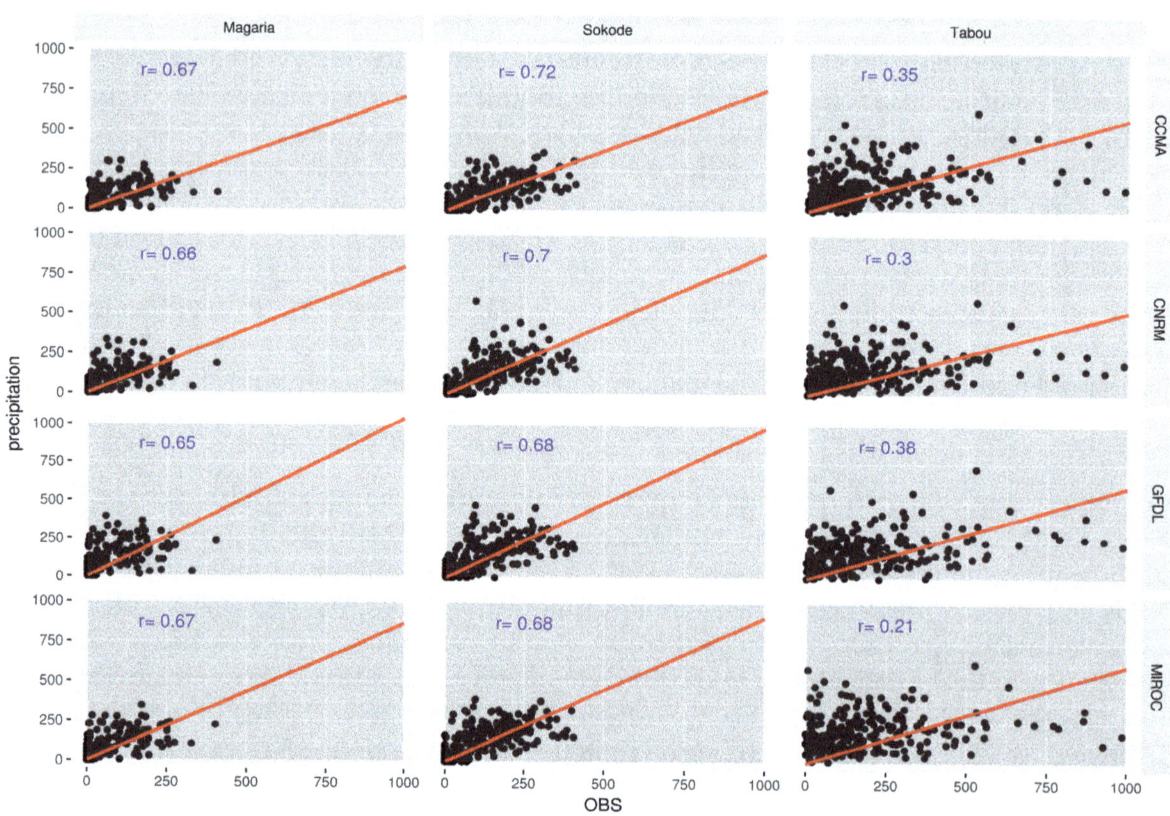

Figure 4. Total monthly rainfall (mm/month) as depicted by station observations and statistically downscaled CMIP5 GCMs (CCMA, CNRM5, GFDL, and MIROC) across the three agro ecological

zones, Tabou, Sokode, and Magaria, of West Africa for the period 1980–2000. The top right corner r-values in each panel represent the correlations between the simulated and observed mean monthly temperature.

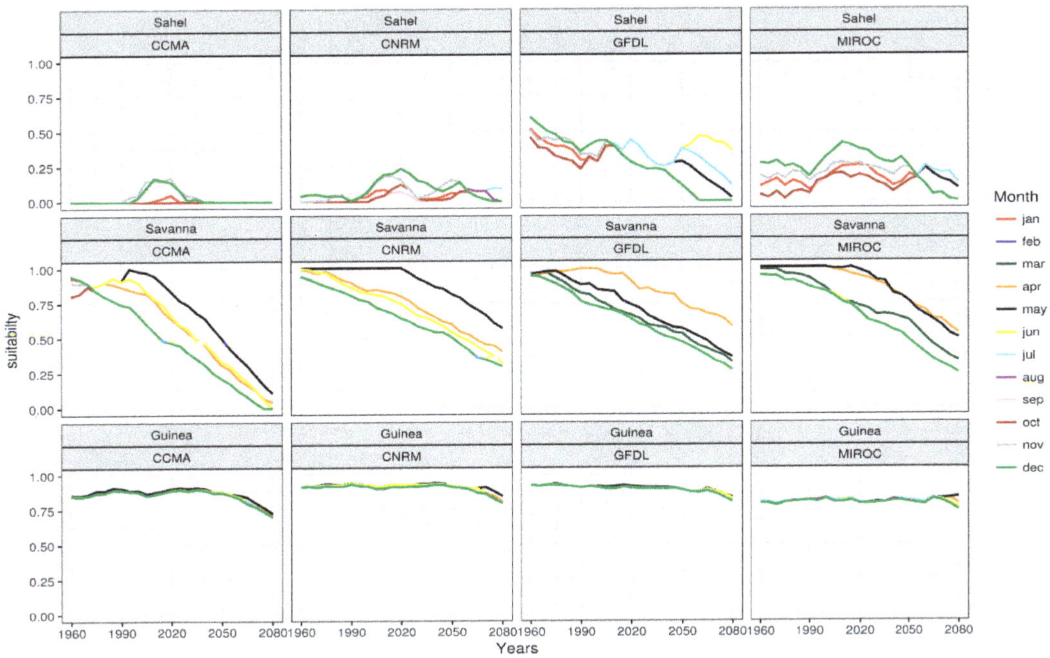

Figure 5. Cassava planting month suitability plots in the Guinea, Sahel, and Savanna as simulated by the four GCMs (CCMA, CNRM, GFDL, and MIROC) used as climate inputs into the Ecocrop suitability model.

Variability in the suitability of the month of planting for cassava crops in response to both AEZ and time increment is observed across the GCMs (Figure 5). The Guinea AEZ is currently the most suitable AEZ in which to grow cassava and is predicted to remain so. Suitable planting months in the Savanna AEZ as identified by all four models includes April–June and December, however a notable decline is observed, consistent for all four GCMs. Suitability declines from just below 1.0 to below 0.5 by 2050 in most cases. Conditions in the Sahel are presently and remain of low suitability. The simulation of the cassava crop growing season and period of planting across the three AEZs: Guinea (January–July and September–December), Savanna (April–November), and the Sahel (May/June–November) corroborates with previous findings with respect to the planting period and growing season in West Africa [54–56].

The concept of crop–climate departure allows for a consolidation of climate outputs from an ensemble of four GCMs into simulated crop suitability indices. Despite the marginal scale differences, the four GCMs consistently represent the unsuitability of cassava in the Sahel AEZ, its fast-declining suitability in the Savanna AEZ, as well as the high suitability of the Guinea AEZ. The cassava growing season is 12 months. Due to the predicted decline in the suitability of growth conditions, it is expected that it will become seasonal i.e., the suitability remains subject to appropriate seasonal planting, but conditions will then become unsuitable in the Savanna AEZ. Farmers will be required to adapt their practices in this AEZ. There is no reason at this stage to prefer one over the other GCMs, thus we use GCMs ensemble data as future climate scenario to simulate crop suitability in the subsequent sections of the paper plots of crop suitability from the Eco-GCMs. As seen from Figure 5, the ensemble suitability plots give a good representation of the Eco-GCMs model simulated suitability across the AEZs over West Africa. It shows the non-suitability of cassava in the Sahel, the fast-declining suitability and the observed seasonality in the Savanna AEZ, and high suitability in the Guinea zone. Thus, the crops ensemble suitability simulations are used in the results and discussion in the subsequent sections of this paper. A summary of crop suitability index values are given in Table 3.

Table 3. Ecocrop simulated crop Suitability Index Value (SIV) for the eight (8) different crops across the three Agro-ecological zones (AEZs) of West Africa.

Years	**1960**			**1990**			**2020**			**2050**			**2080**		
Crops/AEZs	GUI	SAV	SAH	GUI	SAV	SAH	GUI	SAV	SAH	GUI	SAV	SAH	GUI	SAV	SAH
Cassava	>0.75	>0.75	<0.25	>0.75	>0.75	<0.25	>0.75	>0.50	<0.25	>0.75	≤0.50	<0.25	>0.75	<0.50	<0.25
Maize	≤0.50	≥0.50	<0.75	≤0.50	≥0.50	0.50–0.75	<0.50	>0.50	>0.50	<0.50	>0.50	>0.50	<0.50	>0.75	>0.75
Mango	0.50–0.75	1.00	≤0.75	0.50–0.75	1.00	≤0.75	0.50–0.75	>0.75	>0.75	0.50–0.75	>0.75	0.75	0.50–0.75	>0.75	0.75
Orange	>0.75	>0.75	<0.25	>0.75	>0.75	<0.25	>0.75	>0.75	>0.25	>0.75	0.50–0.75	<0.25	>0.75	0.50–0.75	<0.25
Pearl millet	>0.75	>0.50	>0.50	>0.75	>0.50	>0.50	>0.75	>0.50	>0.50	>0.75	>0.50	>0.50	>0.75	>0.50	>0.50
Pineapple	>0.75	>0.75	<0.5	>0.75	<0.50	<0.50	>0.75	>0.75	0.50	>0.75	0.50–0.75	<0.50	>0.75	<0.50	<0.50
Plantain	1.00	>0.75	0.00	1.00	>0.75	0.00	>0.75	<0.75	0.00	>0.75	<0.75	0.00	>0.75	<0.50	0.00
Tomato	>0.75	>0.25	<0.5	>0.75	>0.50	<0.5	>0.75	>0.50	<0.5	<0.75	<0.50	<0.25	<0.75	<0.50	<0.25

GUI—Guinea AEZ, SAV—Savanna AEZ, SAH—Sahel AEZ. Ecocrop suitability Index: 0.0–0.25—Unsuitable/No suitability, 0.25–0.50—Marginally suitable, 0.50–0.75—Suitable, 0.75–1.00—Highly suitable.

3.3. Crop Suitability Response with Past Climate

Past climatic conditions (1960–2010) indicate that a crop growth suitability gradient existed from south to north across the three AEZs for each crop type considered. A Suitability Index Value (SIV) (0.75–1.00) is observed for each crop type in the Guinea and Savanna zones throughout the year (Figures 6 and 7). An exception is the cereal crop maize. Maize is marginally suitable (0.25–0.50) in the Guinea AEZ, suitable (0.50–0.75) for planting in May, September, and December in the Savanna AEZ, and January–February in the Sahel AEZ. The suitability increases (0.75–1.00) in 2050 (Figure 6). In the Sahel AEZ, other cereals and mango are suitable (above 0.50). Crop growth suitability increased for pineapple crops in this AEZ.

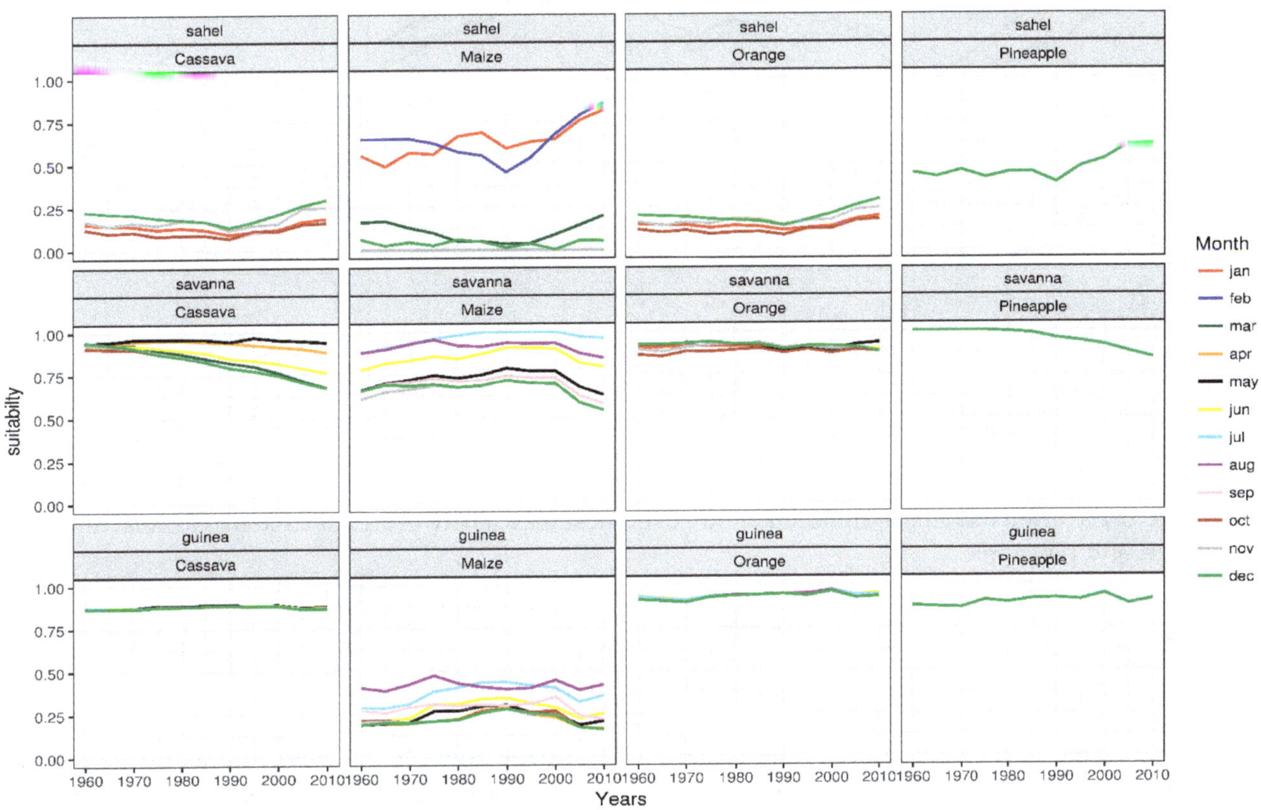

Figure 6. Ensemble crop suitability plots in the past climate (1960–2010) for cassava, maize, orange, and pineapple.

The Ecocrop simulations of crop growth suitability for the period 1960–2010 and the crop types evaluated corroborate previous findings with respect to the type of crops actually grown in the region. Cereals and root and tuber crops were the principal agricultural commodities in the region during this period [57,58]. Cassava and maize crops demonstrated modest yield increases of 6.3 to 10.3 and 1.1 to 1.8 tons ha^{-1}, respectively in the last 40 years [58,59]. The historical suitability of growth conditions for maize across the region (although marginal in the Guinea zone) highlights its importance as a staple crop here, accounting for almost 20% of the calorie intake for the population of West Africa [20,41]. The large area grown and high yield of pearl millet between 1960 and 2010 can be linked to the high suitability indices across the AEZs of West Africa over this period, contributing considerably to the livelihoods and economies of the countries in this region [60,61]. Increased productivity has also been witnessed in crops such as orange, mango, pineapple, and tomatoes in the last 40 years, again correlating with the high crop growth suitability indices identified for these crops [62,63]. Given the importance of these crops in the regional economy, a key question is, how is the projected change in climate predicted to impact on the crop growth suitability of these key crops in West Africa?

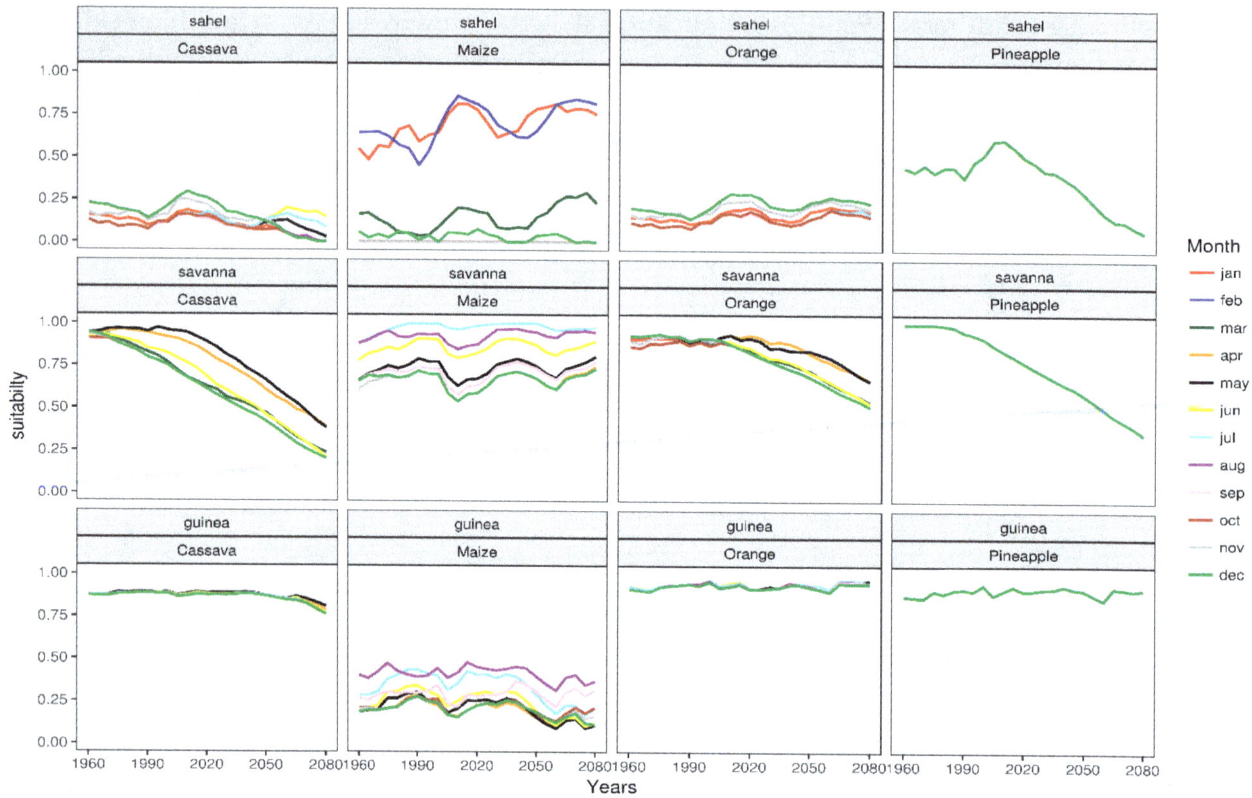

Figure 7. Projected model ensemble suitability over West Africa between 1960–2100 for cassava, maize, orange, and pineapple.

3.4. Projected Changes in Crop Suitability and Time of Planting over West Africa

The projected increase in global temperature is predicted to have a varied impact on crop growth suitability in West Africa (Figures 8 and 9). The Guinea AEZ remains largely unchanged with respect to crop growth suitability, evidently a more resilient area. Drastic declines are predicted for multiple crops in the Savanna AEZ including for cassava, orange, and pineapple. The main staple crop, maize, remains stable with an SIV of 0.5–1.0. It is interesting to note that the SIV for maize in the Sahel AEZ is projected to increase, shifting from suitable in 2020 to highly suitable by 2050 (Figure 7). Conditions for pearl millet will remain highly suitable (Figure 9), although the SIV for mango will decline post 2020.

The impact of future warming will affect crop seasonality, i.e the suitability of the time of planting. For root and tuber crops and cassava, in the months of April and May, they will become marginally suitable for cultivation by mid-century in the Savanna AEZ. Conditions will be unsuitable if planted in March, June, or December, which are currently optimal seasons. No change is predicted for cereal crops mango or orange.

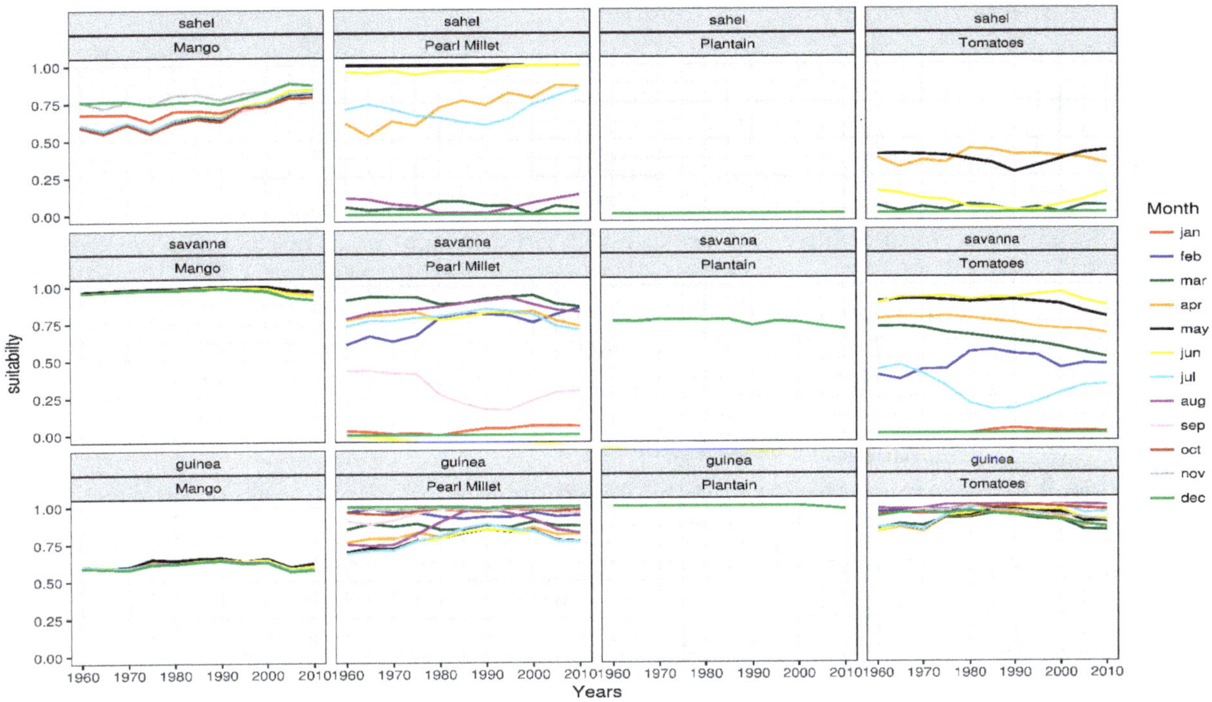

Figure 8. Ensemble crop suitability plots in the past climate (1960–2010) for mango, pearl millet, plantain, and tomatoes.

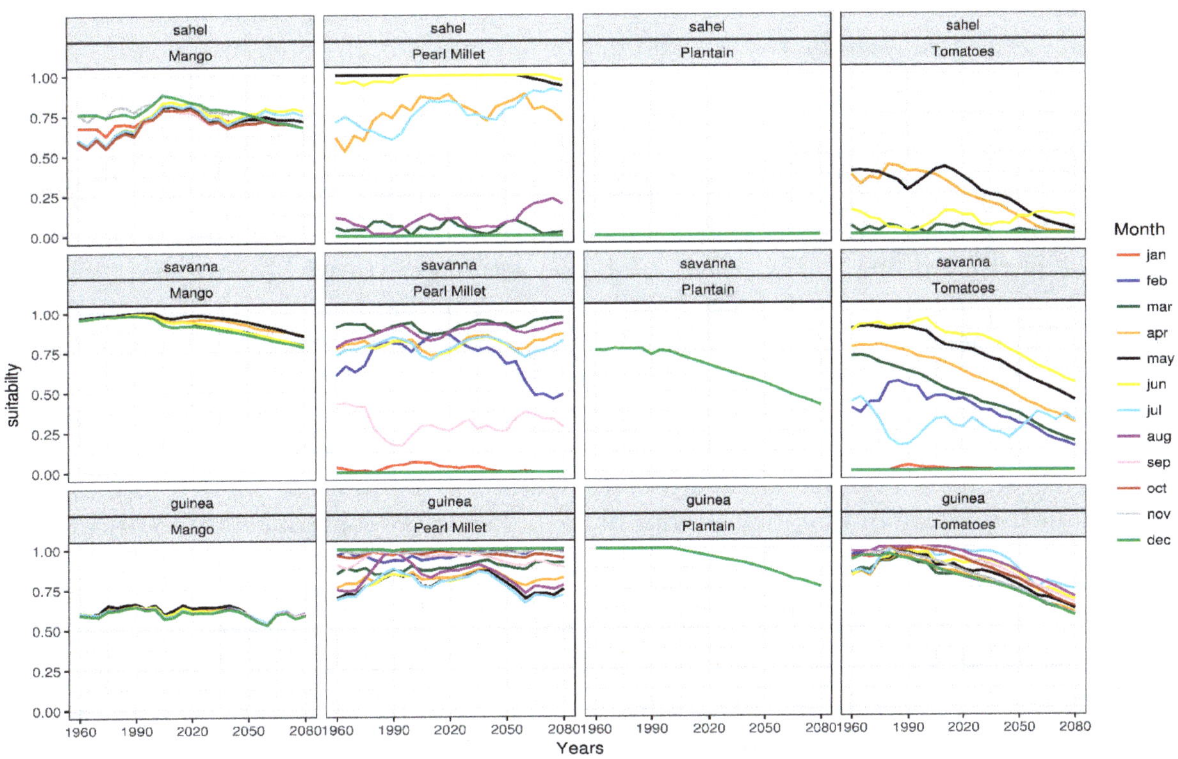

Figure 9. Projected model ensemble suitability over West Africa between 1960–2100 for mango, pearl millet, plantain, and tomatoes.

4. Discussion

4.1. From Climate Departure to Crop–climate Departure

The impact of global warming on the crop–climate relationship in the three AEZs of West Africa varies depending on the crop grown. The combination of a changing climate and crop growth suitability thresholds results in a projected deviation for the SIV from historical data. This may be a predicted increase in SIV, as observed for maize, or a decline as noted for crops such as cassava. A further variable is the AEZ itself. The predicted increase in SIV for maize in the Sahel zone by 2100 results from the projected increase in temperature and precipitation in this location. The decline in SIV forecast for cassava and pineapple in the Savanna AEZ decreases the suitability of the SSA region for these crops, spatially constraining suitable areas to the Guinea AEZ by 2100. The warming climate also influences the suitability of the month of planting of many crops. A potential cause for concern is the timing of crop–climate departure in the Savanna AEZ, which is already evident in root and tuber, cassava, and pineapple crops, and projected for orange crops by 2100. The diversity of crops grown in this AEZ may be diminished in the future.

The projected shifts in suitability and the variation of suitability between different crops and AEZ due to global warming highlights the importance of local climatic conditions in determining the extent of crop growth, development, and yield in response to climate suitability thresholds [31,64]. The above characteristics observed from the crop–climate relationship of different crops in the three AEZs of West Africa supports our proposed crop–climate departure definition of "a departure/shift from a historical crop suitability threshold, whether in terms of variability, mean or both, resulting from climate change whether of radical climatic nature or not due to warming of the climate over a location both in space and time". This definition can be used to inform adaptation responses to impacts resulting from climate change that will influence crop suitability, especially in a vulnerable region with low adaptive capacity like West Africa. Cultivating crops with high suitability as projected by the models such as cassava in the Guinea and maize in the southern Sahel may be the best solution. However, with improved/hybrid seedlings of the crop projected to decline that can withstand the variability in climate, this improved suitability may be considered another option.

4.2. Crop–Climate Departure and the Spatio-Temporal Variability of Crop-Suitability in West Africa

The temporal and spatial targeting of adaptation measures under an increasing global temperature will be crucial in maintaining and improving food security in the future. Identifying the timing and location of changes in crop growth suitability due to climate change can play a potentially key role in addressing the challenge of food production [16,33]. This is particularly relevant for the crops of vital importance in the West African region assessed here. The projected decrease in growth suitability conditions for cassava crops in the Savanna AEZ between 2020 and 2050, for example, depends on the time of planting or season. This projected departure is critical for the Savanna region as it may impact negatively on both the economy and livelihoods within the region. The improved understanding of crop–climate departure timing may permit timely adaptation plans such as modification to crop management regimes that account for this change in crop seasonality. If this were to be included in combination with increased use of key varietal traits, e.g., drought resistance, it will greatly assist in improving the adaptive capacity of such crops and mitigating the future impacts of a warmer climate to increase the resilience of current cropping systems within the region [16,65]. Improvements to the underlying knowledge base can therefore potentially improve both crop yield and crop quality.

In AEZs where the continued growing of given crop types is no longer possible, a further adaptation strategy is a shift to other more resilient crop species [66], or the substitution of existing crops with crops not previously grown within a given AEZ. Maize, for example, is projected to increase in yield by up to 7% in comparison to non-adapted crops under future climate change scenarios in SSA [33]. An increase in the planting area of crops such as maize further northwards into the southern Sahel AEZ due to the projected rainfall increase in this AEZ [31], or a shift of cassava or pineapple cropping into the Guinea AEZ due to reduced rainfall and the crops ability to withstand drought [16,31,65,67] represent further opportunities. A spatio-temporal projection of potential crop growth suitability can help provide information on future opportunities and constraints that will arise from shifts in the location of suitable crop lands within each AEZ [65]. Adaptation measures may then be prioritized for individual countries in response to the predicted changes [31,32], resulting in the maximum utilization of suitable areas for specific crop types, which will greatly assist in mitigating the future impacts from a warmer climate.

5. Summary and Conclusion

In order to improve the understanding of climatic impacts on agriculture, we conceptualized and explored the notion of crop–climate departure from historical variability in West Africa. We used four downscaled CMIP5 GCMs (CCMA CNRM5, GFDL, and MIROC) for the period 1960–2100 under RCP8.5 emission scenario and a crop suitability model, Ecocrop, across three weather stations representative of the Guinea, Savanna, and Sahel AEZs. In summary, all four GCMs correlate with observed weather station data in their simulation of monthly mean temperature and total monthly rainfall in the Savanna and Sahel zones, but moderately over the Guinea zone. It is recommended that future simulations acquire data from additional weather stations and utilize additional CMIP5 GCMs such as CSIRO, ICHEC, HADGEM, IPSL, MPI etc. In terms of crop rotations, the current climate is suitable for maize in the Savanna and Sahel AEZs, while future projections predict a potential for the expansion of maize further into the Sahel zone. The Guinea zone remains less suitable for maize but provides the correct climate both currently and in the future for crops such as cassava and pineapple. The predicted range for pearl millet and mango will remain stable in all three AEZs. Importantly, the Savanna AEZ, given its current cropping regime, is the most sensitive to climate change and shows the least resilience of the three AEZs considered. Climate change adaptation strategies will require prioritization in this zone. The climate-departure concept has been used to characterize crop–climate relationships i.e., crop–climate departure with increased warming and how it can, with appropriately planned adaptation and mitigation strategies, increase food security in the future.

Author Contributions: T.S.E. was responsible for developing the initial content of the manuscript, including literature search and data analysis. O.C. and C.L. were the supervisors for the research and provided guidance in terms of the article structure, data analysis, and finalization of the manuscript.

Acknowledgments: This study was supported with bursaries from the National Research Foundation (NRF, South Africa), Alliance Centre for Climate and Earth Systems Science (ACCESS, South Africa), and the JW Jagger Centenary Scholarship and Sari Johnson scholarship from the Postgraduate Funding Office, University of Cape Town, South Africa. We also acknowledge the anonymous reviewers and editor for their constructive comments which helps improve the quality of the study.

Abbreviations

ACCESS	Alliance for Collaboration for Climate and Earth System Sciences
AEZs	AgroEcological Zones
CCCMA	Canadian Centre for Climate Modelling and Analysis
CIP	Climate Information Portal
CMIP5	Coupled Model Intercomparison Project (Phase 5)
CNRM5	Centre National de Recherches Meteorolo-Giques
CSAG	Climate System Analysis Group
FAO	Food and Agriculture Organisation
GCMs	Global Climate Models
GFDL	Geophysical Fluid Dynamic Laboratory
MIROC	Japan agency for Marine-Earth Science and Technology Model
NRF	National Research Foundation
SSA	sub Saharan Africa

Declarations

Ethics approval and consent to participate	Not applicable
Consent for publication	Not applicable
Availability of data and materials	Not applicable

References

1. Roudier, P.; Sultan, B.; Quirion, P.; Berg, A. The impact of future climate change on West African crop yields: What does the recent literature say? *Glob. Environ. Chang.* **2011**, *21*, 1073–1083. [CrossRef]
2. Diasso, U.; Abiodun, B.J. Drought modes in West Africa and how well CORDEX RCMs simulate them. *Theor. Appl. Climatol.* **2017**, *128*, 223–240. [CrossRef]
3. Omotosho, J.B.; Abiodun, B.J. A numerical study of moisture build-up and rainfall over West Africa. *Meteorol. Appl.* **2007**, *14*, 209–225. [CrossRef]
4. World Bank. Making Development Climate Resilient. A World Bank Strategy for Sub-Saharan Africa. Report No. 46947-AFR. 2009, p. 144. Available online: http://siteresources.worldbank.org/INTAFRICA/Resources/ClimateChange-StrategyReport2010-Full_vNoImages.pdf (accessed on 21 August 2019).
5. Blein, R.; Soulé, B.G.; Dupaigre, B.F.; Yérima, B. *Agricultural Potential of West Africa*. Available online: http://www.fondation-farm.org/IMG/pdf/potentialites_rapport_ang_mp.pdf (accessed on 21 August 2019).
6. McCarthy, J.; Canziani, O.; Leary, N.; Dokken, D.; White, C. Climate Change 2001: Impacts, Adaptation, and Vulnerability. Available online: https://library.harvard.edu/collections/ipcc/docs/27_WGIITAR_FINAL.pdf (accessed on 21 August 2019).
7. World Bank. *World Development Report 2013: Jobs*; License: Creative Commons Attribution CC BY 3.0; World Bank: Washington, DC, USA, 2012. [CrossRef]
8. Benhin, J.K.A. South African crop farming and climate change: An economic assessment of impacts. *Glob. Environ. Chang.* **2008**, *18*, 666–678. [CrossRef]
9. Schlenker, W.; Lobell, D.B. Robust negative impacts of climate change on African agriculture. *Environ. Res. Lett.* **2010**, *5*, 014010. [CrossRef]
10. Slingo, J.M.; Challinor, A.J.; Hoskins, B.J.; Wheeler, T.R. Introduction: Food crops in a changing climate. *Philos. Trans. R. Soc. B Biol. Sci.* **2005**, *360*, 1983–1989. [CrossRef] [PubMed]
11. Knox, J.; Hess, T.; Daccache, A.; Wheeler, T. Climate change impacts on crop productivity in Africa and South Asia. *Environ. Res. Lett.* **2012**, *7*, 034032. [CrossRef]
12. Abatan, A.A. West African Extreme Daily Precipitation in Observations and Stretched-grid Simulations by CAM-EULAG. Available online: https://www.researchgate.net/publication/241809235_West_African_extreme_daily_precipitation_in_observations_and_stretched-grid_simulations_by_CAM-EULAG (accessed on 21 August 2019).
13. Grolle, J. Heavy rainfall, famine, and cultural response in the West African Sahel: The 'Muda' of 1953–54. *GeoJournal* **2014**, *43*, 205–214. [CrossRef]
14. Tarhule, A.; Woo, M.K. Towards an interpretation of historical droughts in northern Nigeria. *Clim. Chang.* **1997**, *37*, 601–616. [CrossRef]
15. Challinor, A.; Wheeler, T.; Garforth, C.; Craufurd, P.; Kassam, A. Assessing the vulnerability of food crop systems in Africa to climate change. *Clim. Chang.* **2007**, *83*, 381–399. [CrossRef]
16. Rippke, U.; Ramirez-Villegas, J.; Jarvis, A.; Vermeulen, S.J.; Parker, L.; Mer, F.; Diekkrüger, B.; Challinor, A.J.; Howden, M. Timescales of transformational climate change adaptation in sub-Saharan African agriculture. *Nat. Clim. Chang.* **2016**, *6*, 605–609. [CrossRef]
17. Paeth, H.; Hall, N.M.; Gaertner, M.A.; Alonso, M.D.; Moumouni, S.; Polcher, J.; Ruti, P.M.; Fink, A.H.; Gosset, M.; Lebel, T.; et al. Progress in regional downscaling of west African precipitation. *Atmos. Sci. Lett.* **2011**, *12*, 75–82. [CrossRef]
18. Hansen, J.; Sato, M.; Ruedy, R. Perception of climate change. *Proc. Natl. Acad. Sci. USA* **2012**, *109*, E2415–E2423. [CrossRef] [PubMed]
19. Pretty, J.N.; Morison, J.I.L.; Hine, R.E. Reducing food poverty by increasing agricultural sustainability in developing countries. *Agric. Ecosyst. Environ.* **2003**, *95*, 217–234. [CrossRef]
20. Sultan, B.; Gaetani, M. Agriculture in West Africa in the Twenty-First Century: Climate Change and Impacts

Scenarios, and Potential for Adaptation. *Front. Plant Sci.* **2016**, *7*, 1–20. [CrossRef] [PubMed]

21. Lobell, D.B.; Burke, M.B.; Tebaldi, C.; Mastrandrea, M.D.; Falcon, W.P.; Naylor, R.L. Prioritizing Climate Change Adaptation Needs for Food Security in 2030 Region. *Science* **2008**, *319*, 607–610. [CrossRef]
22. Zinyengere, N.; Crespo, O.; Hachigonta, S. Crop response to climate change in southern Africa: A comprehensive review. *Glob. Planet. Chang.* **2013**, *111*, 118–126. [CrossRef]
23. IPCC. Summary for Policymakers. In *Climate Change 2013: The Physical Science Basis. Contribution of Working Group I to the Fifth Assessment Report of the Intergovernmental Panel on Climate Change*; Stocker, T.F., Qin, D., Plattner, G.K., Tignor, M., Allen, S.K., Boschung, J., Nauels, A., Xia, Y., Bex, V., Midgley, P.M., Eds.; Cambridge University Press: Cambridge, UK, 2013.
24. World Bank. *World Development Report 2010: Development and Climate Change*; World Bank: Washington, DC, USA, 2010; Available online: https://openknowledge.worldbank.org/handle/10986/4387 (accessed on 30 November 2018).
25. Egbebiyi, T.S. Future Changes in Extreme Rainfall Events and African Easterly Waves over West Africa. Master's Thesis, University of Cape Town, Cape Town, South Africa, 2016.
26. Sui, Y.; Lang, X.; Jiang, D. Time of emergence of climate signals over China under the RCP4.5 scenario. *Nat. Clim. Chang.* **2013**, *39*, 1–5.
27. Vermeulen, S.J.; Challinor, A.J.; Thornton, P.K.; Campbell, B.M.; Eriyagama, N.; Vervoort, J.M.; Kinyangi, J.; Jarvis, A.; Läderach, P.; Ramirez-Villegas, J.; et al. Addressing uncertainty in adaptation planning for agriculture. *Proc. Natl. Acad. Sci. USA* **2013**, *110*, 8357–8362. [CrossRef]
28. Hawkins, E.; Sutton, R. Time of emergence of climate signals. *Geophys. Res. Lett.* **2012**, *39*, 1–6. [CrossRef]
29. Mora, C.; Frazier, A.G.; Longman, R.J.; Dacks, R.S.; Walton, M.M.; Tong, E.J.; Sanchez, J.J.; Kaiser, L.R.; Stender, Y.O.; Anderson, J.M. The projected timing of climate departure from recent variability. *Nature* **2013**, *502*, 183–187. [CrossRef]
30. Thornes, J.E. IPCC, 2001: Climate change 2001: Impacts, adaptation and vulnerability, Contribution of Working Group II to the Third Assessment Report of the Intergovernmental Panel on Climate Change, edited by J. J. McCarthy, O.F.; Canziani, N.A.; Leary, D.J., Dokken, A. *Int. J. Climatol.* **2002**, *22*, 1285–1286. [CrossRef]
31. Porter, J.R.; Semenov, M.A. Crop responses to climatic variation. *Philos. Trans. R. Soc. B Biol. Sci.* **2005**, *360*, 2021–2035. [CrossRef]
32. Lobell, D.B.; Schlenker, W.; Costa-Roberts, J. Climate trends and global crop production since 1980—Supporting Online Material. *Science* **2011**, *333*, 616–620. [CrossRef]
33. Challinor, A.J.; Watson, J.; Lobell, D.B.; Howden, S.M.; Smith, D.R.; Chhetri, N. A meta-analysis of crop yield under climate change and adaptation. *Nat. Clim. Chang.* **2014**, *4*, 287–291. [CrossRef]
34. Abiodun, B.J.; Adeyewa, Z.D.; Oguntunde, P.G.; Salami, A.T.; Ajayi, V.O. Modeling the impacts of reforestation on future climate in West Africa. *Theor. Appl. Climatol.* **2012**, *110*, 77–96. [CrossRef]
35. Klutse, N.A.; Ajayi, V.O.; Gbobaniyi, E.O.; Egbebiyi, T.S.; Kouadio, K.; Nkrumah, F.; Quagraine, K.A.; Olusegun, C.; Diasso, U.; Abiodun, B.J. Potential impact of 1.5 °C and 2 °C global warming on consecutive dry and wet days over West Africa. *Environ. Res. Lett.* **2018**, *13*, 055013. [CrossRef]
36. Janicot, S.; Caniaux, G.; Chauvin, F.; De Coëtlogon, G.; Fontaine, B.; Hall, N.; Kiladis, G.; Lafore, J.P.; Lavaysse, C.; Lavender, S.; et al. Intraseasonal variability of the West African monsoon. *Atmos. Sci. Lett.* **2011**, *12*, 58–66. [CrossRef]
37. Nicholson, S.E. The West African Sahel: A Review of Recent Studies on the Rainfall Regime and Its Interannual Variability. *ISRN Meteorol.* **2013**, *2013*, 1–32. [CrossRef]
38. Paeth, H.; Capo-Chichi, A.; Endlicher, W. Climate Change and Food Security in Tropical West Africa—A Dynamic-Statistical Modelling Approach. *Erdkunde* **2008**, *2*, 101–115. [CrossRef]
39. Nelson, G.C.; Van Der Mensbrugghe, D.; Ahammad, H.; Blanc, E.; Calvin, K.; Hasegawa, T.; Havlik, P.; Heyhoe, E.; Kyle, P.; Lotze-Campen, H.; et al. Agriculture and climate change in global scenarios: Why don't the models agree. *Agric. Econ.* **2014**, *45*, 85–101. [CrossRef]
40. Jarvis, A.; Ramirez-Villegas, J.; Campo, B.V.H.; Navarro-Racines, C. Is Cassava the Answer to African Climate Change Adaptation? *Trop. Plant Biol.* **2012**, *5*, 9–29. [CrossRef]
41. FAOSTAT. *FAO Statistical Yearbook 2014*. 2014. Available online: http://www.fao.org/3/a-i3590e.pdf (accessed on 21 August 2019).
42. Srivastava, A.K.; Gaiser, T.; Ewert, F. Climate change impact and potential adaptation strategies under alternate climate scenarios for yam production in the sub-humid savannah zone of West Africa. *Mitig. Adapt. Strateg. Glob. Chang.* **2016**, *21*, 955–968. [CrossRef]

43. Hijmans, R.J.; Guarino, L.; Cruz, M.; Rojas, E. Computer tools for spatial analysis of plant genetic resources data: 1. DIVA-GIS. *Plant Genet. Resour. Newsletter* **2001**, *127*, 15–19.
44. Cong, R.G.; Brady, M. The interdependence between rainfall and temperature: Copula analyses. *Sci. World J.* **2012**, *12*. [CrossRef]
45. IPCC. IPCC Expert Meeting on Climate Change, Food, and Agriculture. Available online: https://www.ipcc.ch/site/assets/uploads/2018/05/Food-EM_MeetingReport_FINAL-1.pdf (accessed on 21 August 2019).
46. Medori, M.; Michelini, L.; Nogues, I.; Loreto, F.; Calfapietra, C. The impact of root temperature on photosynthesis and isoprene emission in three different plant species. *Sci. World J.* **2012**, *2012*. [CrossRef]
47. Olesen, J.E.; Bindi, M. Consequences of climate change for European agricultural productivity, land use and policy. *Eur. J. Agron.* **2002**, *16*, 239–262. [CrossRef]
48. Cantelaube, P.; Terres, J.M. Seasonal weather forecasts for crop yield modelling in Europe. *Tellus Ser. A Dyn. Meteorol. Oceanogr.* **2005**, *57*, 476–487. [CrossRef]
49. Hewitson, B.C.; Crane, R.G. Gridded area-averaged daily precipitation via conditional interpolation. *J. Clim.* **2005**, *18*, 41–57. [CrossRef]
50. Ramirez-Villegas, J.; Jarvis, A.; Läderach, P. Empirical approaches for assessing impacts of climate change on agriculture: The EcoCrop model and a case study with grain sorghum. *Agric. For. Meteorol.* **2013**, *170*, 67–78. [CrossRef]
51. Hijmans, A.R.J.; Phillips, S.; Leathwick, J.; Elith, J.; Hijmans, M.R.J.; Hijmans, R.J. Species Distribution Modeling. Available online: https://cran.r-project.org/web/packages/dismo/dismo.pdf (accessed on 21 August 2019).
52. Hunter, R.; Crespo, O. *Large Scale Crop Suitability Assessment Under Future Climate Using the Ecocrop Model: The Case of Six Provinces in Angola's Planalto Region*; Springer: Cham, Switzerland, 2018.
53. White, J.W.; Hoogenboom, G.; Kimball, B.A.; Wall, G.W. Methodologies for simulating impacts of climate change on crop production. *F. Crop. Res.* **2011**, *124*, 357–368. [CrossRef]
54. Vrieling, A.; de Leeuw, J.; Said, M.Y. Length of growing period over africa: Variability and trends from 30 years of NDVI time series. *Remote Sens.* **2013**, *5*, 982–1000. [CrossRef]
55. Butt, B.; Turner, M.D.; Singh, A.; Brottem, L. Use of MODIS NDVI to evaluate changing latitudinal gradients of rangeland phenology in Sudano-Sahelian West Africa. *Remote Sens. Environ.* **2011**, *115*, 3367–3376. [CrossRef]
56. Brown, M.E.; de Beurs, K.M. Evaluation of multi-sensor semi-arid crop season parameters based on NDVI and rainfall. *Remote Sens. Environ.* **2008**, *112*, 2261–2271. [CrossRef]
57. Dixon, J.A.; Gibbon, D.P.; Gulliver, A. *Farming Systems and Poverty: Improving Farmers' Livelihoods in a Changing World*; FAO & World Bank: Rome, Italy; Washington, DC, USA, 2001.
58. Alliance for Green Revolution in Africa (AGRA). The Africa Agriculture Status Report 2014: Climate Change and Smallholder Agriculture in Sub-Saharan Africa. Available online: https://ccafs.cgiar.org/publications/africa-agriculture-status-report-2014-climate-change-and-smallholder-agriculture-sub#.XVy0U0G-lPY (accessed on 21 August 2019).
59. Bationo, A.; Hartemink, A.E.; Lungo, O.; Naimi, M.; Okoth, P.; Smaling, E.M.A.; Thiombiano, L. African Soils: Their Productivity and Profitability of Fertilizer Use. Available online: https://library.wur.nl/WebQuery/wurpubs/fulltext/26759 (accessed on 21 August 2019).
60. Mason, S.C.; Maman, N.; Palé, S. Pearl millet production practices in semi-arid West Africa: A review. *Exp. Agric.* **2015**, *51*, 501–521. [CrossRef]
61. Singh, P.; Boote, K.J.; Kadiyala, M.D.; Nedumaran, S.; Gupta, S.K.; Srinivas, K.; Bantilan, M.C. Science of the Total Environment An assessment of yield gains under climate change due to genetic modi fi cation of pearl millet. *Sci. Total Environ.* **2017**, *601–602*, 1226–1237. [CrossRef]
62. Barrett, H.; Browne, A. Export Horticultural Production in Sub-Saharan Africa: The Incorporation of The Gambia. *Geography* **1996**, *81*, 47–56.
63. Takane, T. Smallholders and nontraditional exports under economic liberalization: The case of pineapples in Ghana. *Afr. Study Monogr.* **2004**, *25*, 29–43.
64. IPCC. Summary for Policymakers. In *Climate Change 2007: The Physical Science Basis. Contribution of Working Group I to the Fourth Assessment Report of the Intergovernmental Panel on Climate Change*; Solomon, S., Qin, D., Manning, M., Chen, Z., Marquis, M., Ave, K.B., Eds.; Cambridge University Press: Cambridge, UK; New York, NY, USA, 2007.

65. Wheeler, T.; von Braun, J. Climate change impacts on global food security. *Science* **2013**, *341*, 55–60. [CrossRef]
66. Ramirez-Villegas, J.; Thornton, P.K. *Climate Change Impacts on African Crop Production*; CCAFS Working Paper No. 119; CGIAR Research Program on Climate Change, Agriculture and Food Security (CCAFS): Copenhagen, Denmark, 2015; 27p.
67. Ramirez-Villegas, J.; Lau, C.; Hooker, J.; Jarvis, A.; Ann-Kristin, K.; Arnell, N.; Tom, O. *Climate Analogues: Finding Tomorrow's Agriculture Today CGIAR Research Program on Climate Change*; Agriculture and Food Security (CCAFS): Copenhagen, Denmark, 2011. Available online: https://ccafs.cgiar.org/publications/climate-analogues-finding-tomorrow%E2%80%99s-agriculture-today-0#.XVy3U0G-lPY (accessed on 21 August 2019).

4

Effect of Planting Date on Accumulated Temperature and Maize Growth under Mulched Drip Irrigation in a Middle-Latitude Area with Frequent Chilling Injury

Dan Wang [1], Guangyong Li [1,*], Yan Mo [1,2], Mingkun Cai [1] and Xinyang Bian [3]

1 College of Water Resources & Civil Engineering, China Agriculture University, Beijing 100083, China; wangdan_9090@cau.edu.cn (D.W.); cmk1993@cau.edu.cn (M.C.)

2 Department of Irrigation and Drainage, China Institute of Water Resources and Hydropower Research, Beijing 100091, China; moyanSDI@cau.edu.cn or moyan@iwhr.com

3 Scientific Research Department, Water-Saving and Equipment Limited Company of Kingland Muhe, Chifeng 024000, China; bianxy@mhjsgf.com

* Correspondence: LGYL@cau.edu.cn

Abstract: Given that chilling injury, which involves late spring cold and early autumn freezing, significantly affects maize growth in middle-latitude cold areas, a highly efficient cultivation technique combining suitable planting date (PD) and mulched drip irrigation is being studied to guarantee maize production. A field experiment for medium-mature variety "Xianyu 335" was conducted in 2015 to 2016 in Chifeng, Inner Mongolia, China, to explore the effects of PD on the active accumulated temperature (AAT) distribution and maize growth under mulched drip irrigation. Based on the dates (around May 1) of late spring cold occurring in the area, four PDs were designed, namely, April 20 (MD_1), May 2–3 (MD_2), May 12 (MD_3), and May 22 (MD_4), and a non-film mulching treatment ($NM\text{-}D_2$) was added on the second PD. Results indicated that: (1) the warming effect of film mulching effectively compensated for the lack of heat during the early stages of maize growth. Compared with that in $NM\text{-}D_2$, the soil temperature under mulching in MD_2 for the sowing–emergence and seedling stage increased by 14.3% and 7.6%, respectively, promoting maize emergence 4 days earlier and presenting 5.6% and 9.7% increases in emergence rate and grain yield, respectively; (2) the AAT reduction caused by PD delay was mainly observed in reproductive stage, which reached 96.6 °C for every 10 days of PD delay in this stage; (3) PD markedly affected maize growth process and yield, which were closely related to the chilling injury. The late spring cold slowed down the emergence or jointing for maize (under MD_1 and MD_2), but brought insignificant adverse effect on maize later growth and grain yield (16.1 and 15.9 $Mg{\cdot}ha^{-1}$, respectively). While the maize in both MD_3 and MD_4 treatments suffered from early autumn freezing damage at the anthesis–maturity stages, resulting in shortening in reproductive period by 4–8 days and decrease in grain yield by 11.4–17.3% compared with those in MD_1 and MD_2; and (4) taking the typical date (May 1) of late spring cold occurring as the starting point, the grain yield penalty reached 8.5% for every 10 days of PD delay; for every 100 °C of AAT decrease during reproductive stage, the grain yield decreased by 6.1%. The conclusions offer certain reference values for maize cultivation in the same latitude areas with similar ecological environments.

Keywords: planting date; mulched drip irrigation; chilling injury; active accumulated temperature; maize growth

---

1. Introduction

Global warming significantly affects agricultural production in middle-high latitude areas. An increasing number of studies have indicated that crop varieties with a long growth period (GP)

should be selected for temperature increasing and frost-free season extension, and photo-thermal resources should be fully utilized to promote crop production [1–3]. However, the selection of planting date (PD), which corresponds to the expanded planting of crop varieties with a long GP, remains challenging [4]. In recent years, researchers have conducted extensive discussions on the changing trend of the PDs for rice, maize and other crops [4–7]. The Corn Belt in Northeast China is located in cold middle-latitude regions. Insufficient accumulated temperature caused by chilling injuries (such as late spring cold and early autumn freezing) is a crucial ecological factor that limits grain yield [8]. Therefore, film mulching has become a primary cultivation practice. However, owing to the lack of systematic research on the effects of PD on maize growth and regulation mechanism of heat factors, wide PD intervals have made the effects of improving production in cold weather imperceptible.

Planting date influences crop growth by changing the corresponding relationship between hydrothermal factors and growth stages. The main obstacles that restrict maize growth in different ecological regions should be identified to determine the optimal PD. In cold middle-latitude regions, the limitation of PD delay should be based on the required accumulated temperature for crop growth [9]. Maize growth is accelerated with PD postponement, leading to a shortening growth period and an adverse effect on biomass accumulation [10,11]. Moreover, delayed sowing increases the probability of early autumn freezing occurring at latter stages in maize growth, thereby restraining grain maturity and leading to output reduction [12,13]. Early planting with film mulching is regarded as an effective measure to increase production against the cold, because despite crop exposure to a low-temperature environment at the prophase, the growth point is not endangered; besides, no adverse effect was observed in middle-later stages, and maize can reach physiological maturity before early autumn freezing occurs [14,15].

The U.S. Corn Belt is located in the same latitude as the China Corn Belt, and the appropriate PDs for maize are from late April to early May, which tends to be advanced in recent years [2,16,17]. Maize planting season ranges from late April to mid-May in the China Corn Belt, and several reports showed that the appropriate PDs for medium-late maturing varieties are from April 25 to May 10, but with considerable yield differences [18,19].

Most of the aforementioned conclusions are drawn from studies on traditional surface irrigation. Mulched drip irrigation technology combines the characteristics of film mulching and drip irrigation. The application of this technology for maize is becoming increasingly widespread in Northeast China these years. According to the latest reports, the mulched drip irrigation area for maize in the region increased 135×10^4 ha during 2012–2015, which made a significant contribution to food security [20]. Film mulching can improve soil temperature and reduce water evaporation [21–24], whereas drip irrigation can achieve accurate and efficient management of soil moisture and nutrients [25,26]. The regulation effects of mulched drip irrigation and PD on the microclimate around the crop will perform a certain role in the maize growth process, yield formation and utilization of hydrothermal resources. However, related research under drip irrigation remains rare and unsystematic, and the reported conclusions were mainly concentrated in arid areas, wherein the primary objective was to improve yield- and water-use efficiency [27–29]. Furthermore, current studies on maize PD under mulched drip irrigation are little. Therefore, identifying a reasonable PD that corresponds to mulched drip irrigation in cold middle-latitude areas with different ecological factors under consideration is important.

This study aims to investigate the warming effect and yield-increasing potential of mulched drip irrigation for maize, to analyze maize growth response to chilling injury (that is, late spring cold and early autumn freezing) at certain growth stages caused by different PDs, quantify the accumulated temperature loss and yield reduction with PD delay, and provide a theoretical basis for improving the rational utilization of heat resources and minimizing production loss in cold middle-latitude areas.

2. Materials and Methods

2.1. Site Description

The field experiment was conducted in 2015 to 2016, Chifeng in Inner Mongolia, China (42°56′53′′N, 119°4′20′′E). The area has a semi-arid and continental monsoon climate with an approximately 135-day frost-free period. Early autumn freezing generally occurs from the end of September to the beginning of October, and the late spring cold always occurs around May 1. The effective accumulated temperature (≥10 °C) is 2000–3200 °C, and the annual average rainfall and evaporation amounts are 350–450 and 1500–2300 mm, respectively.

Temperature and precipitation during maize GPs in 2015 and 2016 are shown in Figure 1. The late spring cold occurred in both years (the minimum daily temperature dropped to −1.8 °C in 2015 and to 1.4 °C in 2016) and the dates of the early autumn freezing in 2 years were October 2 and September 28, respectively. The weather conditions during maize GPs in 2015 and 2016 were in line with the environmental climate characteristics of years in the region [30]. Total precipitation values during maize GPs for 2 years were 180 and 250 mm, respectively. Note that the climate feature of "late spring cold" in the region is that the daily average temperature exceeding 10 °C lasts for 5–10 days at least in late Spring, but then the daily minimum temperature dropped to about 0 °C for a sudden strong cold air attack (the cold snap always continue 1–2 days), and then the temperature increases again gradually.

The soil layer of 0–60 cm in the test site belong to silt loam soil with a soil bulk density of 1.49 g·cm^{-3} and a field water-holding rate of 34.45% (volumetric water content). Soil organic matter was 10.6 g·kg^{-1}, mass fraction for total nitrogen was 0.60 g·kg^{-1}, and available potassium and phosphorus were 167 and 7.6 mg·kg^{-1}, respectively.

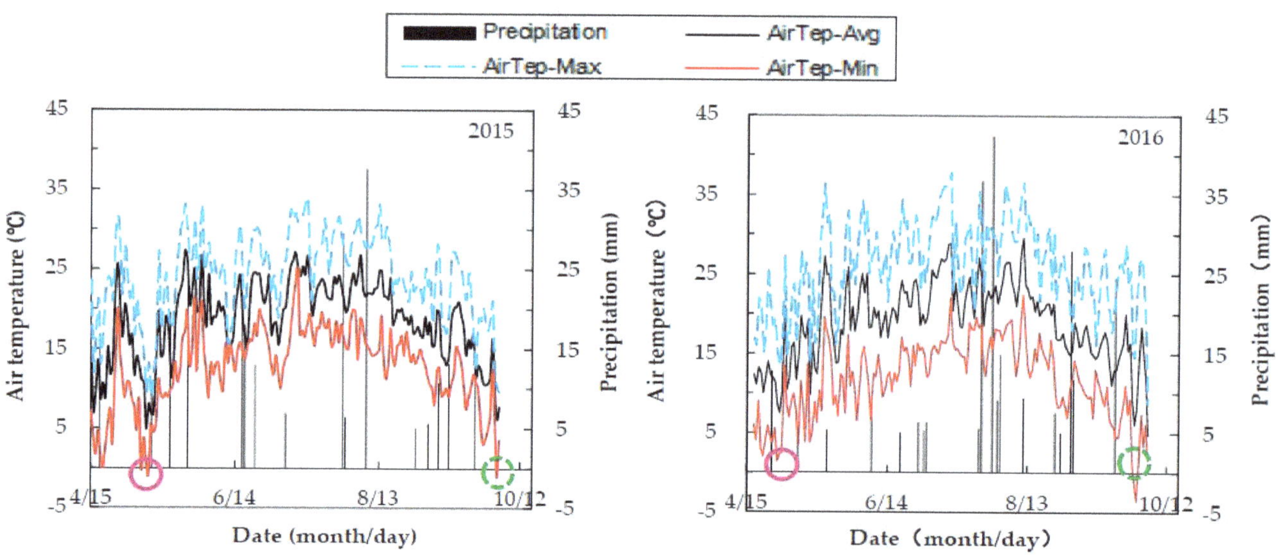

Figure 1. Daily temperature and precipitation during maize growing season in 2015 and 2016. Purple circle with full line marked position for the late spring; green broken circle marked position for the early autumn freezing.

2.2. Field Experiment

The experiment used a completely randomized design for the variable factor of PD, which involved four levels. Based on the date of late spring cold (around May 1), four PDs were set under mulched drip irrigation, namely, MD_1 (April 20), MD_2 (May 2–3), MD_3 (May 12), and MD_4 (May 22). And a treatment (NM-D_2) without mulching was added on the second PD under drip irrigation. Each treatment was performed in triplicate, thus, 15 plots in total were arranged randomly. The size of each plot was 40 m × 6 m.

Maize varieties need to match the light and heat resources of a certain ecological zone otherwise the yield and utilization rate of resources will be reduced [31]. For testing, we selected the medium variety "Xianyu 335", which is a high-yield plant in the Corn Belt of Northeast China [32,33]. Maize was planted in the alternate wide–narrow rows (80–40 cm) with a planting density of 83,330 plants·ha^{-1}. The drip tape with a wall thickness of 0.2 mm, a drip flow of 1.38 L·h^{-1}, and a drip spacing of 40 cm was placed in the middle of the narrow line. The maize cropping pattern and lateral layout of drip tapes under mulched drip irrigation are shown in Figure 2.

The treatments were irrigated according to the lower irrigation limit and crop water requirement, and the lower limit of irrigation at both seedling and mature stages was 70%, whereas that in other growth stages was 75%. The precipitation and irrigation amounts in each GP of maize for all treatments are shown in Figure 3. In treatments of MD_1–MD_4, the total irrigation amounts were 170, 155, 151, and 151 mm, respectively, in 2015 and 135, 135, 125, and 115 mm, respectively, in 2016. The irrigation amount in NM-D_2 treatment was same as that in MD_2 for 2 years. The total precipitation during maize growth season was 180 mm in each treatment in 2015, whereas those in MD_1–MD_4 in 2016 were 251, 242, 232, and 215 mm, respectively.

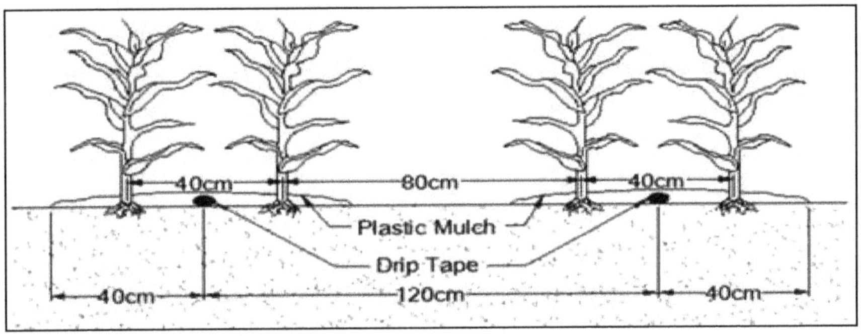

Figure 2. The schematic diagram of cropping pattern and lateral layout of drip tapes under mulch for maize.

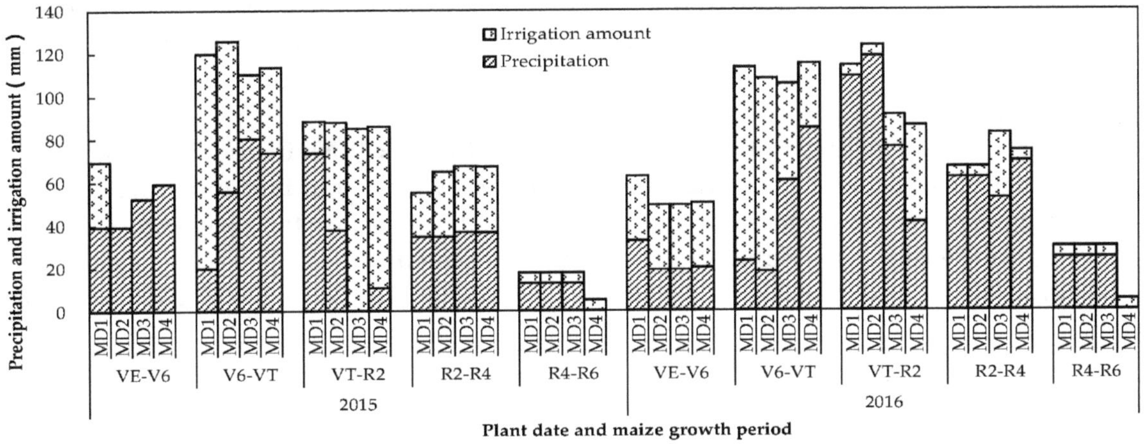

Figure 3. The precipitation and irrigation amount in various growth stages of maize for different treatments. VE—emergence, V6—sixth leaf, VT—tasseling, R2—filling, R4—dough, R6—physiological maturity.

Fertilizer schedule in all treatments was the same for 2 years, and the application amounts of N, P_2O_5, and K_2O were 285, 135, and 135 kg·ha^{-1}, respectively. Fertilizer application percentages in each GP are shown in Table 1. Seed fertilizer, which included urea (N 46%), calcium superphosphate

(P_2O_5 46%), and potassium sulfate were applied to the field by a seeder. The remaining fertilizer was applied in proportion with water by drip tapes, and the soluble fertilizers included urea (N 46%), MAP (monoammonium phosphate, P_2O_5 61%, N 12%), and potassium chloride (K_2O 62%). Other farm tasks were performed as the management in local high-yield farmlands.

Table 1. Fertilizer application percentages in each growth period.

Fertilizer	Sowing	V6–V12	VT	R2	R4
N	20%	45%	15%	15%	5%
P_2O_5	50%	30%	20%	–	–
K_2O	50%	50%	–	–	–

V6—sixth leaf, V12—twelfth leaf, VT—tasseling, R2—filling, R4—dough.

2.3. Observation Indexes and Methods

ET107 automatic meteorological station manufactured by USA Campbell Scientific was used to continuously monitor meteorological data, such as daily temperature and precipitation during the maize growing season.

Soil water content was regularly measured using the gravimetric method. The soil samples were collected by an auger boring in 20 cm increments to a depth of 80 cm, then were dried at 105 °C for 8 h. The measurement range of the electronic scale used is 500 g with an accuracy of 0.01 g. Soil moisture is calculated as the following formula.

$$\omega = \frac{m_w - m_d}{m_d} \times 100\% \quad (1)$$

ω—soil water content, %; m_w—the mass of wet soil, g; m_d—the mass of dry soil, g.

A set of curved tube thermometers were embedded in the narrow lines (under mulching) and wide lines (in bare land) of each treatment to measure the soil temperature at depths of 5, 10, 15, 20 and 25 cm. The observation times were 8:00, 14:00, and 18:00 daily, and the average value of three observations was taken as the daily soil temperature [34,35].

Dates of VE (emergence), V6 (sixth leaf), VT (tasseling), R2 (filling), and R6 (physiological maturity) were recorded according to maize growth traits. Five plants were removed from the ground except for the roots to measure height, stem diameter and dry matter (DM) accumulation (plants were separated into leaves, stems, and ears, which were heat-treated at 105 °C for 30 min and dried at 75 °C to a constant weight).

During harvest season, 10 m of maize from the middle six lines in each plot were harvested separately, grain moisture content was measured, and yield with a standard moisture content of 14% was calculated. Five maize ears from each plot were selected to measure the yield components, such as bald tip length, hundred-grain weight, and grain number per ear.

Statistical analysis and plotting were performed using Microsoft Excel 2010 and IBM SPSS Statistics 17.0.

3. Results

3.1. Effects of Planting Date and Film Mulching on Heat Distribution

3.1.1. Soil Temperature

The dynamic changes of daily soil temperature (DST) in the depth of 5–25 cm under film mulching and in bare land during maize growing season in 2015 and 2016 are shown in Figure 4. The increasing rate of the average DST under film mulching at various GPs, compared with those of bare land in different treatments, are shown in Figure 5.

From MD_1 to MD_4, the accumulated soil temperature (AST) under mulching during maize entire GP for two years averagely increased by 149.9 °C, 136.6 °C, 116.5 °C, and 99.6 °C, respectively, compared with those in bare land. The AST increments of seedling stage in MD_1–MD_4 accounted for 59.7%, 59.1%, 55.0%, and 54.8% of those during the total GP, respectively. Results indicated that the warming effect of film mulching decreased with the PD delay.

For maize plants with different PDs at two stages, namely, sowing–emergence and seeding, the average increasing rates of DST under mulching were 11.2% and 6.7%, respectively. Moreover, the warming effect gradually weakened with continuous GP progress, and the average increasing rates of DST under mulching were 2.3–3.0% during V6–harvest stage with different PDs. The results demonstrated an evident warming effect of film mulching in the prophase of maize.

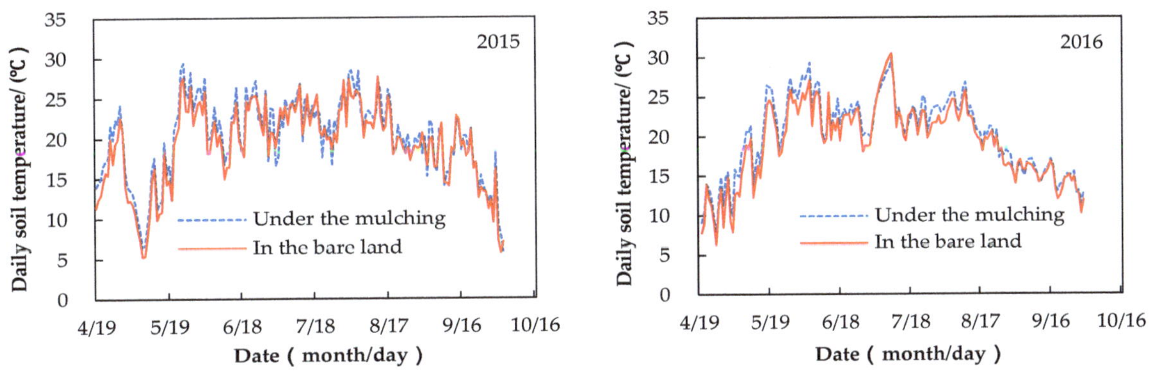

Figure 4. Changes of daily average soil temperature at 5–25 cm depth during maize growing season in 2015 and 2016.

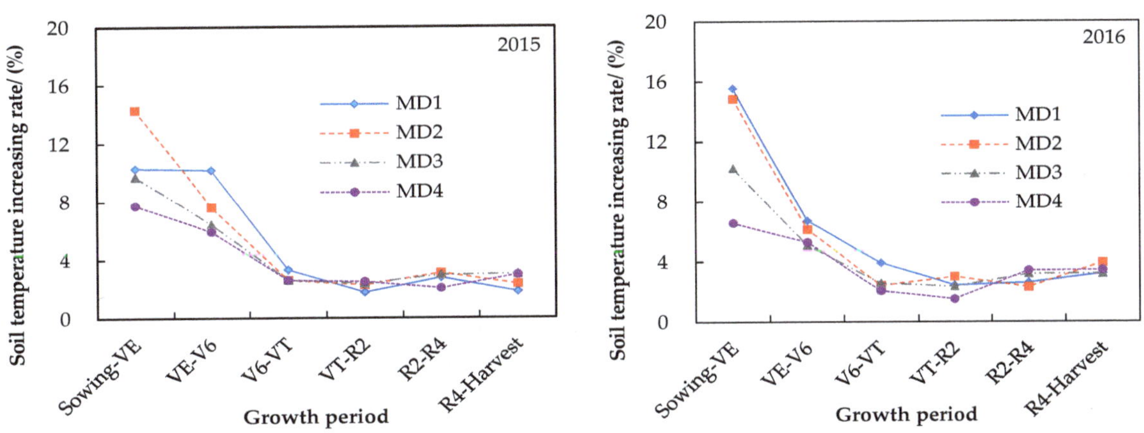

Figure 5. Daily soil temperature increasing rate under mulching compared with that in bare land of various maize growth periods in different treatments. VE—emergence, V6—sixth leaf, VT—tasseling, R—filling, R—dough.

3.1.2. Active Accumulated Temperature

The AATs (≥10 °C) in different treatments for various GPs of maize under mulched drip irrigation are shown in Table 2.

The total AATs exhibited a decreasing trend with the delay in PD, decreasing under MD_2–MD_4 treatments by 73 °C, 226 °C, and 365 °C, respectively, compared with those in MD_1.

The AATs at maize's different GPs were determined based on daily temperature and stage duration. The average AAT required for maize emergence was approximately 208 °C, and the annual variation was caused by fluctuation in the occurrence of late spring cold.

The AATs at both vegetative and reproductive growth stages decreased with the postponement of PD. The AATs at these two stages under MD_1–MD_4 treatments were on the average 1494–1430 °C and 1327–1022 °C, respectively. Compared with those in MD_1, AATs of the maize reproductive stage in MD_2–MD_4 decreased by 89 °C, 198 °C, and 305 °C, respectively, illustrating that PD particularly affected AATs of this stage and insufficient AATs at this stage were unfavorable for grain filling and dehydration.

Table 2. The active accumulated temperature of sowing–harvest, sowing–VE, VE–VT and VT–harvest stages for maize in different treatments (°C).

Year	Treatment	Sowing–Harvest	Sowing–VE	VE–VT	VT–Harvest
2015	MD_1	3050	183	1544	1323
	MD_2	2927	210	1487	1229
	MD_3	2793	235	1435	1122
	MD_4	2677	207	1447	1023
	Mean	2862	209	1478	1174
2016	MD_1	2993	219	1443	1330
	MD_2	2969	214	1508	1247
	MD_3	2798	214	1450	1134
	MD_4	2635	186	1413	1020
	Mean	2849	208	1454	1183
Mean	MD_1	3021	201	1494	1327
	MD_2	2948	212	1498	1238
	MD_3	2796	225	1443	1128
	MD_4	2656	197	1430	1022

VE—emergence, VT—tasseling.

3.2. *Effect of Planting Date on Maize Growth Process*

The durations of various maize GPs under different treatments are shown in Table 3, and no statistically significant differences occurred between years for the average parameter values of all treatments.

Table 3. The durations of sowing–harvest, sowing–VE, VE–VT and VT–harvest stages for maize in different treatments (d).

Year	Treatment	Sowing–Harvest	Sowing–VE	VE–VT	VT–Harvest
2015	MD_1	155d	9a	81b	65c
	MD_2	150c	17c	69a	64c
	MD_3	139b	11b	68a	60b
	MD_4	133a	10a	66a	57a
	F-test	89.9 **	42.9 **	52.0 **	22.7 **
	NM-D_2	153	21	67	65
	Mean	146A	14A	70A	62A
2016	MD_1	154d	18d	69b	67c
	MD_2	149c	14c	69b	66c
	MD_3	141b	12b	67a	62b
	MD_4	131a	10a	64a	57a
	F-test	200.5 **	143.9 **	16.3 **	45.2 **
	NM-D_2	152	18	67	67
	Mean	145A	14A	67A	64A
Mean	MD_1	155d	14ab	75b	66c
	MD_2	150c	16b	69a	65c
	MD_3	140b	12ab	67a	61b
	MD_4	132a	10a	65a	57a
	F-test	200.7 **	4.6 *	8.5 **	51.2 **
	NM-D_2	153	20	67	66

1 The different lowercase letters in the same column show significance at $p < 0.05$. 2 * means significant ($p < 0.05$), ** means extremely significant ($p < 0.01$). 3 Column values for the various years with same uppercase letters are insignificantly different at $p < 0.05$. 4 VE—emergence, VT—tasseling.

Compared with that under NM-D_2 treatment, the duration of total maize GP under MD_2 decreased by 3 days, and that of sowing–VE period shortened by 4 days, whereas the differences in growth durations of middle–late stages were minimal.

For treatments under mulched drip irrigation, the duration of total maize GP decreased gradually with the delay in PD, and the differences among MD_1–MD_4 reached 5–23 days. The total GP duration for maize in both MD_1 and MD_2 exceeded 150 days, which guaranteed the physiological maturity for grain. Whereas, maize in MD_3 and MD_4 experienced early autumn freezing during later growth stages, resulting in evident shortening in GP duration (10–18 days) and failing to reach physiological maturity.

A significant difference reaching 2–8 days was found in sowing–VE duration among MD_1–MD_4 treatments. When sowing was performed on April 20, maize sowing–VE duration was highly unstable with a 9-day difference within 2 years because of the fluctuation in late spring cold. In 2015, maize had entered seedling stage as late spring cold occurred, whereas the occurrence of chilling injury during sowing–VE stage delayed emergence by 9 days in 2016. In MD_2–MD_4 treatments, the sowing–VE duration decreased gradually with PD delay with insignificant inter-annual differences.

The duration of VE–VT for maize showed a decreasing trend with the PD delay. A 12-day difference in this period appeared in MD_1 treatment for 2 years because the seedlings underwent a long period of recovery after acquiring frostbite in 2015; by contrast, no chilling injury occurred during this period in 2016. The inter-annual differences of this period for 2 years were insignificant under MD_2–MD_4 treatments.

The effects of PD on the VT–Harvest duration were highly evident. MD_1 and MD_2 didn't show significant difference during this period, but shortened by 4–8 days when PD delayed to May 12 or later (MD_3 and MD_4) because of early autumn freezing.

3.3. *Effects of Planting Date on Maize Growth Indexes*

Planting date significantly affected maize emergence rate, plant height, stem diameter, and single plant dry weight (DW) and its allocation proportions in each organ under mulched drip irrigation in varying degrees (Table 4). No significant difference occurred in maize growth indexes under MD_1 and MD_2, but the plant height increased significantly and the stem diameter and DW per plant decreased considerably with the continuous delay in PD. And no appreciable differences existed in most growth indexes between years, other than plant height and single plant DW (with greater values in the first year).

The maize emergence rate under treatments of MD_1, MD_3 and MD_4 did not appear to be markedly different, but that in MD_2 decreased by 1.6–2.3%. The reason why maize in MD_2 obtained a lower emergence rate was that the PD (May 2–3) was near the date of late spring cold, the average temperature during maize sowing–VE period was lower than that in other treatments.

A comparison of the distribution ratios of DW in nutritive and reproductive organs of maize under mulched drip irrigation showed that the proportions of DW in stem and leaves exhibited an upward trend with the delay in PD, whereas those in the ears presented a downward trend. No significant difference was found in single-plant DW and its distribution proportion in each organ between MD_1 and MD_2; however, in MD_3 and MD_4 treatments, single-plant DW decreased by 10.0–14.5%, and the proportions of DW in stem and leaves increased by 11.7–15.4% and 16.4–22.8%, respectively, whereas those in ears decreased by 8.6–6.4%. Results indicated that the delay in PD led to "slim" plants, which are not conducive to the translocation of dry matter to ears.

In the comparison of the maize growth indexes between treatments with film mulching and NM-D_2, film mulching improved the maize emergence rate by 5.6–8.1% on average; the single-plant DW in MD_1 and MD_2 increased by 8.0–8.4%, while those in MD_3 and MD_4 decreased by 2.6–7.5%, compared with those in NM-D_2.

Table 4. Maize growth indexes of different treatments in harvest season.

Year	Item	Emergence Rate (%)	Height (cm)	Stem Diameter (cm)	Dry Matter Per Plant (g·Plant^{-1})	Proportion (%) Stem	Proportion (%) Leaves	Proportion (%) Ear
2015	MD_1	94.5b	311.5a	2.6b	434.9b	19.9a	11.9a	68.2b
	MD_2	92.1a	314.2a	2.6b	440.2b	20.2a	12.1a	67.7b
	MD_3	93.5ab	317.3ab	2.4ab	400.0a	23.0ab	14.6b	62.4a
	MD_4	93.9ab	322.0b	2.3a	378.5a	23.9b	15.4b	60.8a
	F-test	6.6 *	4.2 *	6.8 *	11.8 **	11.5 **	16.7 **	18.7 **
	NM-D_2	86.7	316.2	2.4	412.5	22.7	14.1	63.2
	Mean	92.1A	316.2B	2.5A	413.2B	21.9A	13.6A	64.5A
2016	MD_1	93.5a	306.0a	2.5 b	426.1b	19.6a	12.4a	68.0c
	MD_2	93.0a	309.0b	2.5b	423.7b	20.0a	13.1ab	66.9bc
	MD_3	95.2a	314.4c	2.3ab	376.4a	21.5a	14.2bc	64.3ab
	MD_4	95.6a	319.3 c	2.1a	359.3a	22.1a	15.0c	62.9a
	F-test	3.3	6.9 **	15.2 **	39.4 **	1.9	6.6 *	7.3 *
	NM-D_2	88.5	311.5	2.3	384.7	21.6	13.8	64.6
	Mean	93.2A	312.0A	2.3A	394.0A	20.9A	13.7A	65.3A
Mean	MD_1	94.0ab	308.6a	2.6	430.5c	19.8a	12.1a	68.1b
	MD_2	92.5a	311.8a	2.6	432.0c	20.1a	12.6a	67.3b
	MD_3	94.3ab	315.9ab	2.4	388.2b	22.2b	14.4b	63.4a
	MD_4	94.7b	320.7b	2.2	368.9a	23.0b	15.2b	61.8a
	F-test	4.4 *	5.6 **	9.6 **	27.2 **	8.6 **	22.3 **	21.5 **
	NM-D_2	87.6	313.9	2.4	398.6	22.2	14	63.9

1 The different lowercase letters in the same column show significance at $p < 0.05$. 2 * means significant ($p < 0.05$), ** means extremely significant ($p < 0.01$). 3 Column values for the various years with different uppercase letters are significantly different at $p < 0.05$.

3.4. Effects of Planting Date on Maize Yield Indexes

The maize yield indexes under different treatments are shown in Table 5.

A comparison of the maize yield indexes with different PDs under mulched drip irrigation showed that the bald tip length and grain moisture content exhibited an increasing trend with a delay in PD, whereas the hundred-grain weight, grain number per ear (2016), yield, and total DM displayed a decreasing trend. The grain number per ear under MD_2 in 2015 was the largest in terms of relatively lower emergence rate and large ear length. And no statistically evident differences existed in most yield indexes other than the bald tip length and grain number per ear between years, with smaller bald tip length and greater grain number in the first year.

No evident difference was found in the yield indexes between MD_1 and MD_2. Under MD_3 and MD_4, the bald tip length increased by 34.3% and 54.3%, the hundred-grain weight decreased by 4.3% and 9.1%, and grain number per ear decreased by 7.5% and 9.9%, respectively, compared with those in MD_1 and MD_2. The results illustrated that delayed sowing is unfavorable for the formation of yield components. Compared with those in MD_1, under MD_2, MD_3, and MD_4 treatments, DM decreased by 1.2%, 9.5%, and 13.7%, grain moisture content increased by 1.8%, 13.5%, and 17.9%, and yield (14% moisture content) decreased by 1.4%, 11.4%, and 17.3%, respectively. The findings indicated no significant difference in total biomass and economic yield of maize sowed on April 20 and May 2, whereas delayed PD to May 12 or later resulted in a significant yield reduction.

Compared with those yield indexes under non-plastic film mulching (NM-D_2), the bald tip length and grain moisture of maize planted before May 3 (MD_1 and MD_2) averagely decreased by 20.5% and 3.8%, respectively, but the total DM and yield increased by 6.5% and 9.7%, respectively; while under MD_3 and MD_4 treatments, the bald tip length and grain moisture averagely increased by 9.1–22.7% and 8.0–12.3%, respectively, whereas the total DM and yield decreased by 1.5–8.0% and 2.2–6.6%, respectively. The results showed an apparent effect of film mulching on improving yield for the early sowing maize.

Table 5. Maize yield indexes in different treatments.

Year	Treatment	Bald Tip Length (cm)	Hundred- Grain Weight (g)	Grain Number Per Ear	Grain Moisture Content (%)	Yield (14%) ($Mg \cdot ha^{-1}$)	Total Dry Matter Accumulation ($Mg \cdot ha^{-1}$)
2015	MD_1	1.6a	35.5c	592b	24.9a	16.24c	34.25b
	MD_2	1.7a	35.1bc	604c	25.4a	16.02c	33.77b
	MD_3	2.2b	33.4ab	560ab	28.1b	14.45b	31.15a
	MD_4	2.4b	31.8a	548a	29.6b	13.55a	29.60a
	F-test	15.3 **	9.6 **	10.2 **	14.2 **	24.9 **	10.9 **
	NM-D_2	2	34.7	576	26	14.62	31.45
	Mean	2.0A	34.1A	576B	26.8A	15.20A	32.20A
2016	MD_1	1.8a	35.3b	591b	24.8a	15.91b	33.18b
	MD_2	1.9a	35.2b	587b	25.2a	15.68b	32.83b
	MD_3	2.5b	34.1b	538a	28.3b	14.04a	29.86a
	MD_4	3.0b	32.4a	521a	28.9b	13.06a	28.62a
	F-test	48.1 **	6.9 *	29.6 **	14.5 **	9.9 **	28.5 **
	NM-D_2	2.4	34.3	558	26.27	14.29	30.91
	Mean	2.3B	34.3A	559A	26.7A	14.53A	31.04A
Mean	MD_1	1.7a	35.4c	592b	24.8a	16.08c	33.71c
	MD_2	1.8a	35.1c	595b	25.3a	15.85c	33.30c
	MD_3	2.4b	33.7b	549a	28.2b	14.24b	30.50b
	MD_4	2.7c	32.0a	535a	29.3b	13.30a	29.11a
	F-test	30.3 **	41.8 **	21.7 **	30.4 **	59.9 **	26.9 **
	NM-D_2	2.2	34.5	567	26.1	14.46	31.18

1 The different lowercase letters in the same column show significance at $p < 0.05$. 2 * means significant ($p < 0.05$), ** means extremely significant ($p < 0.01$). 3 Column values for the various years with different uppercase letters are significantly different at $p < 0.05$.

According to the significance F-test based on the mean dates of 2 years, illuminating that no evident differences were found in yield indexes between MD_1 and MD_2, whereas significant differences appeared in maize bald tip length, hundred-grain weight, and grain yield among treatments of MD_2, MD_3, and MD_4. We use the regressions to present the relationships of maize bald tip length, hundred-grain weight, and grain yield with the days of PD delay (in Figures 6–8). When the typical date of late spring cold (around May 1) was regarded as the starting point, an evident linear relationship existed between the grain yield indexes and the days of PD delay (MD_2–MD_4). That is, for every 10 days of PD delay, bald tip length increased by 23.2%, whereas hundred-grain weight and grain yield decreased by 4.5% and 8.5%, respectively. According to the regression equation, it concluded that the yield penalty (4.2%) was less than 5.0% with a yield exceeding 15 $Mg \cdot ha^{-1}$ when PD delayed to May 8.

Therefore, obtaining a super high yield is possible under mulched drip irrigation when the medium-mature maize varieties with close-planting are sown from April 20 to May 8 in the area.

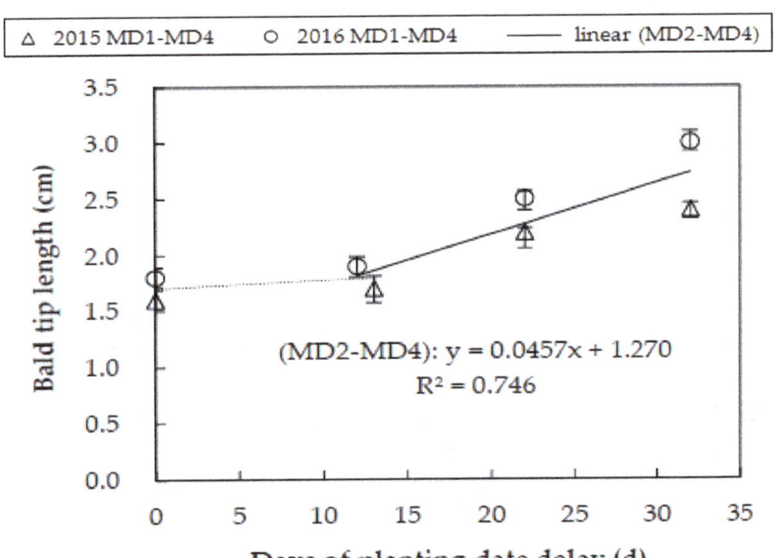

Figure 6. Effect of planting date on Bald tip length.

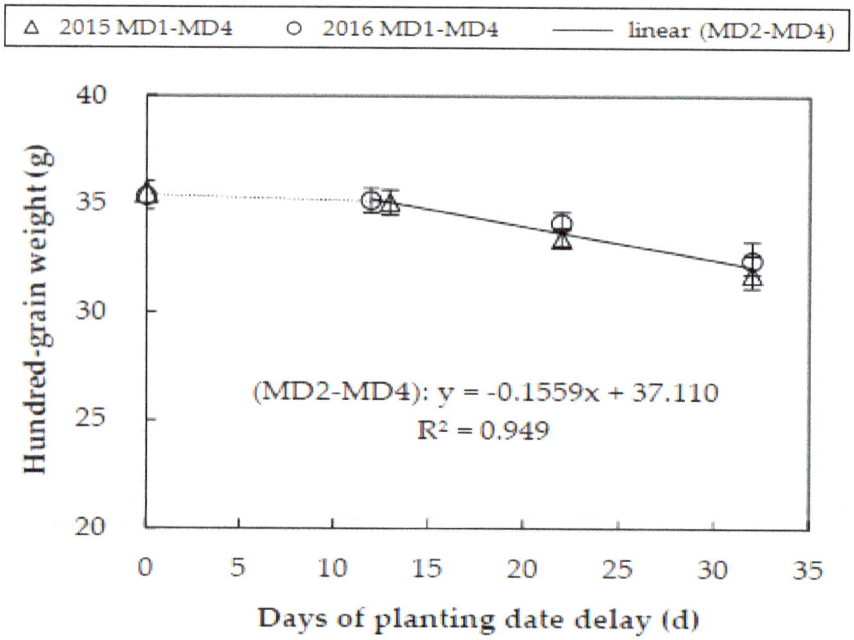

Figure 7. Effect of planting date on hundred-grain weight.

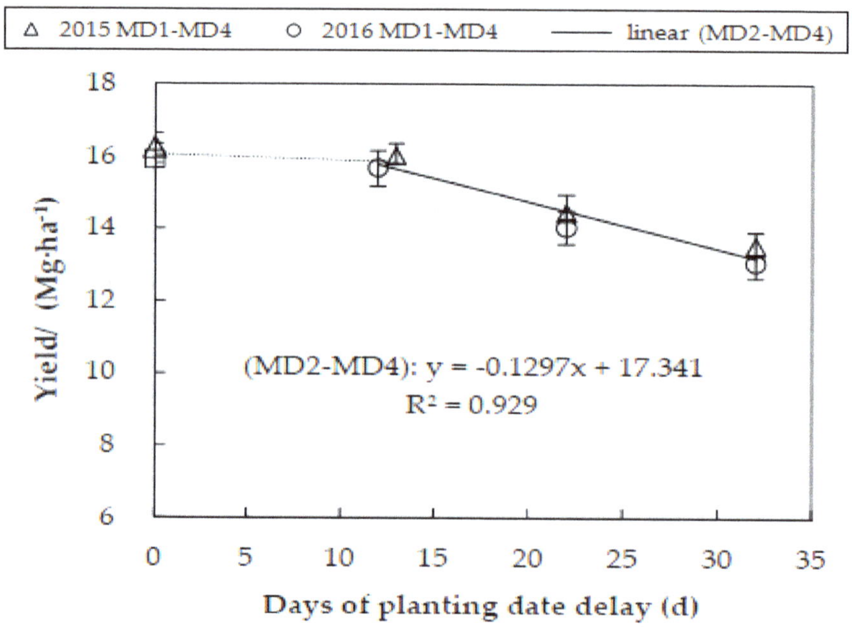

Figure 8. Effect of planting date on maize yield.

4. Discussion

The selection of an appropriate planting time for maize belongs to the category of ecological climate adaptability with regards to the degree of interaction between growth habit and climate condition [36]. Water stress is a key factor that restricts grain yield in areas with limited irrigation, and PD mainly affects yield formation by regulating the reasonable correspondence between rainfall and GP [11]. However, in areas where chilling injury frequently occurs, the selected PD should fully use photo-thermal resources for the limited accumulated temperature [13]. The test was conducted in a cold middle-latitude region with general occurrence of late spring cold from the end of April to early May, along with the regularity of spring drought. Generally, farmers tend to delay maize planting with temperature and soil moisture under consideration. Whereas, the probability of experiencing early autumn freezing (at the end of September) was increased in the later GP of maize, hence resulting in irregular grain yields. In this study, the application of mulched drip irrigation technology provided an environment with appropriate water and fertilizer levels for maize during the entire GP. Therefore, the study mainly analyzed the effects of chilling injury from certain growth stages and AAT differences caused by various PDs on maize growth.

Maize is highly sensitive to chilling injury, which mainly affects crop growth by changing crop root activity, enzyme activity, and photosynthetic rate [37,38]. The chilling injury at the seedling (MD_1) or sowing–VE (MD_2) stages delayed maize jointing or emergence, but the GP (>150 days) met the growth demand and ensured grain physiological maturity. Nevertheless, the maize sowed after May 12 (MD_3 and MD_4) suffered from early autumn freezing at latter growth stages, resulting in significant shortening in GP duration and increase in grain moisture content with failing to reach physiological maturity. The results verified the previous viewpoint that early autumn freezing is detrimental to maize growth [2,39]. On the basis of the maize growth and yield indexes with different PDs, a delay in PD was found to cause weak individual growth and reduced proportion of dry matter in ears (Table 4), which were not conducive to the formation of economic output (Table 5). The results coincided with Dong's viewpoints [40].

Planting date significantly affected the AAT distribution. The AATs in maize's entire GP decreased with the delay in PD under MD_1–MD_4, and the difference mainly manifested during the reproductive growth stage. At this stage, AATs decreased by an average of 96.6 °C for every 10 days of PD delay (Table 3). Whether the AATs in maize reproductive growth stage could meet the growth demand is a key factor in yield formation for the significant effects on grain filling and dehydration process [13,41]. The relationship between maize yield and AATs in this stage is analyzed in Figure 9, which shows that when sowing was performed before May 3 (MD_1 and MD_2), AATs were sufficient to ensure grain physiological maturity with high yield. But with continuous delay in PD (MD_2–MD_4), grain yield exhibited a negative correlation with AATs at this stage, and the amplitude of yield penalty reached 6.14–7.74% for every 100 °C of AAT decrease. Confirming the key measure to guarantee high yield is to ensure grain maturity before early autumn freezing with the maximum use of accumulated temperature [42]. Moreover, the decreasing solar radiation and light interception together with low temperature in latter growth stages exerted adverse effect on yield formation for the late sowing maize [18,43].

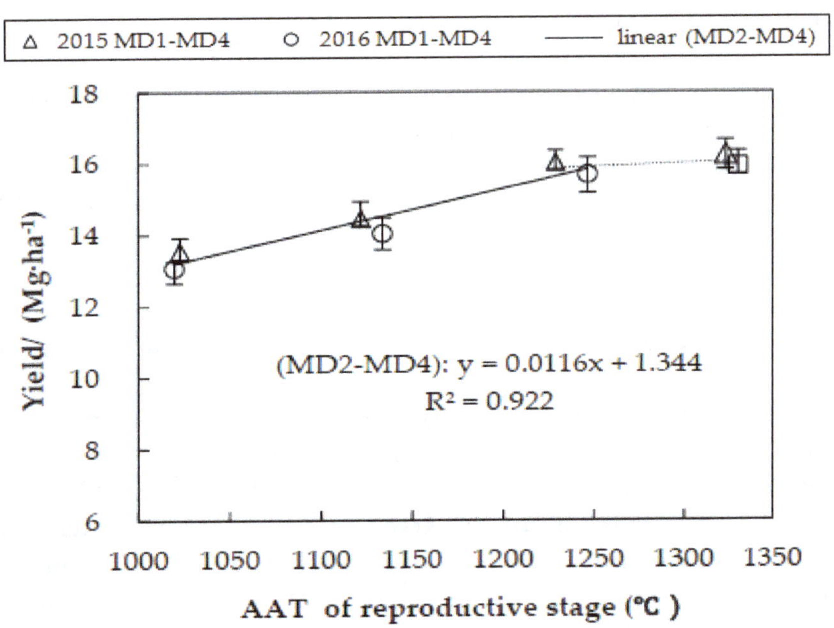

Figure 9. Effect of active accumulated temperature for reproductive stage on maize yield. AAT–active accumulated temperature.

Numerous studies have assumed that chilling injury at each growth stage will lead to varying degrees of yield reduction [13,44,45]. For the early sowing treatments in this study, the late spring cold slowed down maize early growth, but exerted no evident adverse effects on yield benefit. Reasons may be as following: the warming effect of film mulching could partly compensate for the lack of heat during the early stage of maize, thereby guaranteeing high emergence rate and nonfatal hurt from low temperature to seedlings; moreover, sufficient active accumulated temperature and solar radiation during maize growing season promoted the potential productivity. In the test, the yield from NM-D_2 treatment appears to be about the same as the yield from MD_3 treatment but less than that from MD_2 treatment, which showed an evident yield benefit from film mulching. However, early autumn freezing during the reproductive stage led to considerable yield reduction. Therefore, maize should be sown in advance with film mulching in cold middle-latitude regions to avoid early autumn freezing and ensure that GP duration and AATs can meet growth requirements.

In the test region, we suggest that the medium-mature variety of maize is sown from April 20 to May 8 with available irrigation. Compared with the conclusions of Yu et al and Cao et al. (maize cultivation without film mulching) [18,19], appropriate sowing intervals exhibited an increasing trend and were 5–7 days in advance, with 8.3–80.5% increase in grain yield. Appropriate advanced sowing gained a super-high yield in this study, also due to the technology of drip irrigation, which provided in-season supply of water and fertilizers for maize [30,46]. For the same latitude areas with similar ecological environments, it's potential to obtain high yield when the maize sowing date is in range of 10 days before or after the late spring cold, for a medium-mature variety of maize under mulched drip irrigation.

5. Conclusions

Mulched drip irrigation has shown a broad development prospect in middle-latitude spring maize planting area with low temperature and chilling injury. In this study, the effects of environment difference caused by varying PDs on maize growth under mulched drip irrigation were analyzed. The conclusions were as follows:

(1) Film mulching could significantly improve soil temperature, and the warming effect was more evident in maize early growth stages and gradually decreased with the postponement of PDs (based on MD_1). The warming effect of mulching was effective in compensating for the lack of heat caused by the low-temperature environment at the early growth stages of maize, which was beneficial to the improvement of maize emergence rate and economic yield.

(2) Planting date markedly affects AATs and maize growth process. Both AATs and maize GP duration decreased with the delay in PD (based on MD_1). When the PD was within April 20–May 2, the GP and AATs could meet the growth demand, whereas a delay of the PD until May 12 or later failed to guarantee grain physiological maturity.

(3) The effect of PD on maize growth was closely related to chilling injury (the late spring cold and early autumn freezing). The late spring cold slowed down maize early growth process (MD_1 and MD_2), but no significant adverse effect existed on later growth. As to MD_3 and MD_4, the occurrence of early autumn freezing at maize later stages led to significant decrease in AATs during the reproductive stage with adverse effect on grain filing and dehydration, and subsequently resulted in evident yield penalty (11.3–16.8%).

(4) Taking the date (May 1) of late spring cold generally occurring as the starting point, there was a significant positive correlation in maize bald tip length and the days of PD delay, while the changing tendency of hundred-grain weight and grain yield with the days of PD delay showed a linear decreasing trend; the AATs of maize reproductive stage gradually decreased with the postponement of PD, and the grain yield decreased linearly with the decrease in AATs of this stage.

Based on the threat of chilling injury in the middle-latitude area, it seems more reasonable to plant medium-mature maize varieties with close-planting in range of 10 days before and after the late spring cold to obtain a potential high yield (>15 $Mg \cdot ha^{-1}$) under mulched drip irrigation.

Acknowledgments: The authors are greatly indebted to the financial support from the National Support Program of China (2014BAD12B05).

Author Contributions: Dan Wang, Yan Mo and Mingkun Cai are graduate-students of China Agricultural University, they conducted the experiments and collected relevant materials for this research. Xinyang Bian is the key working personnel in the experimental station, he participated in the design and operation of the experiment. Dan Wang analyzed the date and wrote the manuscript. Guangyong Li provided direction of this study and revised the manuscripts. All authors have read and approved the final manuscript.

References

1. Parry, M.; Canzaiani, O.; Palutikof, J.; van der Linden, P.J.; Hanson, C. Climate change 2007: Impacts, adaptation and vulnerability. *J. Environ. Qual.* **2007**, *37*, 2407.
2. Kucharik, C. A multidecadal trend of earlier corn planting in the central USA. *Agron. J.* **2006**, *98*, 1544–1550. [CrossRef]
3. Olesen, J.E.; Trnka, M.; Kersebaum, K.C.; Skjelvåg, A.O.; Seguin, B.; Peltonensainio, P.; Micale, F. Impacts and adaptation of european crop production systems to climate change. *Eur. J. Agron.* **2001**, *34*, 96–112. [CrossRef]
4. Zhao, J.; Yang, X.; Dai, S.; Lv, S.; Wang, J. Increased utilization of lengthening growing season and warming temperatures by adjusting sowing dates and cultivar selection for spring maize in northeast China. *Eur. J. Agron.* **2015**, *67*, 12–19. [CrossRef]
5. Horai, K.; Ishii, A.; Mae, T.; Shimono, H. Effects of early planting on growth and yield of rice cultivars under a cool climate. *Field Crops Res.* **2013**, *144*, 11–18. [CrossRef]
6. Sacks, W.J.; Kucharik, C.J. Crop management and phenology trends in the U.S. Corn Belt: Impacts on yields, evapotranspiration and energy balance. *Agric. For. Meteorol.* **2011**, *151*, 882–894. [CrossRef]
7. Kucharik, C.J. Contribution of planting date trends to increased maize yields in the central united states. *Agron. J.* **2008**, *100*, 328–336. [CrossRef]
8. Ma, S.Q.; Wang, Q.; Wang, C.Y.; Huo, Z.G. The risk division on climate and economic loss of maize chilling damage in Northeast China. *Geogr. Res.* **2008**, *27*, 1169–1177. (In Chinese)
9. Ma, L.L. Response of Maize Growth Process to Sowing Date and the Factors of Light and Temperature. Ph.D. Thesis, Shanxi Agricultural University, Jinzhong, China, 2014. (In Chinese)
10. Hatfield, J.L.; Prueger, J.H. Temperature extremes: Effect on plant growth and development. *Weather Clim. Extrem.* **2015**, *10*, 4–10. [CrossRef]
11. Lu, H.D.; Xue, J.Q.; Guo, D.W. Efficacy of planting date adjustment as a cultivation strategy to cope with drought stress and increase rainfed maize yield and water-use efficiency. *Agric. Water Manag.* **2016**, *179*, 227–235. [CrossRef]
12. Wang, S.Y. *Study on Low Temperature and Chilling Injury for Crops*; Meteorological Press: Beijing, China, 1995.
13. Xu, T.J. Effect of Low Temperature on Maize Senescence and Grain Filling Rate during the Grain Filling Period and PASP-KT-NAA Regulation. Ph.D. Thesis, Chinese Academy of Agricultural Sciences, Beijing, China, 2012.
14. Lauer, J.G.; Carter, P.R.; Wood, T.M.; Diezel, G.; Wiersma, D.W.; Rand, R.E.; Mlynarek, M.J. Corn hybrid response to planting date in the northern corn belt. *Agron. J.* **1999**, *91*, 834–839. [CrossRef]
15. Wu, R.X.; Liu, R.Q.; Lu, C.L.; Lu, Y.L.; Li, H.; Zhang, L.; Lu, X.Z.; Wang, X.H. The optimum sowing Time for Plastic film Corn and the Use of the Two Theories. *Sci. Agric. Sin.* **2001**, *34*, 433–438. (In Chinese)
16. Benson, G.O. Corn replant decisions: A review. *J. Prod. Agric.* **1990**, *3*, 180–184. [CrossRef]
17. Good, D. Can corn and soybean crops overcome late planting? *Corn & Soybean Digest Exclusive Insight*, 6 June 2011; 4.
18. Yu, J.L. Effects of Sowing Date and Density on Matter Production and Yield Formation in Maize. Ph.D. Thesis, Shenyang Agricultural University, Shenyang, China, 2013. (In Chinese)
19. Cao, Q.J.; Yang, F.T.; Chen, X.F.; Lamine, D.; Li, G. Effects of sowing date on growth, yield and quality of spring maize in the central area of Jilin Province. *J. Maize Sci.* **2013**, *21*, 71–75. (In Chinese)
20. Wang, Y.Q.; Gong, S.H.; Xu, D.; Zhang, Y.Q. Several aspects of the research for corn under film drip irrigation in the Northeast of China. *J. Irrig. Drain.* **2015**, *34*, 1–4.
21. Orzolek, M.D.; Murphyj, J.; Giardi, J. The effect of colored polyethylene mulch on the yield of squash. In *Tomato and Auliflower*; College of Agricultural Sciences, Pennsylvania State University: State College, PA, USA, 2003.

22. Kasperbauer, M.J.; Hunt, P.G. Mulch'Surface Color Affects Yield of Fresh-market Tomatoes. *J. Am. Soc. Hortic. Sci.* **1989**, *114*, 216–219.
23. Wang, X.K.; Li, Z.B.; Xing, Y.Y. Effects of mulching and nitrogen on soil temperature, water content, nitrate-n content and maize yield in the loess plateau of China. *Agric. Water Manag.* **2015**, *161*, 53–64.
24. Sun, G.F.; Du, B.; Qu, Z.Y.; Li, C.J.; Zhang, Z.L.; Li, X.H. Changes of soil temperature and response to air temperature under different irrigation modes. *Soils* **2016**, *48*, 581–587. (In Chinese)
25. Lamm, F.R. *Irrigation and Nitrogen Management for Subsurface Drip Irrigated Corn—25 Years of K-State's Efforts;* ASABE Paper, No. 141914980; American Society of Agricultural and Biological Engineers: St. Joseph, MI, USA, 2014.
26. Jayakumar, M.; Janapriya, S.; Surendran, U. Effect of drip fertigation and polythene mulching on growth and productivity of coconut (*Cocos nucifera* L.), water, nutrient use efficiency and economic benefits. *Agric. Water Manag.* **2017**, *182*, 87–93.
27. Sun, Y.L. Effects of Sowing Dates on Spring Corn Female Spike Differentiation and Yield of Xinjiang with High Yield of 15,000 kg/hm^2. Ph.D. Thesis, Shihezi University, Shihezi, China, 2014. (In Chinese)
28. Wei, Y.G.; Chen, L.; Jiang, J.F.; Ding, W.K.; Wang, H.L. Effects of irrigation methods and sowing date on yield and water use efficiency of mulched spring maize. *Chin. Agric. Sci. Bull.* **2014**, *30*, 203–208. (In Chinese)
29. Feyzbakhsh, M.T.; Kamkar, B.; Mokhtarpour, H.; Asadi, M.E. Effect of soil water management and different sowing dates on maize yield and water use efficiency under drip irrigation system. *Arch. Agron. Soil Sci.* **2015**, *61*, 1581–1592. [CrossRef]
30. Li, C.; Wei, X.; Hu, G.J.; Zhang, X.L.; Xu, L.Q. Variation characteristics of first frost dates of Chifeng City in recent 50 years. *Inner Mong. Agric. Sci. Technol.* **2015**, *43*, 93–95. (In Chinese)
31. Bai, C.Y.; Li, S.K.; Bai, J.H.; Zhang, H.B.; Xie, R.Z. Characteristics of accumulated temperature demand and its utilization of maize under different ecological conditions in Northeast China. *Chin. J. Appl. Ecol.* **2011**, *22*, 2337–2342. (In Chinese)
32. Zheng, H.B.; Qi, H.; Liu, W.R.; Zheng, J.Y.; Luo, Y.; Li, R.P.; Li, W.T. Yield characteristics of ZD958 and XY335 under different N application rates. *J. Maize Sci.* **2013**, *21*, 117–119, 126. (In Chinese)
33. Gao, X. Effect on Stalk Lodging-Resistance Characteristics and Photosynthetic Characteristics by Different Types of Maize Cultivar High Density. Master's Thesis, Inner Mongolia Agricultural University, Hohhot, China, 2012. (In Chinese)
34. Liu, Y.; Li, Y.F.; Li, J.S.; Yan, H.J. Effects of mulched drip irrigation on water and heat conditions in field and maize yield in sub-humid region of Northeast China. *Trans. Chin. Soc. Agric. Mach.* **2015**, *46*, 93–104. (In Chinese)
35. Yin, W.; Chen, G.P.; Chai, Q.; Zhao, C.; Feng, F.X.; Yu, A.Z.; Hu, F.L.; Guo, Y. Responses of soil water and temperature to previous wheat straw treatments in plastic film mulching maize field at Hexi Corridor. *Sci. Agric. Sin.* **2016**, *49*, 2898–2908. (In Chinese)
36. Lv, X. Studies on Effects of Ecological Factors on Growth of Maize and Establishment of Climate Ecology Model and Appraisement System. Ph.D. Thesis, Shandong Agricultural University, Jinan, China, 2002. (In Chinese)
37. Zhang, J.L.; Zhou, Y.J.; Hu, M.; Zhu, L.; Zhang, D. The effects of low-temperature stress on chilling resistance ability of maize seedings. *J. Northeast Agric. Univ.* **2004**, *35*, 129–134. (In Chinese)
38. Dong, J.W.; Liu, J.Y.; Tao, F.; Xu, X.L.; Wang, J.B. Spatio-temporal changes in annual accumulated temperature in China and the effects on cropping systems, 1980s to 2000. *Clim. Res.* **2009**, *40*, 37–48. [CrossRef]
39. Duvick, D.N. Possible Genetic Causes of Increased Variability in U.S. Maize Yields. *Plant Breed. Yield Variability* **1989**, 147–156. Available online: http://www.cabdirect.org/abstracts/19901615378.html (accessed on 22 August 2017).
40. Dong, H.F.; Li, H.; Li, A.J.; Yan, X.G.; Zhao, C.M. Relations between delayed sowing date and growth, effective accumulated temperature of maize. *J. Maize Sci.* **2012**, *20*, 97–101. (In Chinese)
41. Tsimba, R.; Edmeades, G.O.; Millner, J.P.; Kemp, P.D. The effect of planting date on maize: Phenology, thermal time durations and growth rates in a cool temperate climate. *Field Crops Res.* **2013**, *150*, 145–155. [CrossRef]
42. Dou, S.J.; Dou, S.Y. Exploration of method to determine the optimum planting date based on accumulated temperature. *Soybean Sci. Technol.* **2009**, *5*, 44–45. (In Chinese)

43. Wilson, D.R.; Muchow, R.C.; Murgatroyd, C.J. Model analysis of temperature and solar radiation limitations to maize potential productivity in a cool climate. *Field Crops Res.* **1995**, *43*, 1–18. [CrossRef]
44. Meng, Y.; Li, M.; Wang, L.M.; Wang, L.Z.; Feng, Y.J. Effects of Chilling Injury on maize and correlative research. *Heilongjiang Agric. Sci.* **2009**, *4*, 150–153. (In Chinese)
45. Martin, M.; Gavazov, K.; Körner, C.; Hättenschwiler, S.; Rixen, C. Reduced early growing season freezing resistance in alpine treeline plants under elevated atmospheric CO_2. *Glob. Chang. Biol.* **2010**, *16*, 1057–1070. [CrossRef]
46. Lv, S.Q. Study on Different Nitrogen Application Schemes under Fertigation for Maize (*Zea mays* L.). Ph.D. Thesis, Chinese Academy of Agricultural Sciences, Beijing, China, 2012. (In Chinese)

5

On-Farm Evaluation of the Potential use of Greenhouse Gas Mitigation Techniques for Rice Cultivation

Noppol Arunrat *, Sukanya Sereenonchai and Nathsuda Pumijumnong

Faculty of Environment and Resource Studies, Mahidol University, Nakhon Pathom 73170, Thailand; sukanya.ser@mahidol.ac.th (S.S.); nathsuda.pum@mahidol.ac.th (N.P.)

* Correspondence: noppol.aru@mahidol.ac.th

Abstract: Environmental and socio-economic evaluations that imply techniques for mitigating greenhouse gas (GHG) emissions from rice cultivation are a challenging and controversial issue. This study was designed to investigate the potential use of mitigation techniques for rice cultivation. Mid-season drainage (MD), using ammonium sulfate instead of urea (AS), and site-specific nutrient management (SSNM) were chosen as mitigation techniques. Data were collected using field surveys and structured questionnaires at the same 156 farms, covering four crop years. The GHG emissions were evaluated based on the concept of the life cycle assessment of the GHG emissions of products. The farmers' assessments of mitigation techniques, with multiple criteria evaluation, were obtained by face-to-face interviews. Opinions on all mitigation techniques were requested two times covering four years with the same 156 farm owners. The multinomial logistic regression model was used to examine the factors influencing the farmers' decisions. The results show that SSNM was evaluated as the highest abatement potential (363.52 $kgCO_2eq\ ha^{-1}$), the negative value of abatement cost (-2565 THB ha^{-1}), and the negative value of the average abatement cost (-14 THB $kgCO_2eq^{-1}$). Among the different techniques, SSNM was perceived as the most suitable one, followed by MD and AS. Highly significant factors influencing decision making consisted of planted area, land size, farmer liability, farmer perception of yield, and GHG emissions. Subsidies or cost-sharing measures to convince farmers to adopt new techniques can enhance their practices, and more support for the development of water systems can increase their availability.

Keywords: rice field; mitigation techniques; greenhouse gas emissions; life cycle assessment; farmer acceptance; incentive measures

1. Introduction

Rice paddies are considered to be one of the most important sources of anthropogenic emissions of greenhouse gases (GHGs), particularly nitrous oxide (N_2O), methane (CH_4), and carbon dioxide (CO_2) [1] and therefore play an important role in climate change [2,3]. Notably, many studies state that N_2O emissions are associated with nitrogen (N) fertilizer application and dry land conditions [4,5], while flooded fields are a significant source of CH_4 and contribute little to N_2O emissions [6–8]. The use of agricultural machines requires the use of fossil fuels, resulting in CO_2 emissions. Projected increases in the demand for rice have raised considerable concerns about increasing greenhouse gas (GHG) emissions [9]. Thus, knowledge about trade-offs between rice yield increases and GHG emission reductions is urgently needed for the development of effective mitigation and adaptation strategies.

Considering possible strategies for mitigating GHG emissions from rice cultivation, those having no effect on rice yield would be the best techniques. Methane emissions vary markedly with water management. In particular, mid-season drainage, with the short-term removal of irrigation water, is one of the most promising strategies for reducing CH_4 emissions [10–12]. Several field measurements indicate that mid-season drainage (MD) significantly reduces CH_4 emissions and exerts a positive impact on rice yields by increasing N mineralization in the soil and increasing rice plant root development [13–17]. However, it also increases N_2O emissions by creating nearly saturated soil conditions, which promote N_2O production [18–20]. Fertilizer management has frequently been suggested as a mitigation option by substituting urea as N fertilizer with ammonium sulfate $(NH_4)_2SO_4$ (inhibits methanogens) and ammonium phosphate (promotes rice plant growth) [21]. Ammonium sulfate has a significant effect on N_2O reduction and slightly depresses CH_4 production by 10–67% [22], because sulfate-reducing bacteria can outcompete CH_4-producing bacteria under these conditions [23]. Moreover, site-specific nutrient management (SSNM) has been suggested as a method to reduce N_2O emissions by controlling the use of fertilizers with synchronization and precise farming techniques, using slow-release nutrients (including nitrification inhibitors) [24,25] and avoiding their overuse [26]. Dobermann and Cassman [27] state that an N recovery of over 70% can be achieved for many cereal crops by using intensive site-specific nutrient management, based on the principles of the 4R nutrient stewardship—the right source at the right rate, time, and place [28]. However, the sources of CH_4 and N_2O from rice fields cannot be reliably identified and discriminated in various areas.

There is an urgent need to quantify the effects and costs of mitigation strategies in rice fields, which, at present, remain difficult to enumerate, and could result as being speculative. A significant problem is that most farmers do not apply these mitigation strategies, for various reasons such as no ownership on farmland [29,30], less education or training on mitigation strategies [30,31], low income and access to credit [30–32], or less farming experience [33]. An evaluation method is therefore required that highlights decision factors and provides insight into the balance between environmental impacts, economic productivity, and social acceptance regarding mitigation strategies. Another significant problem is that the decision-making processes in terms of employing mitigation strategies are complicated by financial incentives and because agricultural activities depend on, and have a large impact on, natural resources [34]. These factors indicate the need to better understand decision making by farmers and the barriers inhibiting the adoption of mitigation and adaptation strategies.

Mitigation and adaptation are two basic, but distinctly different responses. Farmers' attitudes towards these two general responses to tackle changing climate conditions must be understood if scientists, policy makers, and others are to effectively support adaptive and mitigative actions [35,36]. Moreover, integrating mitigation and adaptation are win-win actions because they can mitigate the causes of climate change (mitigation) and adapt to changing climatic conditions (adaptation) [37]. Many studies have investigated farmer behavior and the associated socio-economic characteristics (e.g., [38–40]). Until now, mitigation costs caused by improvements in farming practices have rarely been reported, and information on the socio-economic feasibility of these mitigation techniques are still lacking, while their social acceptance and the minimization of their costs have not been discussed at any length. Therefore, the objectives of this study are: (1) to evaluate the GHG emissions of each mitigation technique for rice cultivation; (2) to clarify the farmers' assessment with multiple criteria evaluation of each mitigation technique; and (3) to examine the factors influencing the farmers' decisions to use a mitigation technique. The knowledge provided by this study can aid policy makers and other related agencies in their efforts to design and compare mitigation policies and reach mitigation goals.

2. Materials and Methods

2.1. Mitigation Technique Selection

Mitigation techniques were selected based on a literature review and on the recommendations of experts, provided in a report by the Office of Agricultural Economics [41], Ministry of Agriculture and Cooperatives, Thailand. Moreover, we expected that any mitigation techniques suggested to government agencies would be likely to be promoted and supported by the government in the near future. Based on these criteria, mid-season drainage (MD), replacement of urea with ammonium sulfate ($(NH_4)_2SO_4$) (AS), and site-specific nutrient management (SSNM) were chosen as mitigation techniques for this study.

2.2. Site Selection

Multi-stage sampling was employed for this study as follows. Firstly, at the provincial level, purposive sampling was used, focusing on farmers who have grown rice. They voluntarily participated and provided their information and opinions. Secondly, at the district and sub-district levels, cluster sampling was used to determine two clusters: irrigated areas and rain-fed areas. Moreover, farmers' average net household incomes (calculated by subtracting expenses from total revenue) for each district and sub-district were set as the criterion, based on the assumption that money is the major factor that can improve their livelihood and is the major factor likely to convince them to change their behavior. The four districts with the highest net incomes (Bang Mun Nak, Taphan Hin, Bueng Na Rang, and Pho Prathap Chang districts) and the four districts with the lowest net incomes (Sam Ngam, Wachira Barami, Wang Sai Phun, and Thap Khlo districts) in Phichit province were selected as samples.

2.3. Data Collection

Data were obtained from participatory observation, in-depth interviews, and a questionnaire survey at the same 156 farms (in irrigated and rain-fed areas of 78 farms, respectively) in four crop years (2012/2013, 2013/2014, 2014/2015 and 2015/2016) to avoid data variation. Data throughout the crop years from each crop, consisting of cultivation practices, agricultural inputs (e.g., fossil fuels, fertilizers, insecticides, herbicides, and water sources), yields, transportation costs, and benefits were collected from the farm owners. Data were also obtained from the record books for the standards for good agricultural practices (GAP) for farm owners, which was disseminated to the farmers by the Department of Agricultural Extension, Ministry of Agriculture and Cooperatives, Thailand.

2.4. Estimation of GHG Emissions

2.4.1. System Boundary and Functional Unit

The concept of the life cycle assessment of the greenhouse gas emissions of products, based on cradle-to-gate, was employed. It is because this approach is widely used for evaluating and comparing the environmental impacts of various products, and also to identify, quantify, and track the sources of GHG emissions throughout production process [42]. System boundary covers raw material production, transport of agricultural inputs (diesel fuel, gasoline fuel, chemical fertilizers, insecticides and herbicides) to the farm, land preparation, planting, harvesting, storing and post-harvest burning of crop residues (Figure 1). The transportation data were considered for two distances: the average distance from the farms to the retailer in the municipality of each sub-district and the average distance from the farms to the retailer in the community of each farm. Burning crop residues in the paddy field were included in this study because it is a common way to eliminate rice residues in Asia, including Thailand [43,44], and GHG emissions from open burning concentrated in the harvest season [45]. It is indicated that emissions from burning crop residues play an important role in the air pollution and climate change [46]. To assess the combined global warming potential (GWP), CH_4, and N_2O were calculated as CO_2 equivalents over a 100-year time scale, using a radiative forcing potential relative to CO_2 of 28 for CH_4 and 265 for N_2O [47]. The functional unit used in assessments was kg CO_2eq ha^{-1} for each technique.

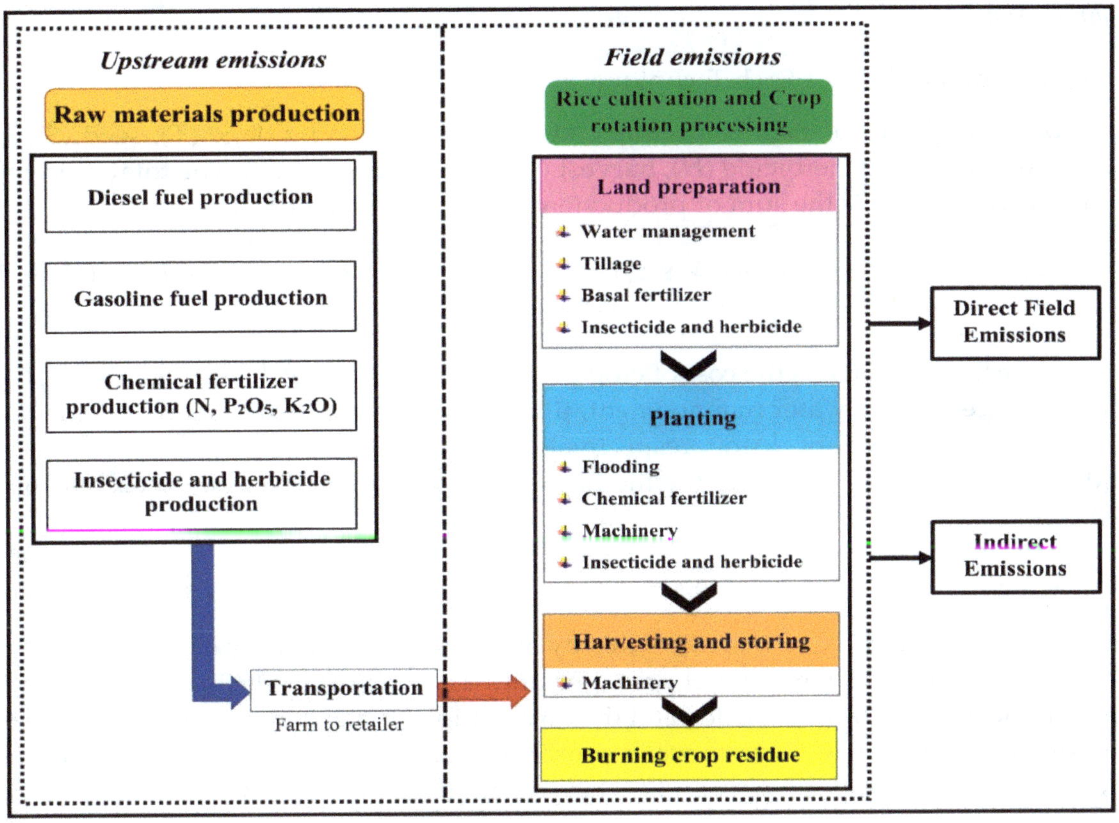

Figure 1. System boundary from cradle to farm gate of the study (adapted from Arunrat et al. [48]).

2.4.2. Calculation of GHG Emissions

The GHG emissions were calculated for each farm using four scenarios, including the business as usual (BAU) case, and the use of MD, AS, and SSNM techniques. Upstream emissions were accounted for in terms of raw material production and the transportation of agricultural inputs to the farm. Fossil fuels, chemical fertilizers, as well as insecticide and herbicide production were estimated using specific emission factors, as characterized in Ecoinvent 3.2 [49]. Emissions from the transportation of agricultural inputs to the farm were estimated based on diesel fuel consumption, using the emission factors from the National Technical Committee on Product Carbon Footprinting (Thailand) [50]. In some cases, specific emission factors for gasoline or insecticides and herbicides were not available in Ecoinvent 3.2, so country-specific emission factors for Thailand from the National Technical Committee on Product Carbon Footprinting (Thailand) [50] were used instead.

Field CH_4 emissions from rice cultivation were used as the model for the calculations, according to the 2006 Intergovernmental Panel on Climate Change (IPCC) Guidelines for National Greenhouse Gas Inventories [51]. The baseline emission factor was taken from Yan et al. [16], who adjusted region-specific emission factors for rice fields in east, southeast, and south Asian countries, and all scaling factors used were derived from the IPCC [51]. Direct and indirect N_2O emissions and CO_2 emissions from urea applications were also estimated using the methodology proposed by the IPCC [50]. The GHG emission calculations and parameters and emission factors for diesel and gasoline usage in stationary combustion were taken from the IPCC [51]. The GHG emissions from the mobile combustion of diesel fuel by farm tractors and harvesters were estimated from the emission factors of Maciel et al. [52], and GHG emissions from gasoline fuel were estimated following the EPA [53]. Figures for insecticides and herbicides were provided by the emission factors from Lal [54]. Equations, parameters, and emission factors for the calculation of GHG emissions are presented in the Supplementary Material by Arunrat et al. [48].

2.5. Economic Analysis

2.5.1. Estimation of the Costs of Each Technique

The production input of each technique consists of water (W), tillage (T), seed (S), labor (L), fertilizer (F), insecticide (P), herbicide (H), harvest (V), and land rental (R). The total production cost $[C(Q_i)]$ for each technique is the sum of production input costs Equation (1).

$$C(Q_i) = (C_W \times W_i) + (C_T \times T_i) + (C_s \times S_i) + (C_L \times L_i) + (C_F \times F_i) + (C_P \times P_i) + (C_H \times H_i) + (C_V \times V_i) + (C_R \times R_i) \quad (1)$$

where i is each technique, $C(Q)$ is the cost of crop production, in Baht ha^{-1}, and C_W, C_T, C_S, C_L, C_F, C_P, C_H, C_V, and C_R are costs of water management, tillage, seed, labor, fertilizer, insecticide, herbicide, harvest, and land rental, in Baht^{-1} unit, respectively.

In addition, the specific details of the methods used to estimate the costs of each technique are described below.

(1) *MD Technique*

The cost of the MD technique was calculated by multiplying the quantity of fuel used for pumping water back into the fields, using the fuel price per unit. The cost of this technique was investigated depending on the distance from the fields and the ownership of the water source by dividing the farms into two groups: (1) those far away from water sources (natural sources or irrigation systems at >100 m or >50 m from the fields, respectively); and (2) farms with their own surface pond or artesian well.

(2) *AS Technique*

The use of ammonium sulfate (21-0-0) instead of urea (46-0-0) requires changes in the quantities of the fertilizers used and their costs. The relevant calculations are as follows: (1) 1 kg of urea contains 0.46 kg N; (2) it takes 2.19 kg of ammonium sulfate to replace 1 kg of urea, providing 0.46 kg of N; (3) the amount of ammonium sulfate used, multiplied by its unit price, is equal to the total cost of the ammonium sulfate used.

(3) *SSNM Technique*

The cost of the SSNM technique was calculated based on the following steps. Firstly, the amount of each fertilizer to be used was calculated based on the instructions provided by the Land Development Department of Thailand after soil factor analysis. For instance, in the Nong Phra sub-district, Wang Sai Phun district, the soil series is Chiang Rai, suitable for growing photosensitive rice varieties. Suggested fertilizers are 31 kg ha^{-1} of 46-0-0, 71 kg ha^{-1} of 16-20-0, and 37 kg ha^{-1} of 0-0-60, to be applied 7–10 days after sowing or 25–30 days after transplanting, and 31 kg ha^{-1} of 46-0-0, to be applied again during the early flowering phase. After the suitable amounts of all fertilizers were established, the cost of each fertilizer used was calculated by multiplying the quantity by the price per unit. Finally, the total fertilizer cost of the SSNM technique was compared to the fertilizer cost of the BAU case.

2.5.2. Average Abatement Cost (AAC)

The AAC was used to assess the economic potential for the reduction of GHG emissions in this study; AAC refers to the cost of implementing a technique to reduce GHG emissions to an anticipated level. Similar to the GHG emission estimations, AAC was estimated using four scenarios comprising the BAU case and the use of the MD, AS, and SSNM techniques. The AAC (THB kgCO_2eq^{-1}) of each technique was calculated by dividing the total abatement cost (THB ha^{-1}) (TAC) by the total abatement potential (kgCO_2eq ha^{-1}) (TAP), and each TAC and TAP were obtained by subtracting the cost under the BAU scenario. Indeed, the reduction of GHG emissions is involved with cropping system, mitigation techniques, and farmers' behavior. Therefore, ACC was then presented to the farmers of each farm during their assessments on each mitigation technique. This is because ACC can help the farmers to visualize about being environmentally friendly and reducing production costs.

2.6. Farmers' Assessment and Analysis Tools

After the last crop year (2015/2016) for data collection, the investigation of the farmers' assessment for each farm was taken place in 2017. A multiple criteria evaluation was developed to assess farmers in the qualitative evaluation of the mitigation techniques. In this study, the criteria applied in the multiple criteria evaluation for farmers' assessment on the three mitigation techniques were as defined in Table 1, adapted from Webb et al. [55]. To reduce the bias and uncertainty from the farmers' assessment, the survey was administered via a face-to-face interview in November 2016 and August 2017, with the same 156 farm owners. The farmers were introduced and explained the purposes of the survey. The farmers' assessment was investigated after calculating the AAC for each scenario and each farm, but the farmers were allowed to choose only one suitable technique to implement. A questionnaire was presented to the farmers to evaluate the rating of each mitigation technique. A four-Likert scale was adopted for the evaluation [56]. The rating scale for the farmers' assessment was: '4' = very good, '3' = good, '2' = poor, and, '1' = very poor. We used a four-point scale to interpret the farmers' response because a mid-point is considered as too ambiguous for decision making [57], which was also mentioned in Webb et al. [55]. The scores of each farmer were summed up from the scores of each criterion for the three mitigation techniques. For instance, 78 farmers gave a score of 4 (very good) to the MD technique on the criteria of effectiveness; the total score was 312 (78 × 4). Moreover, the farmers were asked about their needs for policies and incentives to support their farming.

Table 1. Definitions of the criteria for farmers' assessment (adapted from Webb et al. [55]).

Criteria	Definition
Effective	Evaluates whether or not the mitigation technique reduces GHG emissions
Flexible	Evaluates whether or not the ability of the mitigation technique to enhance opportunity for other cropping systems and places
Economically efficient	Evaluates whether or not implementing the mitigation technique reduces production cost and increases household income
Easy to implement	Evaluations whether a mitigation technique is easy to implement by farmers with technical and managerial ease
Ability to trial	Evaluates whether a mitigation technique can be easily trialed or tested before full implementation
Institutional compatibility	Evaluates whether a mitigation technique is consistent with the current management framework, laws, regulations and will be promoted and supported by the government in the near future

2.7. Estimating the Determinants of Mitigation Techniques and Socio-Economic Variables

Factors that might influence the farmers' decision to adopt or reject the mitigation techniques were examined using the multinomial logistic regression (MNL) model. The MNL model is an extension of logistic regression, which is generally effective when the dependent variable is composed of a polytomous category with multiple choices. Explanatory variables included in the MNL model were defined as two types: dichotomous and continuous variables, as detailed below (Table 2). The model was estimated using the following specification:

$$\begin{aligned} Y = \; & \beta_0 + \beta_1 AREA + \beta_2 EXP + \beta_3 OWN + \beta_4 SIZE + \beta_5 INC + \beta_6 LIB \\ & + \beta_7 LABOR + \beta_8 MEM + \beta_9 PYIELD + \beta_{10} PGHG + \beta_{11} MEA \\ & + \beta_{12} TRAIN + \beta_{13} DOUB + \beta_{14} TRI + u \end{aligned} \quad (2)$$

where *Y* is the acceptability of the mitigation technique; *AREA* is the planted area; *EXP* is the experience; *OWN* is the land owner; *SIZE* is the land size; *INC* is the farmer´s income; *LIB* is liability; *LABOR* is the amount of labor; *MEM* is the membership of the environment group; *PYIELD* is the perception of yield; *PGHG* is the perception of GHG emissions; *MEA* represents government measures; *TRAIN* represents attendance at training; *DOUB* is the double cropping system; *TRI* is the triple cropping system; and μ is the error term.

Table 2. Definition and descriptive statistics of variables used in the MNL model.

Variable	Description
Planted area	Dummy, 1 if the farm is located in a rain fed area; 0 irrigated area
Experience	Continuous, rice cultivation experience of farmer (years)
Land owner	Dummy, 1 if the farmer is a land owner; 0 otherwise
Land size	Continuous, size of plantation (ha)
Farmer income	Continuous, farmer income from in-farm and off-farm (THB year^{-1} household^{-1})
Farmer liability	Continuous, farmer liability from formal and informal financial institutions (THB household^{-1})
Number of labor	Continuous, number of laborers in the household (persons)
Membership of environment group	Dummy, 1 if the farmer is the member of an environmental group or institution; 0 otherwise
Perception on yield	Dummy, 1 if the farmer's perception is that the mitigation technique will increase the rice yield; 0 otherwise
Perception on GHG emissions	Dummy, 1 if the farmer thinks that the mitigation technique can reduce GHG emissions; 0 otherwise
Perception on measures	Dummy, 1 if the farmer's perception is that the mitigation technique will be supported by government agencies; 0 otherwise
Attendance in training	Dummy, 1 if the farmer had attended the training about the impact of climate change impact on the environment; 0 otherwise
Double cropping system	Dummy, 1 if the farmer practices as usual the double cropping system; 0 otherwise
Triple cropping system	Dummy, 1 if the farmer practices as usual the triple cropping system; 0 otherwise

3. Results and Discussion

3.1. Cost of Rice Production under BAU and Mitigation Techniques

Marked significant differences in costs between irrigated and rain-fed areas were revealed using the t-test ($p < 0.05$). The average production costs under BAU were 27,521 and 24,240 THB ha^{-1} for irrigated and rain-fed areas, respectively. Using cost structure analysis, the average variable cost was 22,375 THB ha^{-1}, consisting of an average labor cost of 11,918 THB ha^{-1} and an average material cost of 10,456 THB ha^{-1}, while the average fixed cost was 4213 THB ha^{-1}. Furthermore, a lack of laborers and water for planting were the outstanding factors increasing the production costs. The average rice yields were 5.58 and 4.58 tons ha^{-1} for irrigated and rain-fed areas, respectively. The net profit in irrigated areas was higher than that in rain-fed areas, being 34,079 and 32,960 THB ha^{-1}, respectively.

This study found that when implementing the MD technique, the average cost of rice production was 30,100 and 29,662 THB ha^{-1} for irrigated and rain-fed areas, respectively. Rain-fed areas were associated with higher average production costs than irrigated areas, about 2840 THB ha^{-1} or double the increase in costs. Comparing the cost of water source distance, farmers who owned their surface pond or artesian well, implementing MD, would face average costs 1946 THB ha^{-1} higher than those for BAU. Meanwhile, at distances of 100 and 50 m from the water sources, the costs would be 6843 and 5584 THB ha^{-1}, respectively. Consequently, this study reflects that the cost of implementing MD is reduced by 28–35% if farmers own their own surface pond or artesian well for cultivation, while the average cost will be higher with increasing distance to the water source.

To implement the AS technique, the average production costs were 28,985 and 25,998 THB ha^{-1} for irrigated and rain fed-areas, respectively. An interesting point is that organic farmers following the AS technique can reduce their costs by about 645 and 863 THB ha^{-1} for irrigated and rain-fed areas, respectively, due to their lower costs for chemical fertilizer application under the BAU case.

Therefore, if organic farmers switch from using urea to ammonium sulfate, their average costs will be reduced as well. A cost-benefit analysis showed that organic rice farming could generate higher net profits than conventional farming, of about 437 and 289 THB ha^{-1} for irrigated and rain-fed areas, respectively. Consequently, to effectively implement the AS technique, organic fertilizer should be applied in combination to further reduce costs and increase net profit while not affecting rice yields.

For SSNM, the average production costs were 26,450 and 23,354 THB ha^{-1} for irrigated and rain-fed areas, respectively. Following this technique, farmers could achieve reductions in the average production cost compared with BAU of 1068 and 885 THB ha^{-1} for irrigated and rain-fed areas, respectively. The average production costs in irrigated areas were about 182 THB ha^{-1} lower than those in rain-fed areas, as lower amounts of chemical fertilizer were applied under BAU conditions.

Comparing the cost of BAU and using mitigation techniques for both irrigated and rain-fed areas, performing SSNM can reduce the average production costs compared with BAU. However, MD and AS resulted in higher production costs than BAU. Overall, the average production costs were higher in irrigated areas than in rain-fed areas. This result reflects that the average production costs are higher when farmers own more land for growing rice, but this higher average cost tends to decrease when farmers adapt their rice cultivation behavior by adopting the option that has lower costs than BAU, without reducing the rice yields.

3.2. *GHG Emissions, Abatement Potential, and AAC Under BAU and Mitigation Techniques*

The results of estimates of GHG emissions, abatement potential, and AAC between BAU and the different mitigation techniques are presented in Table 3 and Figures 2 and 3. There were highly significant differences in the first and second cultivations between irrigated and rain-fed areas and for each technique. These results reflect the fact that MD is more appropriate for implementation in irrigated rather than rain-fed areas and more appropriate for the second rice cultivation than for the first cultivation. The AS technique led to a higher abatement potential for the second rice cultivation than for the first one. Meanwhile, SSNM generated a 42.6% higher abatement potential for the second rice cultivation than for the first one, with a 9.8% lower AAC for irrigated than rain-fed areas. However, among all techniques, SSNM was the most appropriate one because its AAC was lower than that for BAU, and it had a 60.2 and 58.1% higher abatement potential than MD and AS, respectively.

Table 3. Average abatement cost (AAC) using different mitigation techniques (Authors own calculation).

	GHG Emissions under BAU ($kgCO_2eq\ ha^{-1}$)	GHG Emissions under Mitigation Technique ($kgCO_2eq\ ha^{-1}$)	Abatement Potential ($kgCO_2eq\ ha^{-1}$)	Abatement Cost (THB ha^{-1})	AAC (THB $kgCO_2eq^{-1}$)
MD technique					
1st rice					
Irrigated	3549	3411	138	7372	53
Rain-fed	3214	3089	125	8975	71
2nd rice					
Irrigated	2767	2590	176	7960	45
Rain-fed	2185	2046	139	9663	69
AS technique					
1st rice					
Irrigated	3549	3403	146	3405	23
Rain-fed	3214	3062	151	3002	19
2nd rice					
Irrigated	2767	2618	148	3641	24
Rain-fed	2185	2022	163	3499	21
SSNM technique					
1st rice					
Irrigated	3549	3276	273	−4718	−17
Rain-fed	3214	2888	326	−3747	−11
2nd rice					
Irrigated	2767	2269	497	−6600	−13
Rain-fed	2185	1828	357	−5738	−15

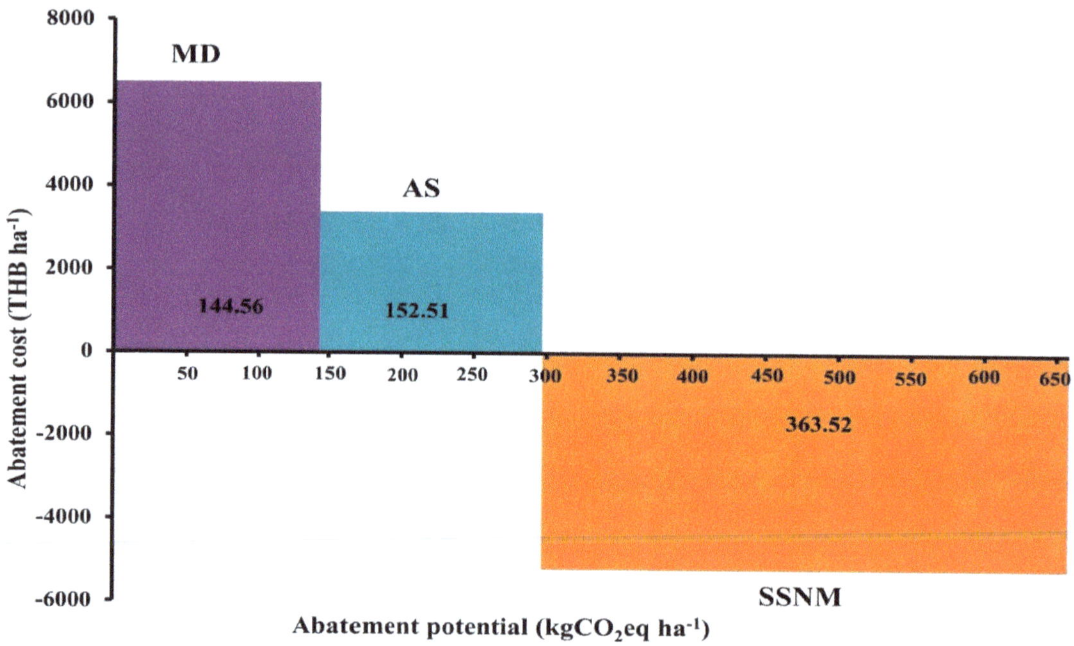

Figure 2. Comparison between abatement cost and abatement potential for each mitigation technique (Authors own calculation).

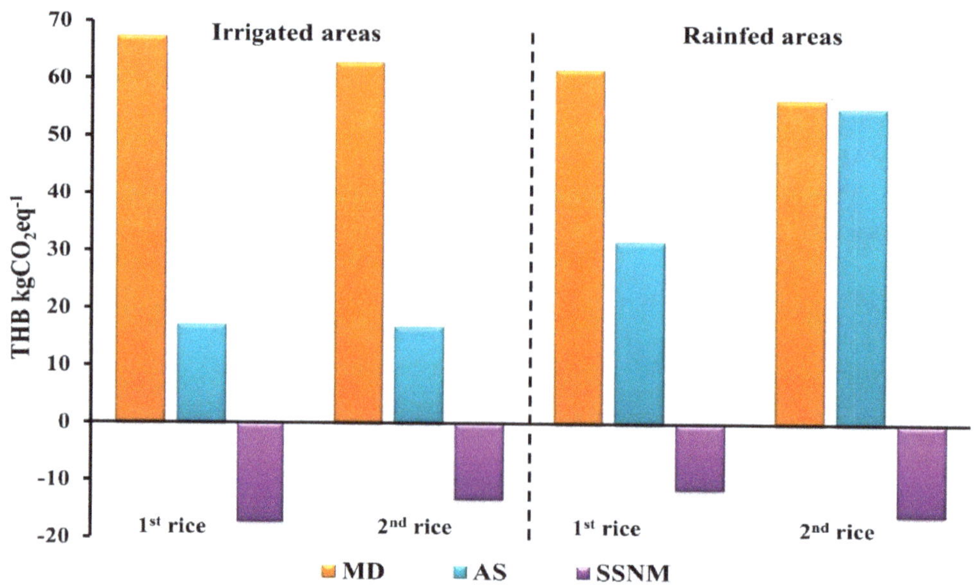

Figure 3. Average abatement cost (AAC) under BAU and using mitigation techniques (Authors own calculation).

3.3. Farmers' Assessment on Mitigation Techniques and Barriers

In the survey, farmers were requested to indicate their opinion on all mitigation techniques. Farmers' assessments across multiple criteria and the total score of each mitigation technique are provided in Table 4. As a result, the SSNM technique was the most favored one and presented the highest score, followed by MD and AS, respectively. The criteria of effectiveness, flexibility, economic efficiency, and institutional compatibility indicated the highest score regarding the SSNM technique. This is in line with Dobermann et al. [58], who reported that the higher benefit for farmers from the implementation of nutrient management strategies can increase the profitability of rice cropping, enhance socio-economic conditions, and mitigate labor shortage. Moreover, efficient nutrient management can also result in environmental benefits through a reduction of chemical fertilizers without a reduction in yield [59]. The criteria "easy to implement" and "ability to trial" were implementing the MD technique because it is easy to drain the water out of the rice field, but farmers need reliable control over irrigation water to implement this technique, otherwise rice yields are impacted. On the other hand, the AS technique obtained the lowest scores for the criteria "economic efficiency", "easy to implement", and "institutional compatibility".

Table 4. Summary of farmers' assessment with multiple criteria evaluation of each mitigation technique (Authors own calculation).

Assessment Criteria	Mitigation Techniques		
	MD	AS	SSNM
Effectiveness	542	393	588
Flexibility	317	446	565
Economic efficiency	376	201	603
Farmer implementability	496	233	468
Ability to trial	510	420	464
Institutional compatibility	495	233	570
Total score	**2736**	**1926**	**3258**

The scale used for scoring is presented in Table 4; green reflects low scores, while red reflects high scores.

The percentage of farmers ranking the mitigation techniques for each criterion, indicating the level of agreement, across the survey is provided in Table 5. The SSNM technique was the technique most favored by the farmers, with 86.5% indicating that they strongly agreed with the highest economic efficiency compared with other mitigation techniques, while only 13.5% of farmers indicated that they strongly agreed that this technique is easy to implement. Indeed, 4.5% of the farmers considered its "ability to trial" as very poor. Similarly, Chinese farmers willing to adopt low-carbon technology when the expenses of required inputs increase less after application [60]. In terms of the MD technique, 50% of the farmers strongly agreed with "effectiveness", followed by "institutional compatibility" (43.6%), "farmer implementability" (41.7%), and "ability to trial" (40.4%). However, 87.8% and 17.9% of farmers considered "flexibility" as poor and "economic efficiency" as very poor, respectively. Further, 32.1 and 10.9% of farmers evaluating the AS technique selected very good in terms of "flexibility" and "ability to trial". On the other hand, 71.2% of the farmers considered "economic efficiency" of the AS technique as very poor.

Table 5. The percentage of farmers showing a score of the level of agreement for each criteria (Authors own calculation).

Criteria/Rank	Mitigation Techniques											
	MD				AS				SSNM			
	1	2	3	4	1	2	3	4	1	2	3	4
Effectiveness	0	2.6	47.4	50.0	10.9	32.1	51.3	5.8	0	0	23.1	76.9
Flexibility	4.5	87.8	7.7	0	3.8	38.5	25.6	32.1	0	2.6	39.1	58.3
Economic efficiency	17.9	41.0	23.1	17.9	71.2	28.8	0	0	0	0	13.5	86.5
Farmer implementability	9.0	5.8	43.6	41.7	50.6	49.4	0	0	0	4.5	82.1	13.5
Ability to trial	1.3	10.9	47.4	40.4	2.6	36.5	50.0	10.9	4.5	16.7	17.9	60.9
Institutional compatibility	0	26.3	30.1	43.6	34.0	54.5	11.5	0	0	0	34.6	65.4

The scale used for scoring is presented in Table 5; green reflects low scores, while red reflects high scores.

When the farmers were asked to select one technique, 58.87% of the respondents were willing to implement SSNM, 29.29% AS, and 11.84% MD. Farmers in irrigated areas were most willing to perform SSNM, followed by AS and MD. In contrast, farmers in rain-fed areas were most willing to operate via SSNM, followed by AS, similar to those in irrigated areas, but no farmers were willing to implement

MD. As a result, we suggest that state policies should encourage SSNM in both irrigated and rain-fed areas as a practice that can result in lower fertilizer use. However, the relative willingness, beliefs, attitudes, and perceptions concerning such choices are indicators of the future likelihood to adopt a certain practice, which have also been described by McCown [61], Morton [62], and Jones et al. [63].

The reasons for the unwillingness to implement MD were water shortage, fear of increased weeds and pests, worries about nutrient losses, potential declines in rice yield, and a perception of MD being time-consuming, labor-consuming, and requiring more investment. Concerning the AS technique, farmers were worried about lower yields when not using urea, as they believe that urea contributes to greater yields, and there was a lack of knowledge about implementing the use of ammonium sulfate. Farmers unwilling to implement SSNM were concerned about yield decrease and felt that SSNM is time-consuming and complex. They also reported a lack of knowledge to support the use of soil analysis and high expenditures on soil analysis as matters of concern.

3.4. Factors Determining Farmers' Decisions

The results of the MNL model are presented in Table 6. The variables that were highly significant in the allocation of the farmers' decisions concerning each mitigation technique were as follows: (i) planted area; (ii) land size; (iii) farmer liability; (iv) farmer's perception of yield; and (v) farmer's perception of GHG emissions. Multicollinearity was checked among independent variables. The variance inflation factor (VIF) for all independent variables ranged from 1.108 to 1.265 (VIF < 5), which means that multicollinearity should not be a serious concern in this regression ($p < 0.01$).

Table 6. Estimated marginal effects of the farmers' decision to use the mitigation technique.

Variable	Mitigation Technique		
	MD	AS	SSNM
Planted area	−0.246 ** (0.0732)	−1.082 *** (0.153)	0.381 *** (0.022)
Experience	0.00384 (0.00492)	0.00376 (0.00348)	0.00743 (0.00315)
Land owner	0.00485 (0.0105)	0.0255 * (0.00503)	0.0466 * (0.0062)
Land size	−1.208 *** (0.0632)	−0.00478 (0.00255)	0.050 * (0.0260)
Farmer income	0.164 ** (0.00478)	0.403 ** (0.00455)	0.365 ** (0.00173)
Farmer liability	−0.411 * (0.00251)	−0.548 *** (0.000751)	0.332 *** (0.000177)
Number of labor	0.0301 (0.0137)	0.00428 (0.00199)	0.0676 (0.00295)
Membership of environment group	0.0446 (0.00662)	0.0507 (0.00227)	0.215 ** (0.00351)
Perception on yield	−0.0643 * (0.0338)	0.0661 (0.0255)	0.332 *** (0.00708)
Perception on GHG emissions	−0.0162 (0.0582)	−0.314 ** (0.0122)	−0.209 *** (0.00314)
Perception on measures	0.00944 (0.0132)	0.0407 (0.00671)	0.00194 (0.0118)
Attendance in training	0.0552 (0.00831)	0.0253 (0.0448)	0.0158 ** (0.00257)
Double cropping system	0.0308 (0.000744)	−0.206 ** (0.00678)	0.0321 (0.0186)
Triple cropping system	−0.0269 (0.00731)	0.0316 (0.0733)	−0.00736 (0.00228)
Constant	122461.72 ** (13562.15)	140939.82 ** (18953.05)	−159005.10 *** (10535.43)
Observations	156	156	156

* $p < 0.1$; ** $p < 0.05$; *** $p < 0.01$; SE in parentheses.

In the area studied, a great number of rice fields are located in rain-fed areas. The negative coefficient for rain-fed areas for MD and AS implies that these techniques are considerably less likely to be implemented in rain-fed areas compared with the irrigated areas, or not implemented at all. The reason is that when implementing MD in rain-fed areas, it is difficult to drain water into rice fields after it has been drained out, resulting in higher costs. Similarly, in terms of the AS technique, the farmers felt unaccustomed to the use of ammonium sulfate fertilizers. If adopting AS, farmers face higher costs as more ammonium sulfate fertilizer is required to maintain the same level of nutrients while possibly achieving lower yields. On the other hand, SSNM has a positive and significant influence when implemented, and it is highly likely that farmers will implement this technique.

Land size is an important factor influencing farmers' decisions in terms of various mitigation techniques. Land size had a negative and significant influence on MD, which probably means that the larger the land, the less likely the farmers are to implement MD. The same is true for AS, which can generate higher production costs in water and chemical fertilizer management. In contrast, farmers who owned more land were interested in SSNM because of its obvious cost savings. However, farmers with large areas of land were also worried about high expenses for soil characteristics analysis.

Of the significant variables, farmer liability had a positive influence favoring SSNM, while having a negative influence towards AS. Therefore, farmers with greater liabilities were interested in low-cost techniques and may reject high-cost techniques.

The effect on rice yield of each mitigation technique was the priority of the farmers. Consequently, farmers' perception of yield was one of the significant variables influencing their decision making. The results show that farmers' perception of yield had a positive and significant influence favoring SSNM. It can be inferred that farmers perceived that implementing SSNM could increase their yields, so they decided to use it.

Farmers' perception of GHG emissions had a negative and significant influence favoring SSNM and AS, meaning that farmers perceived that implementing SSNM and AS techniques would reduce GHG emissions, which was particularly the case for SSNM. Likewise, MD had a negative but non-significant influence, which might be because most farmers still do not have sufficient knowledge about the mitigation potential of each technique. It should be noted that relevant and responsible organizations should encourage and provide knowledge on GHG reduction techniques. Sources of information, including extensions, workshops, and training can enhance the adoption of a certain technology [30]. However, there are several farmers who have less chances for training, probably due to a limitation of time and budget. Therefore, participatory action research should receive more attention both from research-funding organizations and researchers to support collaborations among academicians, local authorities/leaders, and farmers [64]. This would increase the effectiveness of transferring knowledge, the sharing of knowledge and experiences, and could serve as a means to raise awareness about the positive effects of mitigation techniques.

3.5. *Prioritizing Incentive Measures for the Adoption of Mitigation Techniques*

Understanding farmers' decision-making behavior regarding their current practices is important and must be based on the knowledge of why farmers reject or accept different techniques [65]. Based on the results of the field survey and the in-depth interviews, three incentive measures were important from the point of the view of farmers: (1) cash incentives from governmental agencies to convince farmers to adapt their practices; (2) assistance for cost reduction—seed support and soil property analysis; and (3) support for water system development for agricultural activities—digging ponds and drilling wells near rice fields. The classification of farmers' characteristics for prioritizing supporting measures were identified as follows.

3.5.1. Planted Area

Farmers in irrigated areas rated cash incentive measures as the highest priority, while farmers in rain-fed areas were more concerned about supports for water system development.

3.5.2. Land Size

According to land tenure, farmers could be grouped as: (1) small land owners (1.3–6.5 ha); (2) medium land owners (6.6–11.6 ha); and (3) large land owners (11.7–16.8 ha). Medium land owners rated supports for water system development as the highest priority, while small and large land owners rated assistance with cost reduction as their major concern.

3.5.3. Farmer Income

Farmers could be categorized into three groups based on their income: (1) low-income farmers (52,800–128,000 THB year^{-1} household^{-1}); (2) medium-income farmers (128,001–203,200 THB year^{-1} household^{-1}); and (3) high-income farmers (203,201–278,400 THB year^{-1} household^{-1}). Farmers with medium and high incomes rated support for water system development as the first priority, followed by assistance with cost reduction and cash incentive measures. For farmers with low income, cash incentive measures were most important, because this measure had a direct impact on their income and expenses for implementing GHG mitigation techniques.

3.5.4. Farmer Liability

Regarding the levels of liability, there were three groups of farmers: low liability (58,400–538,933 THB household^{-1}), medium liability (538,934–1,019,467 THB household^{-1}), and high liability (1,019,468–1,500,000 THB household^{-1}). Low liability farmers mainly highlighted support for water system development, while medium liability farmers stressed assistance with cost reduction. High liability farmers highly valued cash incentive measures due to their direct and immediate impact on income. Farmers with low or medium liability gave higher priority to investment in their land (seeds, soil property analysis, and water sources).

3.5.5. Number of Laborers in a Household

According to the number of household members, farms were grouped into low-labor households (1–3 persons) and high-labor households (3–5 persons). Low-labor households made seed support a higher priority than high-labor households. This was because most low-labor households conducted their agricultural activities on smaller areas, so seed support and soil property analysis could greatly help to reduce their production costs. High-labor households prioritized support for water system development, because potential improvements in their water systems could allow them to increase their agricultural activities and gain more income.

3.5.6. Cropping System Pattern

Farmers using a double cropping system preferred support for water system measures, followed by cash incentives and assistance for cost reduction measures. This was because although the farmers' way of making a living in Thailand was based on rice cultivation, these farmers had limited water sources, so they selected crop rotation, which requires less water during the dry season. This could also reduce the cost of water management for agricultural activities. Among farmers using a triple cropping system, assistance for cost reduction measures was the first priority as it reduces the costs of seeds and soil property analysis.

The outstanding point was that cash incentives can be appropriate for low-income farmers or small land owners, who have fewer opportunities to increase their income and need more assistance. These farmers obviously considered subsidies are the priority. Besides, small land owners also placed emphasis on developing their land to be more appropriate for agricultural activities, as their main income relies on their land. On the other hand, high-income farmers and large land owners were aware of other alternatives to increase their income, whether from rice grain or crop rotation. Farmers with medium incomes or medium land owners were more concerned about water system development for agricultural activities than the other groups, because having enough water could lead to greater income and increased crop production efficiency [66]. For farmers with high liabilities, subsidies were of greater concern than for farmers with low or medium liabilities due to their direct and immediate effect on income.

4. Conclusions

Site-specific nutrient management (SSNM) was evaluated as the highest abatement potential (363.52 kgCO_2eq ha^{-1}), the negative value of abatement cost (−2565 THB ha^{-1}), and the negative value of the average abatement cost (−14 THB kgCO_2eq^{-1}). Based on farmers' assessment to be a mitigation technique for rice cultivation, SSNM reached the highest score for effectiveness, flexibility, economic efficiency, and institutional compatibility. This indicated that SSNM was obviously preferable and presented the highest scores for farmer acceptability, followed by the replacement of urea with ammonium sulfate ($(NH_4)_2SO_4$) and mid-season drainage. Irrigation systems, land size, farmers' liability, and perception of yield and GHG emissions were found as the main factors affecting the farmers' decision to accept the mitigation techniques. Therefore, incentive measures, such as subsidies or cost-sharing measures can convince farmers to adopt new techniques and enhance their practices. More support of water system development can increase their availability.

Author Contributions: N.A. collected data and preformed life cycle assessment and average abatement cost analyses, wrote, and revised the manuscript. S.S. collected data, analyzed farmers' assessment, and collaborated in discussion. N.P. provided the discussion and advice to this study.

Acknowledgments: This article was supported by the Thailand Research Fund (TRF) (grant number: TRG 5880123), and National Research Council of Thailand (NRCT) (KO-BO-NGO/2557-68). Furthermore, the author would like to thank the reviewers for their helpful comments to improve the manuscript.

References

1. Smith, P.; Martino, D.; Cai, Z. Agriculture. In *Climate Change 2007: Mitigation. Contribution of Working Group III to the Fourth Assessment Report of the Intergovernmental Panel on Climate Change*; Metz, B., Davidson, O.R., Bosch, P.R., Dave, R., Meyer, L.A., Eds.; Cambridge University Press: Cambridge, UK; New York, NY, USA, 2007; pp. 497–540.
2. Intergovernmental Panel on Climate Change (IPCC). *The Supplementary Report to IPCC Scientific Assessment*; Houghton, J.T., Callander, B.A., Varney, S.K., Eds.; Cambridge University Press: Cambridge, UK, 1992.
3. IPCC. *Climate Change 2007: The Physical Science Basis. Contribution of Working Group I to the Fourth Assessment Report of the Intergovernmental Panel on Climate Change*; Solomon, S., Qin, D., Manning, M., Chen, Z., Marquis, M., Averyt, K.B., Tignor, M., Miller, H.L., Eds.; Cambridge University Press: Cambridge, UK; New York, NY, USA, 2007.
4. Adviento-Borbe, M.A.A.; Haddix, M.L.; Binder, D.L.; Walters, D.T.; Dobermann, A. Soil greenhouse gas fluxes and global warming potential in four high-yielding maize systems. *Glob. Chang. Biol.* **2007**, *13*, 1972–1988. [CrossRef]
5. Zou, J.; Huang, Y.; Zheng, X.; Wang, Y. Quantifying direct N_2O emissions in paddy fields during rice growing season in mainland China: Dependence on water regime. *Atmos. Environ.* **2007**, *41*, 8030–8042. [CrossRef]
6. Shang, Q.; Yang, X.; Gao, C.; Wu, P.; Liu, J.; Xu, Y.; Shen, Q.; Zou, J.; Guo, S. Net annual global warming potential and greenhouse gas intensity in Chinese double rice-cropping systems: A 3-year field measurement in long-term fertilizer experiments. *Glob. Chang. Biol.* **2011**, *17*, 2196–2210. [CrossRef]
7. Wang, W.; Dalal, R.; Reeves, S.; Butterbach-Bahl, K.; Kiese, R. Greenhouse gas fluxes from an Australian subtropical cropland under long-term contrasting management regimes. *Glob. Chang. Biol.* **2011**, *17*, 3089–3101. [CrossRef]
8. Yao, Z.; Zheng, X.; Dong, H.; Wang, R.; Mei, B.; Zhu, J. A 3-year record of N_2O and CH_4 emissions from a sandy loam paddy during rice seasons as affected by different nitrogen application rates. *Agric. Ecosyst. Environ.* **2012**, *152*, 1–9. [CrossRef]
9. Van Groenigen, K.J.; van Kessel, C.; Hungate, B.A. Increased greenhouse-gas intensity of rice production under future atmospheric conditions. *Nat. Clim. Chang.* **2013**, *3*, 288–291. [CrossRef]
10. Yagi, K.; Tsuruta, H.; Minami, K. Possible options for mitigating methane emission from rice cultivation. *Nutr. Cycl. Agroecosyst.* **1997**, *49*, 213–220. [CrossRef]
11. Towprayoon, S.; Smakgahn, K.; Poonkaew, S. Mitigation of methane and nitrous oxide emissions from drained irrigated rice fields. *Chemosphere* **2005**, *59*, 1547–1556. [CrossRef] [PubMed]
12. Nayak, D.; Saetnan, E.; Cheng, K.; Wang, W.; Koslowski, F.; Cheng, Y.-F.; Zhu, W.Y.; Wang, J.-K.; Liu, J.-X.; Moran, D.; et al. Management opportunities to mitigate greenhouse gas emission from Chinese agriculture. *Agric. Ecosyst. Environ.* **2015**, *209*, 108–124. [CrossRef]
13. Cai, Z.; Yan, X.; Yan, G.; Xu, H.; Tsuruta, H.; Yagi, K.; Minami, K. Methane and nitrous oxide emissions from rice paddy fields as affected by nitrogen fertilisers and water management. *Plant Soil* **1997**, *196*, 7–14. [CrossRef]
14. Wassmann, R.; Neue, H.U.; Lantin, R.S.; Makarim, K.; Chareonsilp, N.; Buendia, L.V.; Rennenberg, H. Characterization of methane emissions from rice fields in Asia: II. Differences among irrigation, rainfed, and deepwater rice. *Nutr. Cycl. Agroecosyst.* **2000**, *58*, 13–22. [CrossRef]
15. Lu, W.F.; Chen, W.; Duan, B.W.; Guo, W.M.; Lu, Y.; Lantin, R.S.; Wassmann, R.; Neue, H.U. Methane emissions and mitigation options in irrigated rice fields in southern China. *Nutr. Cycl. Agroecosyst.* **2000**, *58*, 65–73. [CrossRef]

16. Yan, X.; Ohara, T.; Akimoto, H. Development of region-specific emission factors and estimation of methane emission from rice fields in the East, Southeast and South Asian countries. *Glob. Chang. Biol.* **2003**, *9*, 237–254. [CrossRef]
17. Zou, J.; Huang, Y.; Jiang, J.; Zheng, X.; Sass, R.L. A 3-year field measurement of methane and nitrous oxide emissions from rice paddies in China: Effects of water regime, crop residue, and fertilizer application. *Glob. Biogeochem. Cycl.* **2005**, *19*, GB2021. [CrossRef]
18. Zheng, X.H.; Wang, M.X.; Wang, Y.S.; Shen, R.X.; Shangguan, X.J.; Heyer, J.; Kögge, M.; Papen, H.; Jin, J.S.; Li, L.T. CH_4 and N_2O emissions from rice paddies in southeast China. *Chin. J. Atmos. Sci.* **1997**, *21*, 167–174.
19. Zheng, X.; Wang, M.; Wang, Y.; Shen, R.; Gou, J.; Li, J.; Jin, J.; Li, L. Impacts of soil moisture on nitrous oxide emission from croplands: A case study on the rice-based agro-ecosystem in South-east China. *Chemosphere Glob. Chang. Sci.* **2000**, *2*, 207–224. [CrossRef]
20. Cai, Z.C.; Xing, G.X.; Shen, G.Y.; Xu, H.; Yan, X.Y.; Tsuruta, H. Measurements of CH_4 and N_2O emissions from rice paddies in Fengqiu, China. *Soil Sci. Plant Nutr.* **1999**, *45*, 1–13. [CrossRef]
21. Towprayoon, S. Greenhouse Gas Mitigation Options from Rice Field. 2004. Available online: https://unfccc.int/files/meetings/workshops/other_meetings/application/vnd.ms-powerpoint/towprayoon.ppt (accessed on 30 June 2014).
22. Wassmann, R.; Lantin, R.S.; Neue, H.U.; Buendia, L.V.; Corton, T.M.; Lu, Y. Characterization of methane emissions from rice fields in Asia. III. Mitigation options and future research needs. *Nutr. Cycl. Agroecosyst.* **2000**, *58*, 23–36. [CrossRef]
23. Sela-Adler, M.; Ronen, Z.; Herut, B.; Antler, G.; Vigderovich, H.; Eckert, W.; Sivan, O. Co-existence of Methanogenesis and Sulfate Reduction with Common Substrates in Sulfate-Rich Estuarine Sediments. *Front. Microbiol.* **2017**, *8*, 766. [CrossRef] [PubMed]
24. Akiyama, H.; Yan, X.; Yagi, K. Evaluation of effectiveness of enhanced-efficiency fertilizers as mitigation options for N_2O and NO emissions from agricultural soils: Meta-analysis. *Glob. Chang. Biol.* **2009**, *16*, 1837–1846. [CrossRef]
25. Edmeades, D.C. *Nitrification and Urease Inhibitors—A Review of the National and International Literature on Their Effects on Nitrate Leaching, Greenhouse Gas Emissions and Ammonia Volatilisation from Temperate Legume-Based Pastoral Systems*; Technical Report 2004/22; Environment Waikato: Hamilton, New Zealand, 2004.
26. Xiang, Y.; Jin, J.Y.; Ping, H.E.; Liang, M.Z. Recent advances on the technologies to increase fertilizer use efficiency. *Agric. Sci. China* **2008**, *7*, 469–479.
27. Dobermann, A.; Cassman, K.G. Plant nutrient management for enhanced productivity in intensive grain production systems of the United States and Asia. *Plant Soil* **2002**, *247*, 153–175. [CrossRef]
28. Bruulsema, T.W.; Witt, C.; Garcia, F.; Li, S.; Rao, T.N.; Chen, F.; Ivanova, S. A global framework for fertilizer BMPs. *Better Crops* **2008**, *92*, 13–15.
29. Iheke, O.R.; Echebiri, R.N. Rural Land tenancy and Resource Use Efficiency of cassava farmer in South Eastern Nigeria. *J. Food Fibre Prod.* **2010**, *3*, 455–465.
30. Arunrat, N.; Wang, C.; Pumijumnong, N.; Sereenonchai, S.; Cai, W. Farmers' intention and decision to adapt to climate change: A case study in the Yom and Nan basins, Phichit province of Thailand. *J. Clean. Prod.* **2017**, *143*, 672–685. [CrossRef]
31. Sani, S.; Chalchisa, T. Farmers' Perception, Impact and Adaptation Strategies to Climate Change among Smallholder Farmers in Sub-Saharan Africa: A Systematic Review. *J. Ressour. Dev. Manag.* **2016**, *26*, 1–8.
32. Krause, M.A.; Deuson, R.R.; Baker, T.G.; Preckel, P.V.; Lowenberg-DeBoer, J.; Reddy, K.C.; Maliki, K. Risk-Sharing Versus Low-cost Credit Systems for International Development. *Am. J. Econ.* **1990**, *72*, 911–922. [CrossRef]
33. Tazeze, A.; Haji, J.; Ketema, M. Climate Change Adaptation Strategies of Smallholder Farmers: The Case of Babilie District, East Harerghe Zone of Oromia Regional State of Ethiopia. *J. Econ. Sustain. Dev.* **2012**, *3*, 1–13.
34. Sasaki, H. Farmer Behaviour, Agricultural Management and Climate Change. Available online: http://www20.iadb.org/intal/catalogo/pe/2012/09774.pdf (accessed on 18 June 2016).
35. Howden, S.M.; Soussana, J.F.; Tubiello, F.N.; Chhetri, N.; Dunlop, M.; Meinke, H. Adapting Agriculture to Climate Change. *Proc. Natl. Acad. Sci. USA* **2007**, *104*, 19691–19696. [CrossRef] [PubMed]
36. McCarl, B.A. Analysis of climate change implications for agriculture and forestry: An interdisciplinary effort. *Clim. Chang.* **2010**, *100*, 119–124. [CrossRef]
37. Grafakos, S.; Pacteau, C.; Delgado, M.; Landauer, M.; Lucon, O.; Driscoll, P.; Wilk, D.; Zambrano, C.;

O'Donoghue, S.H.; Roberts, D. Integrating Mitigation and Adaptation as Win-Win Actions. In *Climate Change and Cities: Second Assessment Report of the Urban Climate Change Research Network. Summary for City Leaders, Chapter: 4, Publisher: Urban Climate Change Research Network (UCCRN)*; Rosenzweig, C.W., Solecki, P., Romero-Lankao, S., Mehrotra, S., Dhakal, T., Bowman, S., Ibrahim, A., Eds.; Columbia University: New York, NY, USA, 2015; Volume 6, pp. 1–24.

38. Vanslembrouck, I.; Van Huylenbroeck, G.; Verbeke, W. Determinants of the Willingness of Belgian Farmers to Participate in Agri-environmental Measures. *J. Agric. Econ.* **2002**, *53*, 489–511. [CrossRef]
39. Maddison, D. *The Perception and Adaptation to Climate Change in Africa*; CEEPA Discussion Paper No. 10; Centre for Environmental Economics and Policy in Africa; University of Pretoria: Pretoria, South Africa, 2006.
40. Ahnstrom, J.; Francis, C.; Hockert, J.; Skelton, P.; Bergea, H.; Hallgren, L. Farmers and nature conservation: What is known about attitudes, context factors and actions affecting conservation? *Renew. Agric. Food Syst.* **2009**, *24*, 38–47. [CrossRef]
41. OAE (Office of Agricultural Economics). *Mitigation Abatement Cost of Greenhouse Gases from Rice Production*; Ministry of Natural Resources and Environment: Bangkok, Thailand, 2012; 355p. (In Thai)
42. Lattanzio, R.K. *Life-Cycle Greenhouse Gas Assessment of Coal and Natural Gas in the Power Sector*; Congressional Research Service: Washington, DC, USA, 2015.
43. Gadde, B.; Bonnet, S.; Menke, C.; Garivait, S. Air pollutant emissions from rice straw open field burning in India, Thailand and the Philippines. *Environ. Pollut.* **2009**, *157*, 1554–1558. [CrossRef] [PubMed]
44. Kanokkanjana, K.; Cheewaphongphan, P.; Garivait, S. Black Carbon Emission from Paddy Field Open Burning in Thailand. In *Proceedings of the 2011 2nd International Conference on Environmental Science and Technology, IPCBEE (6), Singapore, 26–28 February 2011*; IACSIT Press: Singapore, 2011.
45. Niu, H.; Cheng, W.; Hu, W.; Pian, W. Characteristics of individual particles in a severe short-period haze episode induced by biomass burning in Beijing. *Atmos. Pollut. Res.* **2016**, *7*, 1072–1081. [CrossRef]
46. Zhang, X.; Lu, Y.; Wang, Q.; Qian, X. A high-resolution inventory of air pollutant emissions from crop residue burning in China. *Atmos. Chem. Phys. Discuss* **2018**. [CrossRef]
47. IPCC. *The Physical Science Basis: Working Group I Contribution to the Fifth Assessment Report of the Intergovernmental Panel on Climate Change*; Cambridge University Press: Cambridge, UK; New York, NY, USA, 2013.
48. Arunrat, N.; Wang, C.; Pumijumnong, N. Alternative cropping systems for greenhouse gases mitigation in rice field: A case study in Phichit province of Thailand. *J. Clean. Prod.* **2016**, *133*, 657–671. [CrossRef]
49. Ecoinvent Centre. *Ecoinvent Database*, version 3.2; Swiss Centre for Life Cycle Inventories: Duebendorf, Switzerland, 2005. Available online: http://www.ecoinvent.org/ (accessed on 25 August 2016).
50. The National Technical Committee on Product Carbon Footprinting. *The National Guideline on Product Carbon Footprint*, 3rd ed.; Amarin Publishing: Bangkok, Thailand, 2011.
51. IPCC. Agriculture, forestry and other land use. In *2006 IPCC Guidelines for National Greenhouse Gas Inventories*; Eggleston, H.S., Buendia, L., Miwa, K., Ngara, T., Tanabe, K., Eds.; Institute for Global Environmental Strategies (IGES): Hayama, Japan, 2006; Volume 4.
52. Maciel, V.G.; Zortea, R.B.; Silva da, W.M.; Cybis, L.F.A.; Einloft, S.; Seferin, M. Life Cycle Inventory for the agricultural stages of soybean production in the state of Rio Grande do Sul, Brazil. *J. Clean. Prod.* **2015**, *93*, 65–74. [CrossRef]
53. EPA. *Emission Factors for Greenhouse Gas Inventories*; United States Environmental Protection Agency: Washington, DC, USA, 2014. Available online: https://www.epa.gov/sites/production/files/2015-07/documents/emission-factors_2014.pdf (accessed on 17 February 2016).
54. Lal, R. Carbon emission from farm operations. *Environ. Int.* **2004**, *30*, 981–990. [CrossRef] [PubMed]
55. Webb, N.P.; Stokes, C.J.; Marshall, N.A. Integrating biophysical and socio-economic evaluations to improve the efficacy of adaptation assessments for agriculture. *Glob. Environ. Chang.* **2013**, *23*, 1164–1177. [CrossRef]
56. Likert, R. A technique for the measurement of attitudes. *Arch. Psychol.* **1931**, *22*, 1–55.
57. Marshall, N.A. Understanding social resilience to climate variability in primary enterprises and industries. *Glob. Environ. Chang.* **2010**, *20*, 36–43. [CrossRef]
58. Dobermann, A.C.; Witt, D.; Dawe, S.; Abdulrachman, H.C.; Gines, R.; Nagarajan, S.; Satawathananont, T.T.; Son, P.S.; Tan, G.H.; Wang, N.V.; et al. Adviento Site-specific nutrient management for intensive rice cropping systems in Asia. *Field Crops Res.* **2002**, *74*, 37–66. [CrossRef]
59. Wang, G.H.; Dobermann, C.; Witt, Q.; Sun, Q.Z.; Fu, R.X. Performance of site-specific nutrient management for irrigated rice in Southeast China. *Agron. J.* **2001**, *93*, 869–878. [CrossRef]

60. Zhu, H.J.; Tian, Z.H. Analysis on the rice farmer's willingness to use low-carbon technologies measures: Based on the survey of the rice-producing areas in Southern China. *J. Agric. Econ.* **2013**, *215*, 62–71.
61. McCown, R.L. New thinking about farmer decision makers. In *The Farmer's Decision*; Hatfield, J.L., Ed.; Soil and Water Conservation Society: Ankeny, Lowa, 2005; pp. 11–44.
62. Morton, L.W. Citizen involvement. In *Pathways for Getting to Better Water Quality: The Citizen Effect*; Morton, L.W., Brown, S.S., Eds.; Springer: New York, NY, USA, 2011.
63. Jones, A.K.; Jones, D.; Edwards-Jones, G.; Cross, P. Informing decision making in agricultural greenhouse gas mitigation policy: A Best–Worst Scaling survey of expert and farmer opinion in the sheep industry. *Environ. Sci. Policy* **2013**, *29*, 46–56. [CrossRef]
64. Abebaw, D.; Belay, K. Factors influencing adoption of high yielding maize varieties in South-western Ethiopia. An application of logit. *Q. J. Int. Agric.* **2001**, *40*, 149–167.
65. Nowak, P. Why farmers adopt production technology. *J. Soil Water Conserv.* **1992**, *47*, 14–16.
66. Richard, T.; Shively, E.G. Technical Change and Productive Efficiency: Irrigated Rice in the Philippines. *Asian Econ. J.* **2007**, *21*, 155–168.

6

Evaluation of the Agronomic Impacts on Yield-Scaled N_2O Emission from Wheat and Maize Fields in China

Wenling Gao and Xinmin Bian *

College of Resources and Environmental Sciences, Nanjing Agricultural University, 1st Weigang Road, Xuanwu District, Nanjing 210095, China; 2007203003@njau.edu.cn

* Correspondence: bxm@njau.edu.cn

Abstract: Contemporary crop production faces dual challenges of increasing crop yield while simultaneously reducing greenhouse gas emission. An integrated evaluation of the mitigation potential of yield-scaled nitrous oxide (N_2O) emission by adjusting cropping practices can benefit the innovation of climate smart cropping. This study conducted a meta-analysis to assess the impact of cropping systems and soil management practices on area- and yield-scaled N_2O emissions during wheat and maize growing seasons in China. Results showed that the yield-scaled N_2O emissions of winter wheat-upland crops rotation and single spring maize systems were respectively 64.6% and 40.2% lower than that of winter wheat-rice and summer maize-upland crops rotation systems. Compared to conventional N fertilizer, application of nitrification inhibitors and controlled-release fertilizers significantly decreased yield-scaled N_2O emission by 41.7% and 22.0%, respectively. Crop straw returning showed no significant impacts on area- and yield-scaled N_2O emissions. The effect of manure on yield-scaled N_2O emission highly depended on its application mode. No tillage significantly increased the yield-scaled N_2O emission as compared to conventional tillage. The above findings demonstrate that there is great potential to increase wheat and maize yields with lower N_2O emissions through innovative cropping technique in China.

Keywords: climate change; food security; cropping system; soil management; greenhouse gas emission

---

1. Introduction

Nitrous oxide (N_2O) is a long-lasting greenhouse gas that significantly contributes to stratospheric ozone depletion and global warming. It is estimated that about 60% of total anthropogenic N_2O is emitted from agricultural soil, which is mainly produced by nitrification and denitrification processes of reactive nitrogen (N) in soil [1,2]. Reducing the N_2O emission from soil is urgent in contemporary crop production for the mitigation of global warming. However, global crop production is also facing a great challenge of growing by 70~100% by 2050 to meet an expected 34% increase in world population [3,4]. Meeting this goal will result in increased pressure to use more N fertilizer, thereby potentially increasing N_2O emission [5–7]. Maize (*Zea mays* L.) and wheat (*Triticum aestivum* L.) account for the largest and second largest global consumption of all fertilizer N in major cereal crops [8]. Therefore, it is necessary and urgent to study how to increase maize and wheat yields with lower N_2O emissions in the future.

Agronomic practices such as cropping systems and soil management options are the primary factors regulating N_2O emission from cropland soil. Improving these practices (e.g., reducing inorganic N fertilizer, use of enhanced-efficiency fertilizers and no tillage) has the potential to reduce N_2O emission from soil [9]. However, changes to agronomical practices often simultaneously affect crop yield. It is still early to decide which option is optimal for the balance of mitigating N_2O emission and increasing crop yield. A particular practice beneficial to reducing N_2O emissions may or may not favor crop yield enhancement. For example, replacing N fertilizer with manure can mitigate N_2O emission but could decrease crop yields compared to inorganic N fertilizer application only [10]. Application of enhanced-efficiency N fertilizers can reduce N_2O emissions but can either increase [11,12] or decrease crop yields [13,14]. Therefore, integrating assessment on both N_2O emissions and crop yield is essential in optimizing cropping practices. Although many studies have evaluated the impact and mitigation potential of cropping practices on N_2O emission [15–22], few studies have been linked to crop yield [23–25]. Recent studies suggest that comprehensive assessments of cropping practices per unit yield (yield-scaled) rather than land area (area-scaled) could benefit sustainable intensification of cropping practices and policy selection with a trade-off of N_2O emission mitigation and food security [23,24,26].

China takes the first and the second positions, respectively, in global wheat and maize production. Wheat and maize production in China was 121.7 and 217.8 million tons in 2013, approximately 17.1% and 21.4% of global output, respectively [27]. Meanwhile, N_2O emissions from croplands in China occur mostly during wheat and maize growing seasons in China [28]. As a result, mitigating yield-scaled N_2O emissions during these growing seasons in China plays an important role in the sustainable development of global cereal crop production. Using meta-analysis, this study integrated the results of field measurements to assess the mitigation potential of major agronomic practices on yield-scaled N_2O emissions from croplands during wheat and maize growing seasons in China.

2. Materials and Methods

2.1. Data Selection

A literature review of English and Chinese language peer-reviewed studies on N_2O emissions from Chinese wheat and maize fields prior to January 2017 was conducted using Thomson Reuters' ISI-Web of Science research database (http://thomsonreuters.com/thomson-reuters-web-of-science) and the China Knowledge Resource Integrated Database (www.cnki.net), the largest Chinese academic journal database. The 52 studies including 186 wheat and 167 maize measurements were selected based on the following criteria: (1) measurements were conducted under field conditions; (2) N_2O flux rates were measured during an entire crop growth period using the static chamber method; (3) N_2O emission and grain yield were determined simultaneously. (See Supplementary Materials Table S1 for details).

2.2. Data Analysis

For every study, the value of N_2O emission was converted to global warming potential (GWP) using a 100-year radiative forcing potential coefficient of 298 [29]. Area- and yield-scaled N_2O emissions were calculated in GWP of N_2O emission per unit of cropland and yield, respectively. In some studies, measurements were taken during more than one year; the mean value of the results measured in different years was calculated as a single observation.

Based on the field experiments conducted in selected studies, two kinds of major agronomic practices (cropping system and soil management practices) were assessed in the current study. Cropping systems were divided into four groups: winter wheat-upland crops rotation (W-U), winter wheat-rice rotation (W-R), single spring maize (M) and summer maize-upland crops rotation (M-U). W-U is mostly practiced in the semi-arid regions of northern China such as Shandong, Henan and Hebei provinces, where about 60% of China's wheat supply is produced. The winter wheat in W-U is usually planted between late September and early October and harvested between late May and early June. W-R is practiced in the humid regions along the Yangtze River of southern China; it accounts for about 28% of China's wheat production. The winter wheat in W-R is usually planted between late October and mid-November and harvested from late May to early June. M is mostly practiced in the northeast and northwest regions of China and accounts for about 44% of China's maize production.

It is usually planted between late April and early May and harvested from late September to early October. M-U is mostly practiced in the semi-arid or arid regions of eastern China and accounts for about 33% of China's maize production. It is summer maize and is usually planted in late June after the harvest of previous crops; it is harvested during similar periods to spring maize.

Weighted mean values of area-scaled N_2O emission, crop yield and yield-scaled N_2O emission were used as effect size indexes in current study to compare the difference between the four cropping systems. The equations used were as follows [24,30]:

$$Mean = \sum(y_i \times wt_i)/\sum wt_i \tag{1}$$

$$wt_i = n \times f/o \tag{2}$$

The details of these formulas can be found in Feng et al. [30]. Briefly, Equation (1) was used to calculate the weighted mean values of cropping systems. *Mean* is the mean value of area-scaled N_2O emission, crop yield and yield-scaled N_2O emission. Whereas y_i is the observation of area-scaled N_2O emission, crop yield and yield-scaled N_2O emission at the ith site, respectively. wt_i is the weight of the observations from the ith site and was calculated using Equation (2), in which, n is the number of replicates in the field experiment. f is the number of N_2O flux measurements per month and o is the total number of observations from the ith site. This weighting approach assigned more weight to the field measurements that were well replicated and in which more precise fluxes were estimated. The approach adjusted the weights according to total number of observations from one site to avoid dominating the dataset with studies with many observations from one site.

Four types of soil management practices were assessed in the study. These included inorganic N fertilizer application, enhanced-efficiency N fertilizers application, organic amendments and soil tillage. Their impact on area-scaled N_2O emission, crop yield, and yield-scaled N_2O emission were evaluated by the response ratio (Rr) [31]. Only studies including side-by-side comparisons were selected in the analysis of soil management practices. The rates of inorganic N fertilizer were empirically divided into six levels ($N < 100$, $100 \leq N < 150$, $150 \leq N < 200$, $200 \leq N < 250$, $250 \leq N < 300$ and $N > 300$ kg N ha^{-1} per season). The enhanced-efficiency fertilizers were categorized into two groups: nitrification inhibitors (NI) and controlled-release fertilizers (CRF). The organic amendments were classified as crop straw retention and three modes of manure application: (1) equal inorganic N fertilizer as the control with additional manure application (Equal IN + manure), (2) reduced inorganic N fertilizer with additional manure application (Reduced IN + manure), and (3) manure only application with N amount equal to the control (Manure alone). The mean retention amount of crop straw was 5768 kg ha^{-1} in selected studies. As in the three modes of manure application, the mean input rates of inorganic N fertilizer and manure were 175 kg and 194 kg N ha^{-1} for Equal IN + manure, 95 and 74 kg N ha^{-1} for Reduced IN + manure, and 0 and 154 kg N ha^{-1} for Manure alone, respectively, in selected studies. Finally, two groups of soil tillage practices (no tillage and reduced tillage) were analyzed.

The response ratio (Rr) of each management practice was calculated using Equation (3):

$$\ln \mathrm{Rr} = \ln(x_t/x_c) \tag{3}$$

where, x_t and x_c are the measurements for treatments and controls, respectively. The controls were non-fertilization, conventional N fertilizer, non-organic amendments and conventional tillage, respectively, which corresponded to inorganic N fertilizer application, enhanced-efficiency N fertilizers application, organic amendments and conservational tillage.

In addition, the mean of the response ratios was calculated from lnRr of individual studies using Equation (4):

$$\mathrm{Mr} = EXP(\sum[\ln r(i) \times wr(i)]/\sum wr(i)) \tag{4}$$

In Equation (4), $w(i)$ is the weighting factor and is estimated by Equation (5):

$$wr(i) = n \times f \tag{5}$$

where, n is the number of experiment replicates and f is the number of N_2O flux measurements per month.

Additionally, we further analyzed the effects of cropping systems and soil management practices under different aridity regions. Aridity is an integrated indicator of rainfall and potential evapotranspiration. Following the generalized climate classification scheme for Global-Aridity values, study sites with an aridity index < 0.65 were classified as "arid"; whereas study sites with a higher index (>0.65) were classified as "humid" [25].

The meta-analysis was performed using MetaWin 2.1 (Sinauer Associates Inc., Sunderland, UK) [32]. Mean effect sizes were estimated using the random-effects model. The 95% confidence intervals (CIs) of the mean effect sizes were calculated using the bootstrapping with 4999 iterations [24,32].

3. Results and Discussion

3.1. Mitigation Potential of Cropping Systems

As shown in Figure 1, there were significant differences in area-scaled N_2O emission, crop yield and yield-scaled N_2O emission between the cropping systems. Area-scaled N_2O emission during the wheat season of W-R was significantly higher (256%) than that of W-U (Figure 1a), although average N application amounts were similar (W-R, 171.4 kg N ha^{-1}; W-U, 161.7 kg N ha^{-1}). There are two possible reasons that might explain this. Firstly, continuous flooding during the rice season of W-R could have provided more substrate and favorable soil conditions for N_2O production in the following wheat season [33]. As a result, W-R stimulated more N_2O emission during the following wheat season compared to W-U. Secondly, W-R and W-U were respectively located in the humid subtropical and semi-arid temperate regions of China. The mean annual temperature and precipitation were higher for W-R (16–24 °C, 1000–2000 mm) compared to W-U (9–15 °C, 520–980 mm) [34], A relatively higher temperature and precipitation might have increased the N_2O emission during the wheat season of W-R [35].

However, wheat yields did not significantly differ between W-U and W-R (Figure 1b). The yield-scaled N_2O emission during the wheat season of W-U was 107.8 kg CO_2 eq Mg^{-1}, which was close to the estimation of N_2O emission of global wheat production [24]. The yield-scaled N_2O emission of W-R was 304.7 kg CO_2 eq Mg^{-1}, which was significantly higher than that of W-U. Thus, increasing the planting area of W-U and reducing W-R could reduce the yield-scaled N_2O emission by 64.6% (196.9 kg CO_2 eq Mg^{-1}) without wheat yield loss.

There was no significant difference in area-scaled N_2O emissions between M and M-U during the maize season (Figure 1d). However, the maize yield of M was significantly higher than that of M-U by 25.7% (Figure 1e). In China, spring maize is usually planted between late April and early May and harvested from late September to early October [36], while summer maize is usually planted in late June after harvesting previous crops and harvested at the same time as spring maize [10]. As a result, the longer growth period of spring maize contributed to the relatively higher yield.

The yield-scaled N_2O emission of M-U was 144.1 kg CO_2 eq Mg^{-1} (Figure 1f), which was also close to the estimation of N_2O emission of global maize production [24]. But the yield-scaled N_2O emission of M (86.1 kg CO_2 eq Mg^{-1}) was significantly lower than that of M-U. Although increasing M did not reduce the N_2O emission per unit of cropland, the N_2O emission per unit of maize yield could be mitigated by 40.2% (58.0 kg CO_2 eq Mg^{-1}) due to the relatively higher yield.

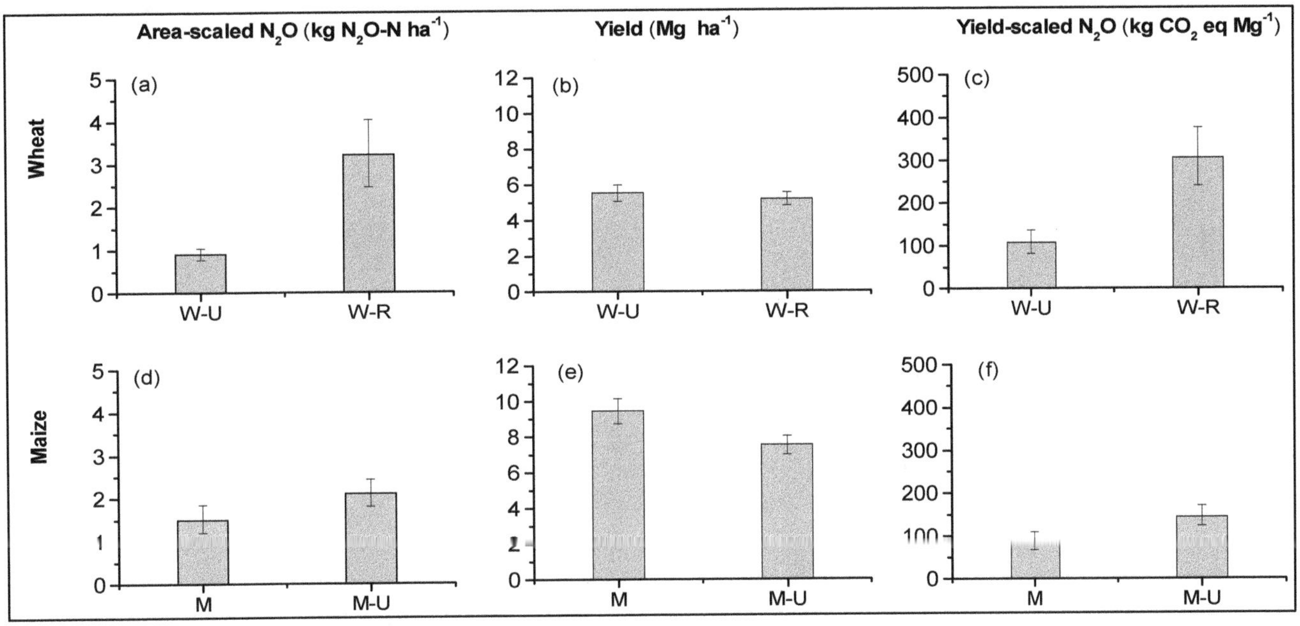

Figure 1. Impacts of cropping systems on area-scaled N_2O emission, crop yield, and yield-scaled N_2O emission during wheat and maize growing seasons ((**a**): area-scaled N_2O of wheat; (**b**) yield of wheat; (**c**): yield-scaled N_2O of wheat; (**d**): area-scaled N_2O of maize; (**e**) yield of maize; (**f**) yield-scaled N_2O of maize). The observations for winter wheat-upland crops rotation system (W-U), winter wheat-rice rotation system (W-R), single spring maize (M), and summer maize-upland crops rotation system (M-U) were 104, 76, 45, and 124, respectively. The error bars represent 95% confidence intervals.

We further analyzed the effects of aridity on the performance of the cropping system on the area-scaled N_2O emission, crop yield and yield-scaled N_2O emission. In China, W-U, M and M-U were located in both arid and humid regions, while W-R was mainly located in humid regions. So, we analyzed the differences of W-U, M and M-U in arid and humid regions (Figure 2). Though the mean N rate for M in arid area (201 kg N ha^{-1}) was higher than that in humid region (182 kg N ha^{-1}); the mean area- and yield-scaled N_2O emissions for M was significantly lower in arid than humid regions. As for W-U and M-U, the mean N rates were also higher in arid (172 kg and 175 kg N ha^{-1} for W-U and M-U,) than humid regions (119 kg and 126 kg N ha^{-1} for W-U and M-U); however, the higher N rate raised both the area-scaled N_2O emission and crop yield in arid than humid regions, resulting in no significant difference in yield-scaled N_2O emission between arid and humid regions. These results indicated that an arid climate was favorable for wheat and maize to control the yield-scaled N_2O emissions. This was possible because that low soil moisture inhibited N_2O production [25].

These results suggest that adjusting cropping systems had great potential in the mitigation of yield-scaled N_2O emission. Replacing W-R and M-U with W-U and M was the recommend strategy to mitigate yield-scaled N_2O emissions in national wheat and maize productions, especially in arid regions. During the past 20 years, the planting areas of W-R and W-U have been respectively reduced by 17.3% and 8.1%, which has been effective in mitigating yield-scaled N_2O emission. These changes are mostly affected by the comparative profits and consumption of wheat and maize in different agro-eco regions [37]. However, there has been almost no attention placed on the mitigation of N_2O emissions. Therefore, a national-scale plan is needed to balance the N_2O emission mitigation and food security by adjusting cropping systems in future wheat and maize production.

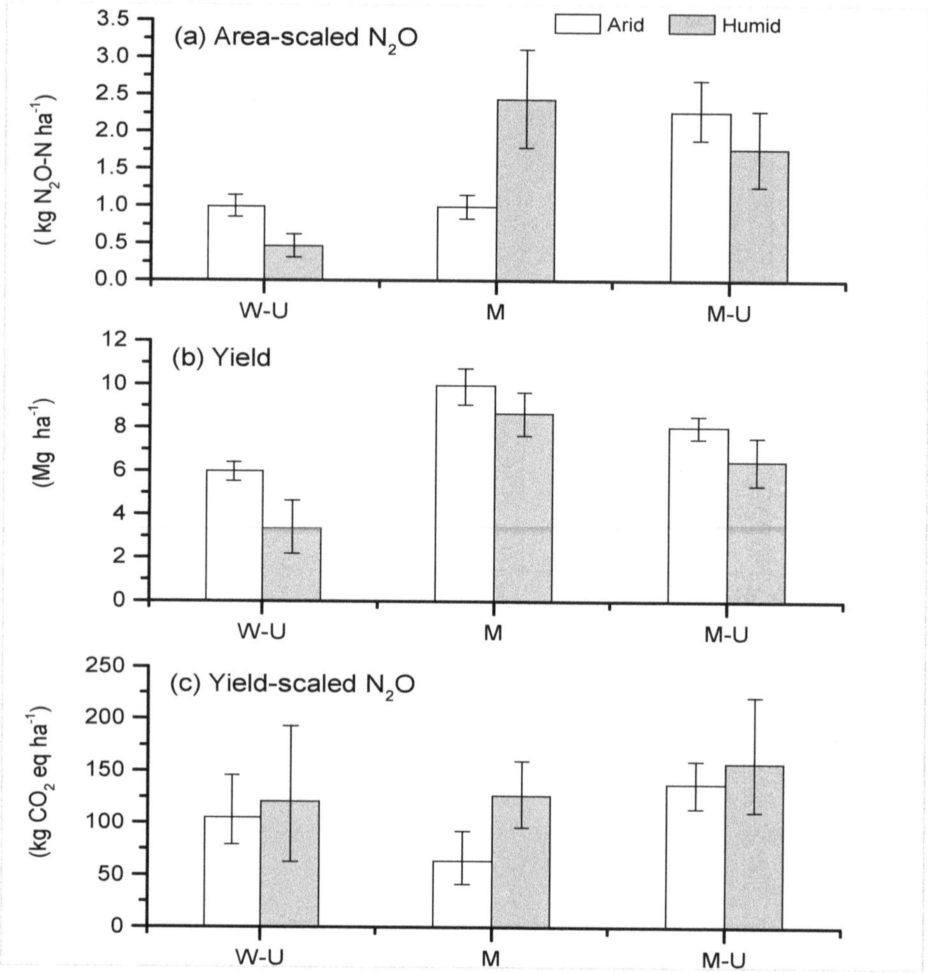

Figure 2. The impacts of aridity on the area-scaled N_2O emission (**a**), yield (**b**) and yield-scaled N_2O emission (**c**) of three cropping system.

3.2. Mitigation Potential of Inorganic N Fertilizer

The application of inorganic N fertilizer is essential for high crop production; however, it also directly provides the substrate for N_2O production. Comparing its contribution to crop yield and N_2O emission is essential for deciding the optimal N rate to mitigating yield-scaled N_2O emission. Results showed that the response ratios of N_2O emission to N addition were higher than that of crop yield at all N levels (Figure 3a), indicating that the application of inorganic N fertilizer could stimulate more N_2O emission than crop yield compared to no N fertilizer. In addition, the differences in response ratios between N_2O emission and crop yield increased with N input rates. This result was inconsistent with that of paddy fields, which showed that N fertilizer application raised more rice yield than total GWP of CH_4 and N_2O emissions [30]. In paddy fields, CH_4 emission contributed more than 80% of total GWP. When the N application rate was above 140 kg ha^{-1}, the inorganic N fertilizer began to inhibit CH_4 emission [38].

Therefore, it was difficult to obtain an optimal N rate that increased more crop yield than N_2O emission. Reducing the N application rate is the most promising option for mitigating N_2O emissions; however, it can affect crop yield. Based on this, the level that can achieve maximum economical returns or N uptake efficiency is usually suggested as the optimal N rate for the balance of crop yield enhancement and N_2O emission mitigation, because the addition of N beyond this level only slightly increases crop yield but produces far more N_2O emissions [23,39]. In our results, when the N fertilizer application rate was below 211 kg N ha^{-1}, the response ratios of N_2O emission and crop yield showed insignificant differences. However, when the N addition rate increased to 282 kg N ha^{-1}, the response ratio of N_2O emission became significantly higher (233%) than that of crop yield and the response ratio of yield-scaled N_2O emission increased significantly (Figure 3b). Thus, the suggested N rate for the balance of N_2O emission and crop yield was below 211 kg N ha^{-1}.

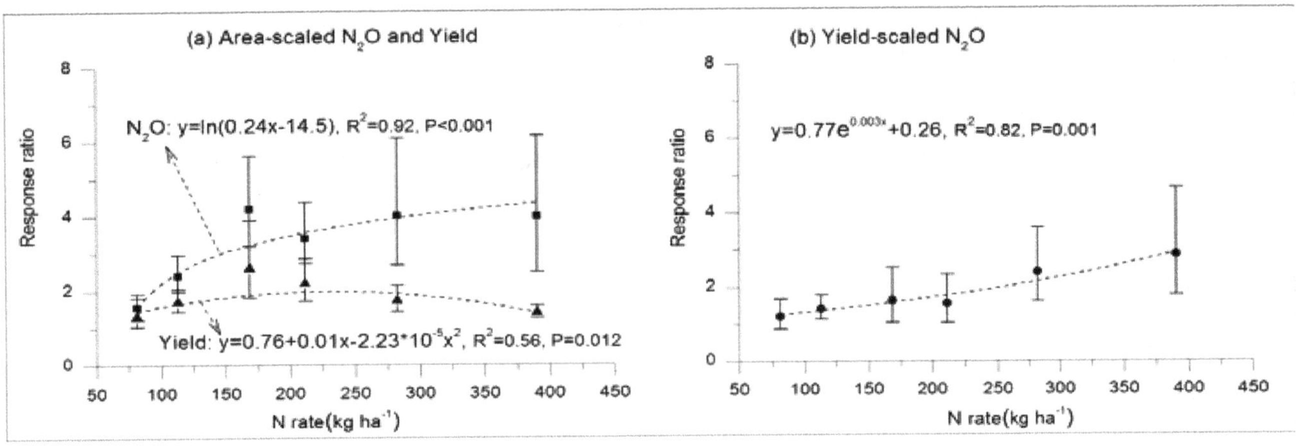

Figure 3. The relationship between N application rates and response ratios of area-scaled N_2O emission (**a**), crop yield (**a**), and yield-scaled N_2O emission (**b**). The data is expressed as mean response ratios of six N levels ($N < 100$, $100 \le N < 150$, $150 \le N < 200$, $200 \le N < 250$, $250 \le N < 300$ and $N > 300$ kg N ha^{-1} per season) with 95% confidence intervals. The observations for six N levels are 5, 12, 21, 14, 13 and 6, respectively. (Note: Only the subgroups of $100 \le N < 150$ and $150 \le N < 200$ have enough observations to differentiate the effects of N application under arid or humid areas; the study sites of other subgroups were all located in arid or humid regions. The results of the subgroups of $100 \le N < 150$ and $150 \le N < 200$ under different aridity regions were listed in the Supplementary Materials (Figure S1)).

In order to improve use efficiency of inorganic N fertilizer and reduce environmental impact, a more precise inorganic N application scheme had been recommended in major cereal crops planting regions in China since 2013 based on a national project of soil testing and fertilizer recommendation [40]. The suggested inorganic N application rates were 103–127 kg N ha^{-1}, 144–209 kg N ha^{-1} and 236–258 kg N ha^{-1} for low-yield (<6 Mg ha^{-1}), medium-yield (6–9 Mg ha^{-1}) and high-yield (>9 Mg ha^{-1}) croplands of wheat production respectively, and 105–167 kg N ha^{-1}, 136–206 kg N ha^{-1}, and 190–235 kg N ha^{-1} for low-yield (<7.5 Mg ha^{-1}), medium-yield (7.5–10.5 Mg ha^{-1}) and high-yield (>10.5 Mg ha^{-1}) croplands of maize production respectively. Only the N rates for high-yield croplands of wheat and maize production exceeded 211 kg N ha^{-1}. Thus, reducing N application rates in the high-yield croplands is essential for the mitigation of N_2O emission. However, reducing the N rate in high-yield croplands could decrease crop production and farmer's profits, because the output of high-yield croplands makes up a large part of farmers' profits. Therefore, financial incentives might be required to compensate farmers for reducing N application rates. Additionally, more work is needed to optimize the application options of inorganic N fertilizer (such as N source, placement and application time) that allow for N-rate reductions to better match crop growth demand and mitigate N_2O emissions without yield loss in high-yield croplands [41]. Improving these options could lessen the need for financial compensation [42].

3.3. Mitigation Potential of Enhanced-Efficiency N Fertilizers

Enhanced-efficiency N fertilizers have been developed to increase crop N use efficiency and decrease N loss to the environment. Our results (Figure 4) showed that, compared to conventional N fertilizer, NI significantly reduced N_2O emissions by 34.2%, which was similar to the report by [18]. NI can delay the bacterial oxidation of ammonium to nitrite and subsequently reduce the denitrification, which is an important process of N_2O production in upland soil [43]. So the application of NI can mitigate N_2O emission from soil. Additionally, the delay of nitrification also provides a better opportunity for the crop to uptake N fertilizer. Our results showed that wheat and maize yield increased 12.9% due to NI application compared to conventional N fertilizer, thereby resulting in a significant reduction in yield-scaled N_2O emission by 41.7% (Figure 4). Aridity did not affect the performance of NI. The effect sizes of NI on area-scaled N_2O emission, crop yield and yield-scaled N_2O did not show significant difference.

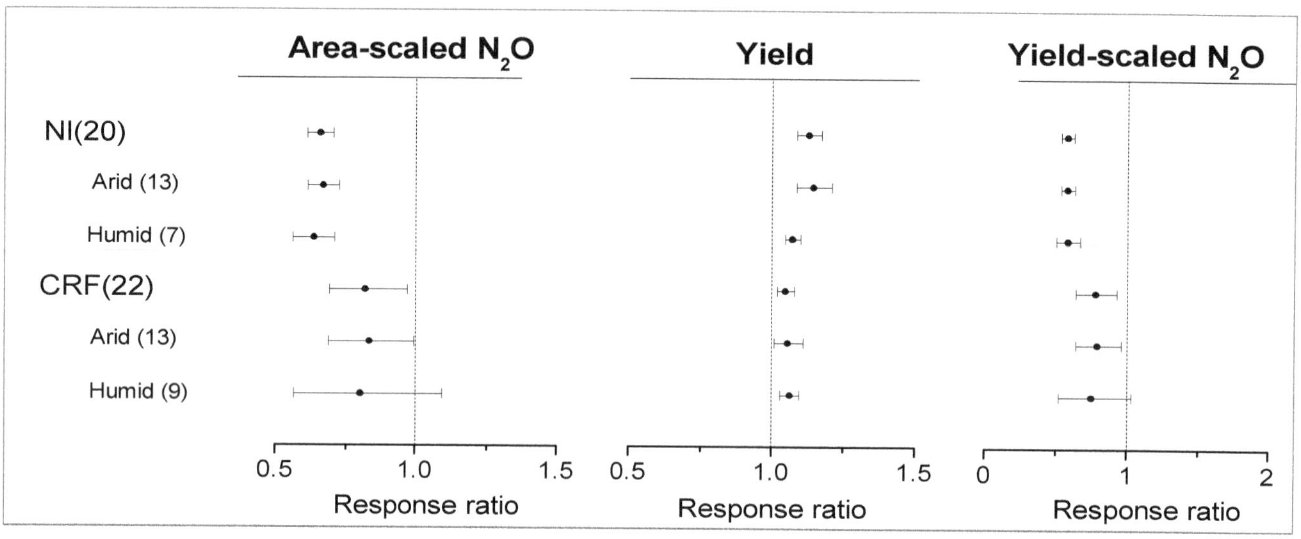

Figure 4. Impacts of enhanced-efficiency N fertilizers on area-scaled N_2O emission, crop yield and yield-scaled N_2O emission. The data is expressed as mean response ratio with 95% confidence intervals. The numbers of observations are indicated in the parentheses.

CRF also showed significant effect size on N_2O emission and crop yield (Figure 4). Compared to conventional inorganic fertilizer, N_2O emission was reduced 18.2% by CRF, which was lower than the report by [18]. In addition, crop yield increased by 4.9%. Yield-scaled N_2O emission was significantly mitigated by 22.0% due to CRF. The effect size of CRF was affected by aridity. CRF performed better in arid than humid regions. Though CRF significantly enhanced the crop yield in humid regions, its effects on area- and yield-scaled N_2O emissions were not significant. In this analysis, the CRF in selected studies was polymer-coated urea; this coating can slow down the release of N and subsequently reduce the loss of N_2O emission [44]. High soil moisture in humid regions may weaken the effect of CRF on controlling N release, and increase the N release from CRF [45], which may raise the N_2O production.

CRF did not perform as well as NI. The mitigation effect of CRF on area- and yield-scaled N_2O emissions was weaker than NI, and showed a greater 95% CI. A possible reason for this was that the release of nitrogen from CRF was easily affected by environmental factors such as soil moisture and temperature [13,46]. If N released from the CRF did not synchronize with crop N demands, the redundant N in favorable environmental conditions could raise the amount of N_2O emissions from denitrification [14,18]. The effect of CRF on N_2O emission might depend on field condition, climate aridity and crop growth.

Both NI and CRF showed a significant ability in mitigating yield-scaled N_2O emissions, and could be recommended to mitigate N_2O emissions without yield loss in wheat and maize production in China. However, NI and CRF are not widely used by farmers in cereal crop production in China since the additional costs of NI and CRF only increase limited crop yields. As a result, additional studies are needed to optimize the management options such as application time and irrigation approaches to improve the effectiveness of enhanced-efficiency fertilizers, especially CRF, on crop productivity [47] and to encourage farmers to use enhanced-efficiency fertilizers in maize and wheat production.

3.4. Mitigation Potential of Organic Amendments

No significant effect of crop straw retention was found on area-scaled N_2O emission, crop yield and yield-scaled N_2O emission (Figure 5). Existing evidence showed that the effect of straw retention can be either positive or negative on N_2O emissions [48,49]. On one hand, straw retention can increase soil temperature and/or moisture, which can stimulate the microbial process of nitrification and denitrification, and thereby raise N_2O emissions [49]. On the other hand, straw with a high C/N ratio can immobilize soil mineral N and decrease soil N availability, consequently leading to a reduction in the substrate N for N_2O production [48,50]. Additionally, the allelochemicals produced from the decomposition of crop straw can reduce the activity of nitrifiers and inhibit N_2O production [51]. The integrated impact of these effects was mostly determined by basal inorganic N application rate, retention timing and straw type [52,53]. As shown in Figure 6, the response ratio of N_2O emissions increased with the application rates of basal inorganic N fertilizer. Field experiments also reported that incorporation of straw with low N application rate could reduce N_2O emission compared to no straw retention [48]. Additionally, the performance of crop straw was affected by climate aridity. Crop straw returning significantly increased the N_2O emission in arid region, but did not affect N_2O emission in humid regions. In arid regions, the positive effect of crop straw, such as increased soil moisture and substrate C, may raise the N_2O emission. As in humid regions, the decomposition of straw possibly intensified the O_2 limitation due to the rapid microbial decomposition, and active the further reduction from N_2O to N_2 [20].

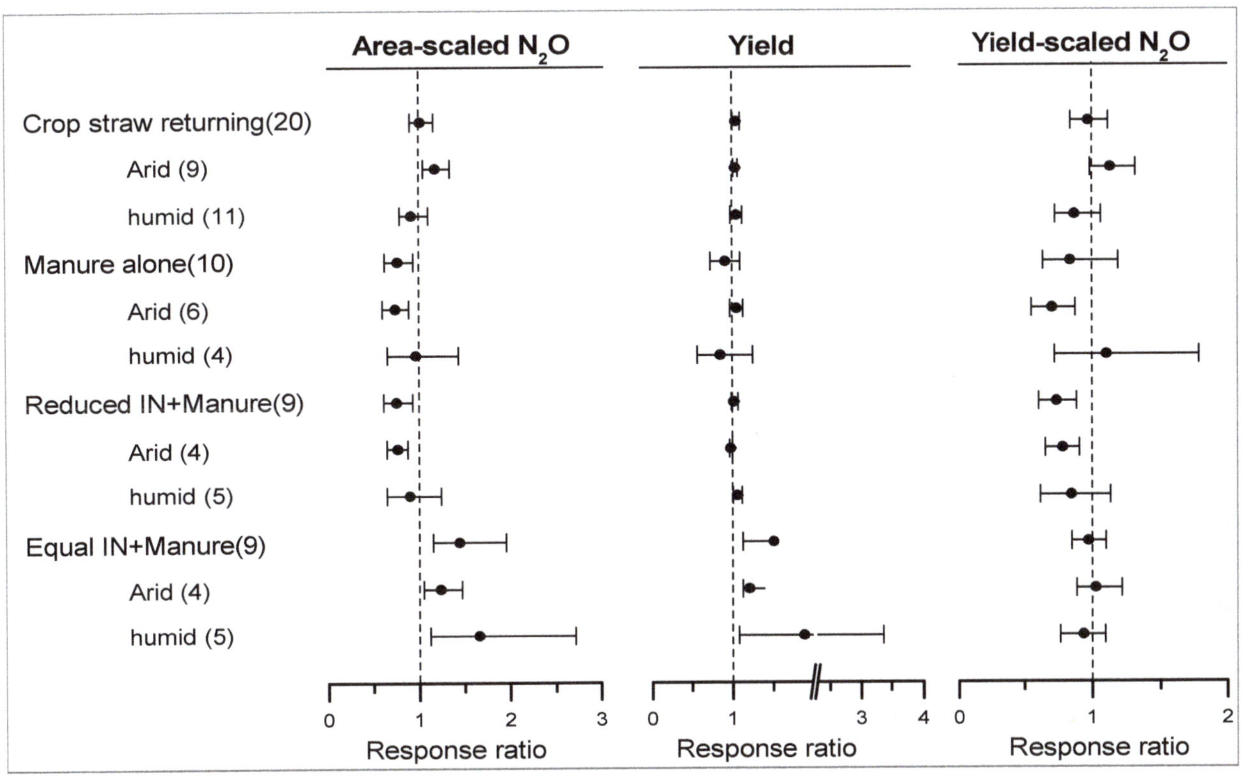

Figure 5. Impacts of organic amendments on area-scaled N_2O emission, crop yield and yield-scaled N_2O emission.

Contradictory effects (either an increase or a reduction) of manure application on N_2O emissions have been demonstrated in previous field experiments [10,54]. Our results showed that the mode of manure application was an important factor influencing the impacts on N_2O emissions (Figure 5). Manure application without inorganic N fertilizer (manure alone) significantly reduced N_2O emissions by 22.8% compared to inorganic N fertilizer. Generally, incorporation of manure in agricultural soil can provide abundant easily decomposable C and cause N_2O to be completely denitrified to N_2 [55]. Therefore, although the total N amount in manure was the same as that in the inorganic N fertilizer control; the N_2O emission was significantly lower under manure alone. However, manure alone did not reduce yield-scaled N_2O emission due to decreases in wheat and maize yields (Figure 5). Aridity affected the effect of manure alone treatment. In arid regions, manure alone significantly mitigated the N_2O emission by 29.2%, which may be primarily due to the enhancement of the crop yield. Soil water was an important factor affecting the crop yield in arid regions. The application of manure could increase the rainfall use efficiency of crop plants by improving soil penetration [56], which provided a benefit to the enhancement of crop yield.

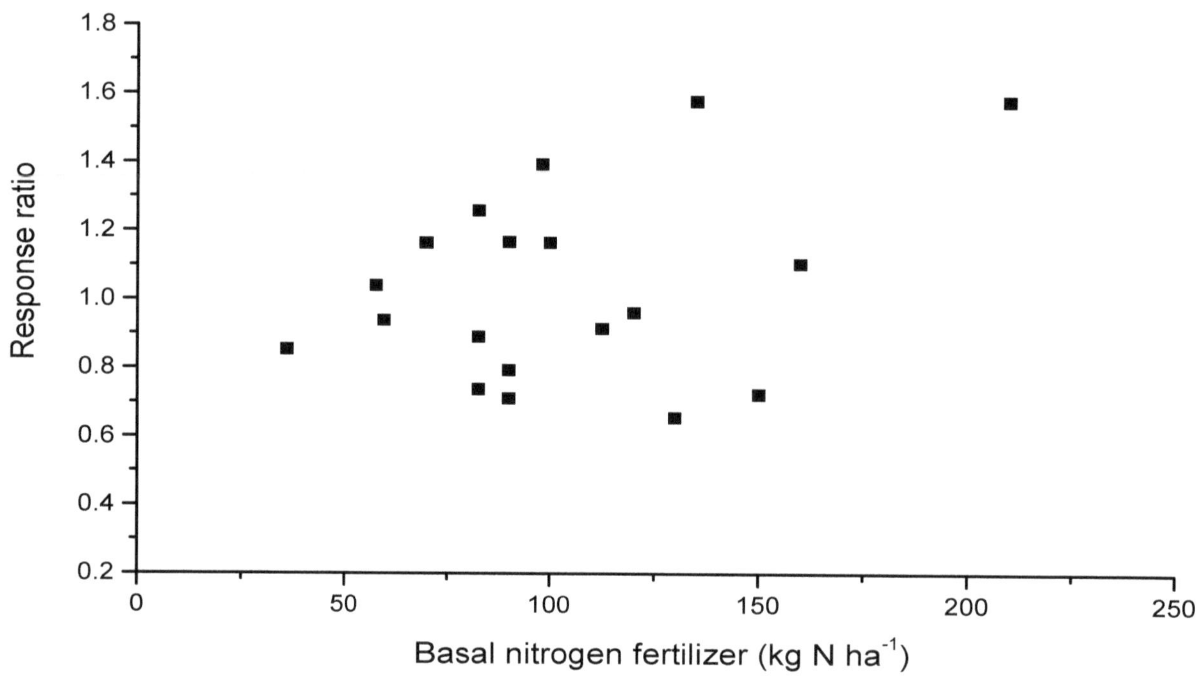

Figure 6. The relationship between inorganic N application rates of basal fertilizer and response ratios of crop straw returning on area-scaled N_2O emission.

Partial substitution of inorganic N with manure in basal fertilizer (mean: 43.6%, range: 22% to 50% in selected studies) (Reduced IN + manure) significantly reduced N_2O emission by 24.0% but had no significant effects on wheat and maize yields. Consequently, the yield-scaled N_2O emission was significantly reduced by 25.8%. However, additional manure application with an equal inorganic N fertilizer amount to the control (Equal IN + manure) significantly increased N_2O emission by 44.6% (Figure 5). Under the same chemical N conditions, manure application can provide additional N and available C for the microbial processes of nitrification and denitrification [57], and thereby significantly stimulate N_2O emission. Although crop yield increased by 50.5% under Equal IN + manure, yield-scaled N_2O emission showed no significant difference between Equal IN + manure and control. Therefore, partial substitution of inorganic N with manure can be suggested as a climate smart practice for balancing crop yield increase and N_2O emission mitigation. Recently, a long-term field experiment in North China also demonstrated that replacing 50% of inorganic N with manure significantly reduced N_2O emission by 41.7% without significant decreases in wheat and maize yields [10].

3.5. Mitigation Potential of Soil Tillage

As shown in Figure 7, no tillage significantly increased N_2O emission (26.6%) compared to conventional tillage. This was due to the fact that no tillage tended to increase soil moisture and bulk density and maintained the N fertilizers on the soil surface [13], consequently resulting in a significant stimulation in N_2O emissions. Wheat and maize yields were lower under no tillage than conventional tillage (Figure 7), which was consistent with a previous report [58]. Therefore, yield-scaled N_2O emission increased significantly by 42.6% as a result of no tillage compared to conventional tillage (Figure 7). As to reduced tillage, no significant effects were found on area-scaled N_2O emission, crop yield and yield-scaled N_2O emission compared to conventional tillage (Figure 7).

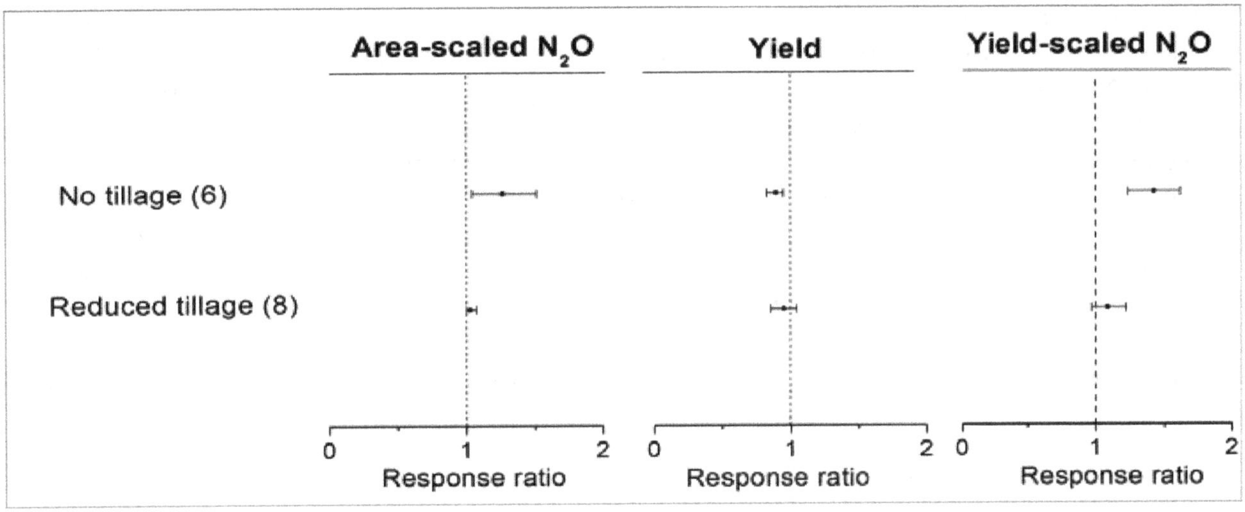

Figure 7. Impacts of reduced and no tillage on area-scaled N_2O emission, crop yield and yield-scaled N_2O emission. (Note: only one observation located in arid region for NT and two observations located in humid area for RT. So, the impact of climate aridity was not analyzed.).

The effect of soil tillage on N_2O emission was affected by N placement, duration and environmental factors [25]. The different effects on N_2O emission between reduced tillage and no tillage could be attributed to the placement depth of N fertilizer. In the select studies of this analysis, the N fertilizer was generally placed on soil surface under no-tillage and incorporated into soil layers (5–10 cm) under reduced tillage. A previous study had reported that tillage interacted with N fertilizer placement depth to regulate N_2O emission; no tillage with surface N placement tends to stimulate N_2O emission compared to reduced tillage [59]. Thus, deep N placement is suggested in no tillage to reduce N_2O emission. In addition, some studies have suggested that tillage duration was an important factor influencing the impact on N_2O emission [16]. Based on a meta-analysis, for example, Kessel et al. [25] reported that reduced and no tillage significantly mitigated the N_2O emission by 14% when experiment durations lasted > 10 years, especially in a dry climate. Recently, a 10-year tillage experiment in the North China Plain reported that reduced tillage (rotary tillage and subsoiling) mitigated N_2O emissions and improved crop productivity in wheat-maize rotation system [60]. However, the field experiments on the effects of tillage on N_2O emissions and crop yields in China were still limited; experiment durations were less than five years in the selected experiments of our meta-analysis. Therefore, additional field experiments are needed to investigate the long-term effects of tillage on area and yield-scaled N_2O emissions, and both short- and long-term effects should be considered in the evaluation of tillage impacts.

4. Conclusions

Agronomic practices affect both crop yield and N_2O emission. Ecological intensification of agronomic practices plays an important role in the sustainable development of future crop production. Our study comprehensively evaluated the impacts of main agronomic practices on area- and yield-scaled N_2O emissions, and analyzed the mitigation potential of N_2O emission by optimizing cropping practices during wheat and maize seasons in China. Results demonstrated that adjusting cropping systems, NI, CRF and reduced IN+ manure were recommend for the mitigation of yield-scaled N_2O emission during wheat and maize growing seasons. Policy options are essential to encourage the application of these strategies. For example, a projected macroscopic plan is needed to adjust cropping systems in national wheat and maize production for the mitigation of N_2O emission. Policies that provided sufficient financial compensation for farmers are required to change the agricultural practices.

Due to limited data, this study did not analyze N_2O emission and crop yield in non-wheat and non-maize growing seasons of four cropping systems. More studies should be conducted to investigate year-round N_2O emissions and crop yield during complete durations of these cropping systems. Additionally, this study only evaluated direct N_2O emissions from soil during wheat and maize growing seasons. In the future, indirect N_2O emissions and carbon cost should be considered in the assessment of the mitigation potential of cropping practices. For example, although no tillage increased direct N_2O emissions from soil, it reduced machine and diesel oil input. As a result, a life-cycle assessment of cropping practices could provide more precise references for the recommendation of management practices.

Acknowledgments: This work was supported by the National Key Technology Support Program of China (2011BAD16B14).

Author Contributions: Wenling Gao collected the data, conducted the data analysis and drafted the paper. Xinmin Bian designed the data analysis and revised the paper.

References

1. Wrage, N.; Velthof, G.L.; Van Beusichem, M.L.; Oenema, O. Role of nitrifier denitrification in the production of nitrous oxide. *Soil Biol. Biochem.* **2001**, *33*, 1723–1732. [CrossRef]
2. Intergovernmental Panel on Climate Change. Summary for Policymakers. In *Climate Change 2013: The Physical Science Basis*; Contribution of Working Group I to the Fifth Assessment Report of the Intergovernmental Panel on Climate Change; Stocker, T.F., Qin, D., Plattner, G.-K., Tignor, M., Allen, S.K., Boschung, J., Nauels, A., Xia, Y., Bex, V., Midgley, P.M., Eds.; Cambridge University Press: Cambridge, UK; New York, NY, USA, 2013.
3. Alexandratos, N. How to feed the world in 2050. In Proceedings of the Technical Meeting of Experts, Rome, Italy, 24–26 June 2009; FAO: Rome, Italy, 2009.
4. Tilman, D.; Balzer, C.; Hill, J.; Befort, B.L. Global food demand and the sustainable intensification of agriculture. *Proc. Natl. Acad. Sci. USA* **2011**, *108*, 20260–20264. [CrossRef] [PubMed]
5. Cai, X.; Zhang, X.; Wang, D. Land availability for biofuel production. *Environ. Sci. Technol.* **2010**, *45*, 334–339. [CrossRef] [PubMed]
6. Popp, A.; Lotze-Campen, H.; Bodirsky, B. Food consumption, diet shifts and associated non-CO_2 greenhouse gases from agricultural production. *Glob. Environ. Chang.* **2010**, *20*, 451–462. [CrossRef]
7. Van Beek, C.L.; Meerburg, B.G.; Schils, R.L.M.; Verhagen, J.; Kuikman, P.J. Feeding the world's increasing population while limiting climate change impacts: Linking N_2O and CH_4 emissions from agriculture to population growth. *Environ. Sci. Policy* **2010**, *13*, 89–96. [CrossRef]
8. Heffer, P. *Assessment of Fertilizer Use by Crop at the Global Level*; International Fertilizer Industry Association (IFA): Paris, France, 2013; Available online: www.fertilizer.org//En/Statistics/Agriculture_Committee_Databases.aspx (accessed on 21 May 2014).
9. Smith, P.; Martino, D.; Cai, Z.; Gwary, D.; Janzen, H.; Kumar, P.; McCarl, B.; Ogle, S.; O'Mara, F.; Rice, C.; et al. Agriculture. In *Climate Change 2007: Mitigation*; Contribution of Working Group III to the Fourth Assessment Report of the Intergovernmental Panel on Climate Change; Metz, B., Davidson, O.R., Bosch, P.R., Dave, R., Meyer, L.A., Eds.; Cambridge University Press: Cambridge, UK; New York, NY, USA, 2007.
10. Cai, Y.; Ding, W.; Luo, J. Nitrous oxide emissions from Chinese maize-wheat rotation systems: A 3-year field measurement. *Atmos. Environ.* **2013**, *65*, 112–122. [CrossRef]

11. Parkin, T.; Hatfield, J. Influence of nitrapyrin on N_2O losses from soil receiving fall-applied anhydrous ammonia. *Agric. Ecosyst. Environ.* **2010**, *136*, 81–86. [CrossRef]
12. Ma, Y.; Sun, L.; Zhang, X.; Yang, B.; Wang, J.; Yin, B.; Yan, X.; Xiong, Z. Mitigation of nitrous oxide emissions from paddy soil under conventional and no-till practices using nitrification inhibitors during the winter wheat-growing season. *Biol. Fert. Soils* **2013**, *49*, 627–635. [CrossRef]
13. Venterea, R.T.; Bijesh, M.; Dolan, M.S. Fertilizer source and tillage effects on yield-scaled nitrous oxide emissions in a corn cropping system. *J. Environ. Qual.* **2011**, *40*, 1521–1531. [CrossRef] [PubMed]
14. Hu, X.; Su, F.; Ju, X.; Gao, B.; Oenema, O.; Christie, P.; Huang, B.; Jiang, R.; Zhang, F. Greenhouse gas emissions from a wheat-maize double cropping system with different nitrogen fertilization regimes. *Environ. Pollut.* **2013**, *176*, 198–207. [CrossRef] [PubMed]
15. Cole, C.V.; Duxbury, J.; Freney, J.; Heinemeyer, O.; Minami, K.; Mosier, A.; Paustian, K.; Rosenberg, N.; Sampson, N.; Sauerbeck, D.; et al. Global estimates of potential mitigation of greenhouse gas emissions by agriculture. *Nutr. Cycl. Agroecosyst.* **1997**, *49*, 221–228. [CrossRef]
16. Six, J.; Ogle, S.M.; Conant, R.T.; Mosier, A.R.; Paustian, K. The potential to mitigate global warming with no-tillage management is only realized when practised in the long term. *Glob. Chang. Biol.* **2004**, *10*, 155–160. [CrossRef]
17. Rochette, P.; Worth, D.E.; Lemke, R.L.; McConkey, B.G.; Pennock, D.J.; Wagner-Riddle, C.; Desjardins, R. Estimation of N_2O emissions from agricultural soils in Canada. I. Development of a country-specific methodology. *Can. J. Soil Sci.* **2008**, *88*, 641–654. [CrossRef]
18. Akiyama, H.; Yan, X.; Yagi, K. Evaluation of effectiveness of enhanced-efficiency fertilizers as mitigation options for N_2O and NO emissions from agricultural soils: Meta-analysis. *Glob. Chang. Biol.* **2010**, *16*, 1837–1846. [CrossRef]
19. Kim, D.G.; Hernandez-Ramirez, G.; Giltrap, D. Linear and nonlinear dependency of direct nitrous oxide emissions on fertilizer nitrogen input: A meta-analysis. *Agric. Ecosyst. Environ.* **2013**, *168*, 53–65. [CrossRef]
20. Chen, H.; Li, X.; Hu, F.; Shi, W. Soil nitrous oxide emissions following crop residue addition: A meta-analysis. *Glob. Chang. Biol.* **2013**, *19*, 2956–2964. [CrossRef] [PubMed]
21. Zhao, X.; Liu, S.; Pu, C.; Zhang, X.; Xue, J.; Zhang, R.; Wang, Y.; Lal, R.; Zhang, H.; Chen, F. Methane and nitrous oxide emissions under no-till farming in China: A meta-analysis. *Glob. Chang. Biol.* **2016**, *22*, 1372–1384. [CrossRef] [PubMed]
22. Xia, L.; Lam, S.K.; Chen, D.; Wang, J.; Tang, Q.; Yan, X. Can knowledge-based N management produce more staple grain with lower greenhouse gas emission and reactive nitrogen pollution? A meta-analysis. *Glob. Chang. Biol.* **2017**, *23*, 1917–1925. [CrossRef] [PubMed]
23. Van Groenigen, J.W.; Velthof, G.L.; Oenema, O.; Van Groenigen, K.J.; Van Kessel, C. Towards an agronomic assessment of N_2O emissions: A case study for arable crops. *Eur. J. Soil Sci.* **2010**, *61*, 903–913. [CrossRef]
24. Linquist, B.; Groenigen, K.J.; Adviento-Borbe, M.A.; Pittelkow, C.; Kessel, C. An agronomic assessment of greenhouse gas emissions from major cereal crops. *Glob. Chang. Biol.* **2012**, *18*, 194–209. [CrossRef]
25. Van Kessel, C.; Venterea, R.; Six, J.; Adviento-Borbe, M.A.; Linquist, B.; Van Groenigen, K.J. Climate, duration, and N placement determine N_2O emissions in reduced tillage systems: A meta-analysis. *Glob. Chang. Biol.* **2013**, *19*, 33–44. [CrossRef] [PubMed]
26. Intergovernmental Panel on Climate Change. Agriculture, Forestry and Other Land Use (AFOLU). In *Climate Change 2014, Mitigation of Climate Change*; Contribution of Working Group III to the Fifth Assessment Report of the Intergovernmental Panel on Climate Change; Edenhofer, O., Pichs-Madruga, R., Sokona, Y., Farahani, E., Kadner, S., Seyboth, K., Adler, A., Baum, I., Brunner, S., Eickemeier, P., Eds.; Cambridge University Press: Cambridge, UK; New York, NY, USA, 2014; Chapter 11.
27. FAOSTAT. Available online: http://faostat.fao.org (accessed on 12 October 2014).
28. Xing, G. N_2O emission from cropland in China. *Nutr. Cycl. Agroecosyst.* **1998**, *52*, 249–254. [CrossRef]
29. Intergovernmental Panel on Climate Change. *Climate Change 2007: The Physical Science Basis*; Solomon, S., Qin, D., Manning, M., Chen, Z., Marquis, M., Averyt, K.B., Tignor, M., Miller, H.L., Eds.; Cambridge University Press: Cambridge, UK; New York, NY, USA, 2007.
30. Feng, J.; Chen, C.; Zhang, Y.; Song, Z.; Deng, A.; Zheng, C.; Zhang, W. Impacts of cropping practices on yield-scaled greenhouse gas emissions from rice fields in China: A meta-analysis. *Agric. Ecosyst. Environ.* **2013**, *164*, 220–228. [CrossRef]

31. Hedges, L.V.; Gurevitch, J.; Curtis, P.S. The Meta-analysis of response ratios in experimental ecology. *Ecology* **1999**, *80*, 1150–1156. [CrossRef]
32. Rosenberg, M.S.; Adams, D.C.; Gurevitch, J. *MetaWin-Statistical Software for Meta-Analysis*; Sinauer Associates Inc.: Sunderland, UK, 2000.
33. Peng, S.; Hou, H.; Xu, J.; Yang, S.; Mao, Z. Lasting effects of controlled irrigation during rice-growing season on nitrous oxide emissions from winter wheat croplands in Southeast China. *Paddy Water Environ.* **2013**, *11*, 583–591. [CrossRef]
34. Zhao, G. Study on Chinese wheat planting regionalization (I). *J. Triticeae Crop.* **2010**, *30*, 886–895. (In Chinese with English abstract).
35. Smith, K.A.; Thomson, P.E.; Clayton, H.; McTaggart, I.P.; Conen, F. Effects of temperature, water content and nitrogen fertilisation on emissions of nitrous oxide by soils. *Atmos. Environ.* **1998**, *32*, 3301–3309. [CrossRef]
36. Hou, P.; Gao, Q.; Xie, R.; Li, S.; Meng, Q.; Kirkby, E.A.; Römheld, V.; Müller, T.; Zhang, F.; Cui, Z.; et al. Grain yields in relation to N requirement: Optimizing nitrogen management for spring maize grown in China. *Field Crop. Res.* **2012**, *129*, 1–6. [CrossRef]
37. Deng, Z.; Feng, Y.; Zhang, J.; Wang, J. Regional pattern change and its influencing factors of cereals crops production in China. *Macroeconomics* **2014**, *3*, 94–100. (In Chinese with English abstract).
38. Banger, K.; Tian, H.; Lu, C. Do nitrogen fertilizers stimulate or inhibit methane emissions from rice fields? *Glob. Chang. Biol.* **2012**, *18*, 3259–3267. [CrossRef]
39. Hoben, J.; Gehl, R.; Millar, N.; Grace, P.; Robertson, G. Nonlinear nitrous oxide (N_2O) response to nitrogen fertilizer in on-farm corn crops of the US Midwest. *Glob. Chang. Biol.* **2011**, *17*, 1140–1152. [CrossRef]
40. Ministry of Agriculture of the People's Republic of China. Fertilizer Recommendation for Rice, Wheat and Maize Production in Major Growing Regions. Available online: http://www.moa.gov.cn/govpublic/ZZYGLS/201307/t20130729_3541508.htm (accessed on 29 July 2013). (In Chinese)
41. Decock, C. Mitigating nitrous oxide emissions from corn cropping systems in the Midwestern U.S.: Potential and data gaps. *Environ. Sci. Technol.* **2014**, *48*, 4247–4256. [CrossRef] [PubMed]
42. Venterea, R.T.; Halvorson, A.D.; Kitchen, N.; Liebig, M.A.; Cavigelli, M.A.; Grosso, S.J.D.; Motavalli, P.P.; Nelson, K.A.; Spokas, K.A.; Singh, B.P.; et al. Challenges and opportunities for mitigating nitrous oxide emissions from fertilized cropping systems. *Front. Ecol. Environ.* **2012**, *10*, 562–570. [CrossRef]
43. Liu, C.; Wang, K.; Zheng, X. Effects of nitrification inhibitors (DCD and DMPP) on nitrous oxide emission, crop yield and nitrogen uptake in a wheat-maize cropping system. *Biogeosciences* **2013**, *10*, 2427–2437. [CrossRef]
44. Ji, Y.; Liu, G.; Ma, J.; Xu, H.; Yagi, K. Effect of controlled-release fertilizer on nitrous oxide emission from a winter wheat field. *Nutr. Cycl. Agroecosyst.* **2012**, *94*, 111–122. [CrossRef]
45. Feng, J.; Li, F.; Deng, A.; Feng, X.; Fang, F.; Zhang, W. Integrated assessment of the impact of enhanced-efficiency nitrogen fertilizer on N_2O emission and crop yield. *Agric. Ecosyst. Environ.* **2016**, *231*, 218–228. [CrossRef]
46. Jiang, J.; Hu, Z.; Sun, W.; Huang, Y. Nitrous oxide emissions from Chinese cropland fertilized with a range of slow-release nitrogen compounds. *Agric. Ecosyst. Environ.* **2010**, *135*, 216–225. [CrossRef]
47. Abalos, D.; Jeffery, S.; Sanz-Cobena, A.; Guardia, G.; Vallejo, A. Meta-analysis of the effect of urease and nitrification inhibitors on crop productivity and nitrogen use efficiency. *Agric. Ecosyst. Environ.* **2014**, *189*, 136–144. [CrossRef]
48. Ma, E.; Zhang, G.; Ma, J.; Xu, H.; Cai, Z.; Yagi, K. Effects of rice straw returning methods on N_2O emission during wheat-growing season. *Nutr. Cycl. Agroecosyst.* **2010**, *88*, 463–469. [CrossRef]
49. Liu, C.; Wang, K.; Meng, S.; Zheng, X.; Zhou, Z.; Han, S.; Chen, D.; Yang, Z. Effects of irrigation, fertilization and crop straw management on nitrous oxide and nitric oxide emissions from a wheat-maize rotation field in northern China. *Agric. Ecosyst. Environ.* **2011**, *140*, 226–233. [CrossRef]
50. McKenney, D.; Wang, S.; Drury, C.; Findlay, W. Dentrification and mineralization in soil amended with legume, grass, and corn residues. *Soil Sci. Soc. Am. J.* **1993**, *57*, 1013–1020. [CrossRef]
51. Huang, Y.; Zhang, F.; Liu, S.; Cao, Q. Effect of allelochemicals on N_2O emission from soil. *Acta Sci. Circumst.* **1999**, *19*, 478–482. (In Chinese with English abstract).
52. Hao, X.; Chang, C.; Carefoot, J.M.; Janzen, H.H.; Ellert, B.H. Nitrous oxide emissions from an irrigated soil as affected by fertilizer and straw management. *Nutr. Cycl. Agroecosyst.* **2001**, *60*, 1–8. [CrossRef]

53. Baggs, E.M.; Stevenson, M.; Pihlatie, M.; Regar, A.; Cook, H.; Cadisch, G. Nitrous oxide emissions following application of residues and fertiliser under zero and conventional tillage. *Plant Soil* **2003**, *254*, 361–370. [CrossRef]
54. Adviento-Borbe, M.; Kaye, J.; Bruns, M.; McDaniel, M.; McCoy, M.; Harkcom, S. Soil greenhouse gas and ammonia emissions in long-term maize-based cropping systems. *Soil Sci. Soc. Am. J.* **2010**, *74*, 1623–1634. [CrossRef]
55. Sánchez-Martín, L.; Vallejo, A.; Dick, J.; Skiba, U. The influence of soluble carbon and fertilizer nitrogen on nitric oxide and nitrous oxide emissions from two contrasting agricultural soils. *Soil Biol. Biochem.* **2008**, *40*, 142–151. [CrossRef]
56. Wang, X.; Jia, Z.; Liang, L.; Yang, B.; Ding, R.; Nie, J.; Wang, J. Impacts of manure application on soil environment, rainfall use efficiency and crop biomass under dryland farming. *Sci. Rep.* **2016**, *6*, 20994. [CrossRef] [PubMed]
57. Van Groenigen, J.W.; Kasper, G.J.; Velthof, G.L.; Van den Pol-van Dasselaar, A.; Kuikman, P.J. Nitrous oxide emissions from silage maize fields under different mineral nitrogen fertilizer and slurry applications. *Plant Soil* **2004**, *263*, 101–111. [CrossRef]
58. Van den Putte, A.; Govers, G.; Diels, J.; Gillijns, K.; Demuzere, M. Assessing the effect of soil tillage on crop growth: A meta-regression analysis on European crop yields under conservation agriculture. *Eur. J. Agron.* **2010**, *33*, 231–241. [CrossRef]
59. Venterea, R.T.; Stanenas, A.J. Profile analysis and modeling of reduced tillage effects on soil nitrous oxide flux. *J. Environ. Qual.* **2008**, *37*, 1360–1367. [CrossRef] [PubMed]
60. Tian, S.; Wang, Y.; Ning, T.; Zhao, H.; Wang, B.; Li, N.; Li, Z.; Chi, S. Greenhouse gas flux and crop productivity after 10 years of reduced and no tillage in a wheat-maize cropping system. *PLoS ONE* **2013**, *8*, e73450. [CrossRef] [PubMed]

7

Climate Change-Induced Impacts on Smallholder Farmers in Selected Districts of Sidama, Southern Ethiopia

Tafesse Matewos

Geography & Environmental Studies & PhD Fellow in Development Studies at Institute of Policy & Development Research (IPDR), Hawassa University, Ethiopia. Po Box 05 Hawassa, Ethiopia; tafessemk@gmail.com

Abstract: Different factors control the types of adaptive strategies and likelihoods of experiencing climate change-induced impacts by smallholder farmers. By using a mixed research method, this study examines the types and determinants of climate change-induced impacts on smallholder rural farmers in drought-prone low lands of Sidama, Southern Ethiopia. Randomly selected (401) households were surveyed on climate change-induced impacts. Longitudinal climatic data were also collected from the Ethiopian National Meteorological Agency to assess the trend of rainfall (RF), temperature and drought incidents. The analyses of the data revealed that RF and temperature had shown decreasing and increasing trends, respectively, during the three decades under consideration (1983–2014). These changes in RF and temperature exposed farmers to climate-related epidemics, drought, harvest loss, and hunger. The logit model results revealed that different factors control the likelihood of exposure to climate change-induced impacts. The findings revealed that literacy level, involving women in family decisions and farmers' involvement in adaptation planning, reduces the likelihood of exposure to climate change-induced hunger. Therefore, there is a need to work on human capital of the farmers through expanding education, strengthening women's participation in family decision-making, and by improving public participation in climate change adaptation undertakings to minimize climate change-induced impacts.

Keywords: climate change-induced impacts; smallholder farmers; drought-prone low lands; rural Sidama; southern Ethiopia

---

1. Introduction

Climate change has emerged as one of the development challenges of the 21st century [1,2]. There is high confidence and agreement among the global scientific community that climate change poses a serious threat to current and future sustainable development. According to the forecasts, climate change-induced impacts will continue to affect people, even if anthropogenic greenhouse gasses emissions stop today [3]. Sub-Saharan Africa (SSA), where smallholder farmers dominate agriculture [4], is one of the global hotspots for climate change-induced impacts [5]. In SSA, agriculture directly employs about 175 million people who cultivate degraded lands where there is no reliable supply of water for irrigation [6]. These smallholder farmers in SSA are among the most vulnerable groups to climate change and variability-induced impacts due to dependency on RF, limited use of irrigation, and weak adaptive capacity. Furthermore, limited human and material capacity, poor infrastructure, fragile environments, political instability, and marginalization contributed to the vulnerability in this region [7–10]. The case is not different in Ethiopia where smallholder farmers, who produce 90% of the total agricultural yield and own 95% of the total cultivated land [11], suffer from different climate change induced-impacts. Most of these subsistent farmers practice rain-fed traditional farming, use little modern agricultural inputs, and have little surplus to sell in local markets. They are not resilient enough to cope with climate change-induced recurrent shocks and long-term impacts [12–14]. Although climate change-induced impacts have been recurrent in Ethiopia and in the study area, little is known about the factors that affect the likelihood of exposure to such impacts in Ethiopia and SSA. Thus, by using mixed research strategy, this study presents the types and determinants of climate change induced-impacts on smallholder rural farmers in drought-prone lowland context.

Climate change induced-impacts refers to effects on natural and human systems due to gradual changes in climate, variations in weather and climatic elements from the average, and that of the impacts of climatic extremes. As IPCC report, climate change-induced impacts refer to impacts on human lives, livelihoods, culture, economies, ecosystems, and material resources due to hazardous climatic events over a period of time [3]. Climate change-induced impacts also include consequences and outcomes of the direct impacts including droughts, hunger, famine, loss of life and property, and sea level rise which can be potential or residual (ibid). The former includes impacts that occur due to a projected change in climate without adaptation. The latter refers to the impacts with adaptation [15]. Therefore, in this research, climate change-induced impacts include the immediate effects as well as outcomes and consequences of the immediate effects on human and natural systems. As global warming increases the likelihood of experiencing severe, persistent, and irreversible impacts on natural and human systems will be stronger [3,16]. Greenhouse gas emissions have adverse effects on biodiversity, ecosystem services, and economic development, which are causing risks for livelihoods and human security [17,18]. The 4th IPCC report argues that ... *by 2020, between 75 million and 250 million people are projected to be exposed to increased water stress due to climate change*[15] (p13). Agricultural crop yield in some African countries could be reduced by 50% by 2020 and this can endanger the existence of many smallholder rural farmers.

Climate change-induced impacts have hampered poverty reduction and sustainable development in the Global South [5,19,20]. This is because most climate-related impacts affect the poor, who are struggling to come out of poverty [21]. Furthermore, livelihood strategies and housing conditions of the poor are more vulnerable to climate change and variability-induced impacts [22,23]. The poor usually have houses made of mud, bamboo, straw, and other inexpensive materials that are the most vulnerable to extreme weather events. The poor also cannot buy climate insurance against climate-related risks. Besides the direct impacts on agriculture-related livelihoods, climate change is also indirectly affecting the health and well-being of the poor through its effect on human and livestock health [24,25]. For example, because of increasing surface temperature, malaria is expected to migrate to higher altitudes and cause further health problems, which affect the poor's income and productivity. According to the World Health Organization (WHO) estimates, global warming causes about 150,000 deaths per year [26]. RF and temperature variability also result in pests and diseases incidents that affect the quantity and the quality of the crop yield [27]. Besides impacts, social strains caused by increased resource scarcity may lead to greater conflict [5,28], with the poor again being the most likely victims.

Erratic RF and higher temperatures characterize the drought-prone lowlands of Sidama. The area is part of the Great East African Rift Valley (GEARV). Climate change-induced impacts such as drought, crop failure, livestock loss, flooding, and water-borne and related diseases (such as diarrhea and malaria) have been recurrent in this area. The problem worsened in 2016 despite various adaptive strategies of the households, communities, and other actors, including the government. This needed a large-scale emergency relief for about 100,000 people, which is highest in recent documented drought history of the area. Despite a few research works on climate change-induced impacts on smallholder farmers, there is a gap in the literature about factors that control the likelihood of exposure to the impacts. Therefore, the main goal of this study is to identify factors that determine the likelihood of exposure to climate-induced impacts on smallholder rural farmers. Identifying such determinants is important for policy-makers, practitioners, and smallholder farmers to identify leverage points to manage the imminent impacts of climate change in the country and beyond.

2. Climate Change-Induced Impacts in Ethiopia

While agriculture is the backbone of Ethiopia's economy, it is RF dependent and dominated by smallholder subsistent farmers. About 80% of the Ethiopian population lives in rural areas making drought-prone agriculture as the primary means of livelihood. This figure is much higher than the SSA average, which stood at 63% in the year 2014 [11,29]. Furthermore, agriculture constitutes 40% of the GDP, supports 80% of the total employment, and is a source of 90% of the export in Ethiopia [11,30,31]. Nonetheless, climate change-induced impacts are challenging the roles of agriculture for the overall growth of the country [12]. In the 20th century alone, twelve extreme droughts happened in Ethiopia that hampered the economic development of the country. The drought incidents claimed the lives of hundreds of thousands and affected the livelihoods of over 50 million people [32]. Drought incidents have been increasing in recent decades in the country. Eight major droughts had occurred in the last three decades in Ethiopia: 1984/85 1987/88 1991 1994 1997 2002/03 2010/11, and in 2015/16 [32,33].

Previous studies revealed that climate change and variability had exerted significant impacts on agriculture and the overall economic growth in Ethiopia [32,34,35]. Specifically, climate change-induced impacts have hampered the country's economic growth and effort to move out of poverty. For instance, it was reported that from 1991–2010, the growth estimates were reduced by 2–9% because of climate change-induced impacts [30]. Under the worst-case scenarios, the economic impacts of climate change may reduce Ethiopia's GDP by up to 8% [36]. On the other hand, another study on climate change impact on agriculture and related sectors revealed a 10% GDP drop from the target [37]. The same study revealed a widening income inequality that could reach up to 20% due to the effect of climate change. In general, with higher vulnerability, and lesser resilience to climate change-induced impacts, the country is projected to experience a 6% decline in total agricultural output in the coming years [38].

Besides the impacts on the national economy, climate change-induced impacts have been affecting health and livelihood status of smallholder farmers in Ethiopia. Some of these impacts include climate-related epidemics, crop failure, flooding, livestock mortality, shortage of water and pasture, migration, and food aid dependency [2,39,40]. Unsustainable adaptation strategies of the farmers, such as selling assets and charcoal production, have resulted in the degradation of assets, environment degradation, and biodiversity losses. Climate change-induced food and water shortages affect the health, educational, and economic status of the households [8]. For instance, during drought events, children are forced to drop out of the school to participate in such household activities as collecting water and fuel-wood from long distances. Malaria affects labor availability at the household level and increases the health expenses of the family.

3. Data and Methods

3.1. Study Area

Located in the Horn of Africa, Ethiopia had a population of 94.35 million in 2017, according to the Ethiopian Central Statistical agency. The country has nine regional states and two city administrations. The Southern Nations Nationalities and Peoples' Regional State (SNNPRS) is one of the nine administrative regions comprising about 20% of the country's total population and 10% of the land area. The Sidama Administration Zone is one of the 14 administrative zones in the SNNPRS. With a population of 3,677,370 in 2014 [41], it is the most populous zone in the region. The zone is located in the central-eastern part of the region bordered by Oromiya in the North, East, and Southeast, Gedeo Zone in the South, and Wolayita Zone in the west. The zone lies between 6°10′ to 7°05′ North latitude and 38°21′ to 39°11′ East longitude. The total area of the zone is 6981.9 km^2 [42].

Three drought-prone districts (Hawassa Zuria, Boricha, and Loka Abaya) were selected for this study (Figure 1). The districts are located in the heart of the East African Rift Valley and they were chosen based on their higher potential risk of exposure to adverse climate change and variability. Variable RF and higher temperatures characterize the districts compared to other districts of the zone. The population of the districts is 576,865 from 113,285 households. About 97% of the population in these districts lives in rural areas where drought-prone agriculture is their source of livelihood. Agro-pastoral communities dominate the southwestern parts of the study area, where climate-related animal diseases are prevalent.

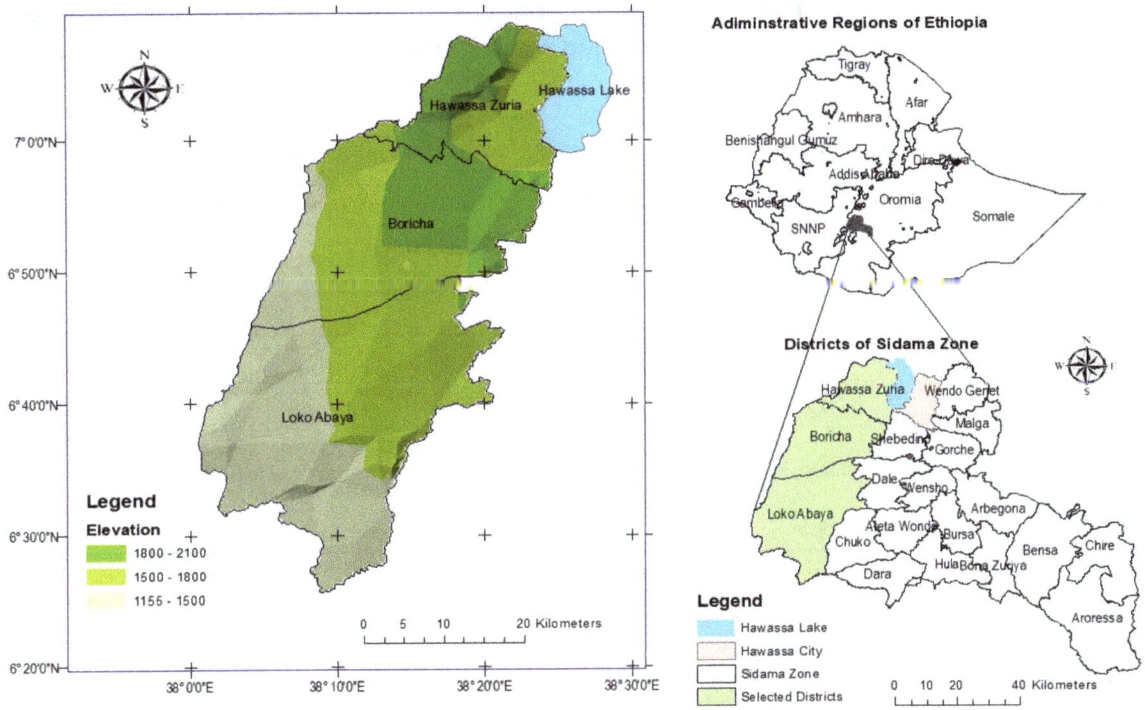

Figure 1. The geographic location of the study area.

Diverse climate characterizes the drought-prone districts of Sidama Zone. Altitude ranges from 560–2300 meters above sea level and the average annual RF ranges between 700 and 1200 mm. The RF pattern in the area is bimodal, which occurs in the summer and spring seasons. The summer (*kiremt*) rains last from June to September while the spring (*belg*) rains usually begin in February and end in late May. Both the *kiremt* and *belg* rains are irregular and unpredictable, resulting in frequent losses of harvest and cattle. The region has a bimodal precipitation distribution and high RF variability due to the seasonal progression of the inter-tropical convergence zone (ITCZ), an atmospheric circulation feature often modified by the El Niño–Southern Oscillation (ENSO). Also, there is severe livestock and human diseases (for example, malaria, cholera, and trypanosomiasis) in the area largely due to the inhospitable climate. The perennial Bilate River, which dissects the Wolayita and Sidama Zones, drains the western part of the area. The eastern part lacks perennial rivers, and households largely depend on artificial ponds and hand-dug shallow wells for water for domestic uses and watering livestock. The area experiences severe water shortages when the ponds run out of water during the December–February dry period.

3.2. Data

This study employed a cross-sectional household survey method. Randomly selected 401 rural household heads took part as respondents for the study. The following sampling procedure was applied. Based on their climatic conditions, accessibility, population size, and population density three kebeles from Boricha, two kebeles from Hawassa Zuria, and two kebeles from Loka Abaya districts (a total of seven kebeles) were selected. Accordingly, Hanja Cafa, Haldada Dela, and Korangoge from

Boricha district, Muticha Gorbe, and Sala Kore kebeles from Loka Abaya district, and Doyo Cala and Doyo Otilicha kebeles from Hawassa Zuria districts were selected for the study. The ideal sample size was calculated based on the Krejcie and Morgan formula; the maximum sample size at a 95% confidence interval and 5% margin of error for 750,000 people is 382 [43]. Considering attritions of the responses, a 5% contingency (19 HHs) gives 401 households, which were selected from the three districts. The 401 sample size was proportionally distributed based on the number of households in each district. So, 189 HHs from Boricha, 93 HHs from Loka Abaya, and 119 HHs from Hawassa Zuria districts were included for this study. The respondents were household heads who were selected by using systematic random sampling. Female-headed households were also included proportionately in all the districts (Table 1).

Table 1. Distribution of rural population and sampled households (HHs) in selected districts.

District Name	Total Rural Population	Total HHs	Kebele Name	Sampled HHs	Total
Boricha	299,175	58,823	Hanja Cafa	66	189
			Haldada Dela	66	
			Korangoge	57	
Loka Abaya	122,445	25,278	Muticha Gorbe	36	93
			Sala Kore	58	
Hawassa Zuriya	155,245	29,184	Doyo Chala	53	119
			Doyo Otilicho	66	
TOTAL	576,865	113,285		401(143)	401

( ) female headed HHs. Source: Own compilation from SNNPRS, BoFED 2015.

The field data collection process took place from April to May 2017 with trained data collectors supervised by the researcher. More data were gathered through focus group discussions, interviews with key informants, and field observation. The study also used RF data collected from the National Meteorological Agency of Ethiopia for the period 1983–2014 to support other data.

The RF data was analyzed by using trend analysis, Mann–Kendall's rank test, and standard RF anomalies (SRA). The SRA refers to standard RF anomaly and Pi and $P\mu$ represent RF of a given year and the mean RF, respectively. Accordingly, the result helps to classify temperature RF condition into different categories (Table 2).

$$\text{SRA} = \frac{P_i - P\mu}{\delta} \tag{1}$$

The collected quantitative data were encoded into STATA software (Version 14.2), and various descriptive and inferential techniques (mean, percent, standard deviation, and logit model) were used to analyze and interpret the results. The logit model analysis was carried out by using the STATA software to identify the determinants of climate change-induced effects on smallholder farmers. The qualitative data were analyzed via systematic thematization to augment the quantitative results.

3.3. Model Specifications

A logit model is a statistical approach used when the dependent variable (DV) is nominal with two or more categories commonly grouped as dummy variables (Table 3). On the other hand, the independent (predictor) variables (IVs) in a logit model can be qualitative (nominal or ordinal level) and interval/ratio level variables. This study used a logit model to identify factors that control the likelihood of households' exposure to climate change-induced impacts. Climate change-induced impacts considered in this study include the likelihoods to exposure to drought, harvest loss, flooding, and hunger. The exposure to hunger is examined vis-à-vis eighteen socio-economic and institutional s (Table 2). The eighteen dependent variables were selected among other socioeconomic and institutional variables after all the variables were tested to have a relationship with the four dependent variables considered for this study. The logit model showed that the selected eighteen variables have some explanatory power on the dependent variables considered.

Table 2. Variables description.

S.N	Variables	Variable Type	Description	Code
1	Sex	Explanatory	Sex of the HHs head	continuous
2	Family size	Explanatory	Number of HH members	continuous
3	Educational status	Explanatory	HHs head no. of years in school)	continuous
4	Total land size	Explanatory	Size of land in hectare	continuous
5	Land use certificate	Explanatory	I have land use right certificate	1) yes 0) no
6	PSNP beneficiary	Explanatory	Beneficiary of PSNP support	1) yes 0) no
7	Produce *enset*	Explanatory	Produce *enset* crop	1) yes 0) no
8	Membership in 1 to 5	Explanatory	Membership in 1 to 5	1) yes 0) no
9	Head of 1 to 5	Explanatory	Head of 1 to 5	1) yes 0) no
10	Communal grazing land	Explanatory	Communal grazing land	1) yes 0) no
11	Communal water pond	Explanatory	Communal water ponds	1) yes 0) no
12	Drought resistant variety	Explanatory	Drought resistant crop variety	1) yes 0) no
13	Fertilizer/urea/NPS	Explanatory	Fertilizer/urea/NPS	1) yes 0) no
14	Training on climate change	Explanatory	Get trained on climate change adaptation	1) yes 0) no
15	Early warning information	Explanatory	Get early climate change induced information	1) yes 0) no
16	Improved animal fodder	Explanatory	Use improved animal fodder	1) yes 0) no
17	HHs participation in adaptive decision	Explanatory	HHs participation in climate decision making	1) yes 0) no
18	Women participation in HHs decision	Explanatory	Women participation in HHs decision making	1) yes 0) no
19	Drought	Dependent	Exposure to the effects of drought	1) yes 0) no
20	Harvest loss	Dependent	Experienced harvest loss	1) yes 0) no
21	Flooding	Dependent	Affected by flooding	1) yes 0) no
22	Famine	Dependent	Experienced hunger	1) yes 0) no

1. Productive Safety Net Program is a government program being implemented in the study area to support poor and vulnerable people. 2. Are groups of 5 people organized by government and used for political, economic, and social purposes. 3. Drought resistant varieties are plant seeds that are supplied by the government and can grow in moisture stress conditions. 4. Urea and NPS are the two types of chemical fertilizers supplied to farmers by the government (usually in loan) NPS refers to Nitrogen-Phosphoric fertilizer containing Sulphur. 5. Improved animal fodder is a plant species (supplied by the government) that is used as a forage to feed animals during a shortage of pasture. 6. HHS are members of a family living in a house.

Table 3. Mann–Kendall's rank test result for inter-annual rainfall (RF).

Coefficicnt	n	Winter RF	Spring RF	Summer RF	Annual RF
Correlation Coefficient	1.000	0.117	0270 *	−0.020	−0.161
Sig. (2-tailed)		0.347	0.030	0.871	0.195
N	32	32	32	32	32

* Correlation is significant at the 0.05 level (2-tailed).

Logistic regression involves a linear function:

$$Z = b_0 + b_1X_1 + b_2X_2 + \ldots + b_nX_n \quad (2)$$

where Z is a dependent variable and $X_i \ldots X_n$ are explanatory variables.

The logistic regression function is thus, given by:

$$p = \frac{e^z}{1+e^z} \quad (3)$$

where p is the probability that an event occurs; and the odds in favor of the occurrence are related according to $p = \frac{\text{Odds}}{1-\text{odds}} = \frac{p}{1-p}$.

$$\log\left(\frac{p}{1-p}\right) = \beta_0 + \beta_1 X \quad (4)$$

Thus, odds is defined for an event with probability p and the logit is the log of the odds, i.e.; Model - $p(x)$ = probability of the event occurring at (Y=(x))

$$p(y/x) = \frac{e^{\beta_0+\beta_1 X_1+\beta_2 X_2+\ldots+\beta_k X_k}}{1+e^{\beta_0+\beta_1 X_1+\beta_2 X_2+\ldots+\beta_k X_k}} \quad (5)$$

A smaller sample size with a large number of predictors creates problems with the analysis, specifically when there are categorical predictors with limited cases in each category. Descriptive statistics were run for each of the predictors to solve the problem. The result showed that all categorical variables have a required number of cases set as a minimum standard [44]. A multicollinearity test was done to check for high intercorrelations among predictor (independent) variables. The test result suggested that the maximum r-value of the independent variables is 0.7, which is within an acceptable limit [44]. The presence of outliers or cases that would not well be explained by the model was also checked to make sure the model fits the data well. The model tried to find out factors that control the HHs likelihood of exposure to climate-induced impacts such as the exposure to the effects of drought, harvest loss, flooding, and hunger in the study area. R^2 and adjusted R^2 were also used to discuss the amount of variation explained by the independent variables. The former supposes that every independent variable in the model explains the variation in the dependent variable by explaining the variation in percent as if all independent variables in the model affect the dependent variable. On the other hand, the adjusted R^2 gives the percentage of variation explained by independent variables that in reality, affect the dependent variable. In this context, it is also called marginal effect after logit.

4. Results and Discussion

4.1. RF Condition and Drought Incidents

Studies on RF over Ethiopia revealed high variability and drought incidents [32,35,45]. Similarly, the analyses of RF data have shown declining RF trends and incidents of drought during the years under consideration (1983–2014). Annual and growing season (*belg* and *kiremt*) RF amounts had shown declining trends in the study area. The *belg* season (February, March, April, and May) RF accounts for 40% of the annual RF in the study area. It is an important season not only to grow short-growing season crops, but also it provides water for livestock and helps the growth of pasture. Though the current study does not check daily RF data of the growing seasons, the experts and focus group discussion participants reported that problems related to its onset, duration, and offset have serious consequence on food security in the study area. However, despite its huge role for livelihood security, it has shown a significant ($p \leq 0.05$) declining trend in the last three decades (Figure 2a, Table 3). The *kiremt* season (June, July, August, and September) is the main rain growing season in the study area which accounts for 41% of the total annual RF. *Kiremt* RF has also shown a declining trend over the years under consideration (Figure 2b), though the trend is not statistically significant (Table 3). The trend analysis of annual RF data has also revealed a declining trend, though the result is statistically insignificant (Table 3). The annual RF had varied from its average amount, which amounts to 1100 mm. The year 2009 was the driest year among the years under consideration with the annual amount of 796.37 mm. The years 1984, 1999, 2000, 2002, 2003, 2004, and 2009 were the drier with annual RF of 950.44 mm, 870.1 mm, 934.02 mm, 960.09 mm, 994.71 mm, and 925.08 mm, respectively.

The years 1983 and 1996 were the wettest years with annual RF amounts of 1420.42 mm and 1368.41 mm, respectively (Figure 3). On the other hand, the analysis of minimum, maximum, and mean monthly temperature had shown increasing trends, all of which are statistically significant

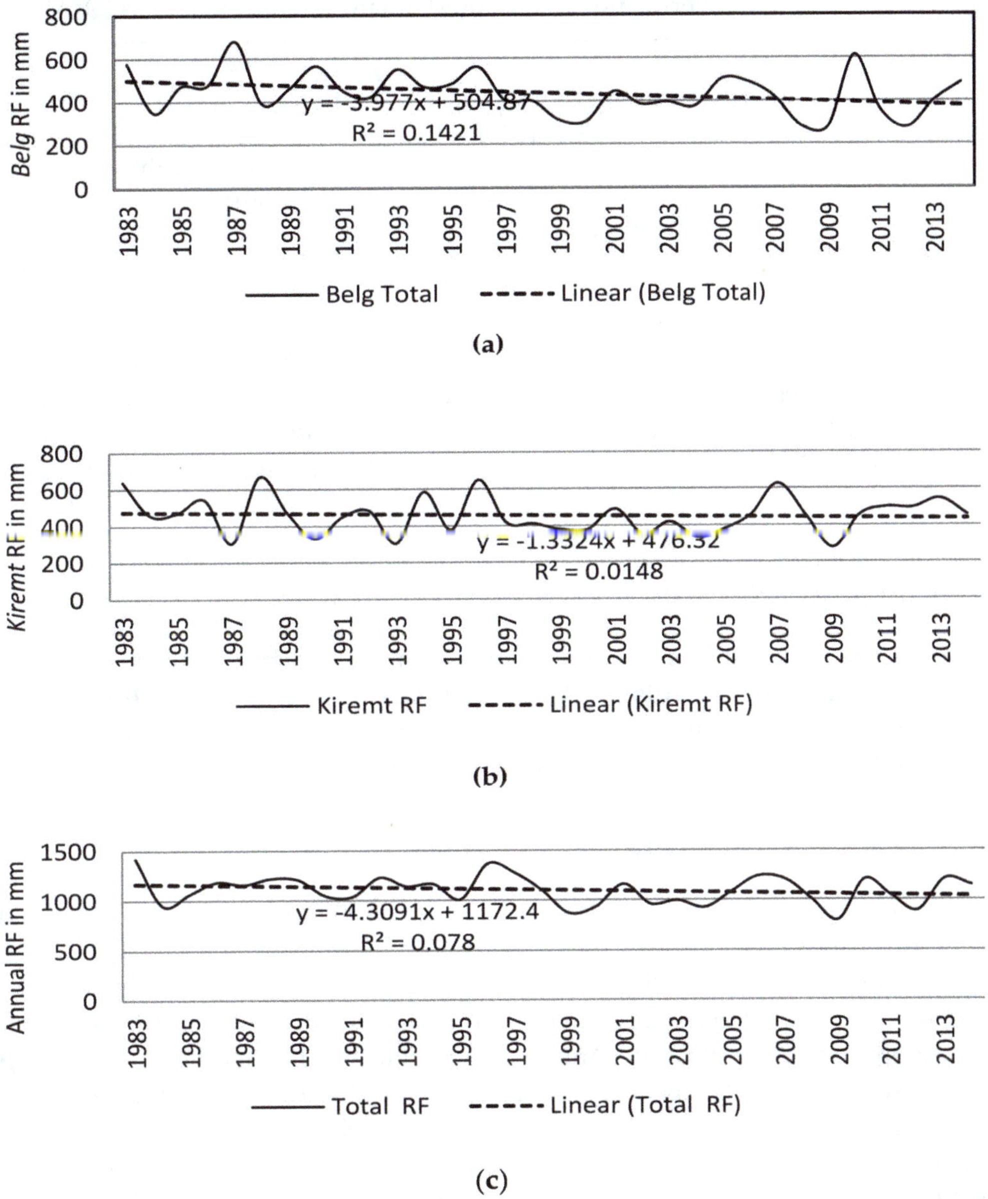

Figure 2. Rainfall trend in the study area (1983–2014). Source: Own computation from Ethiopian National Meteorological Agency data.

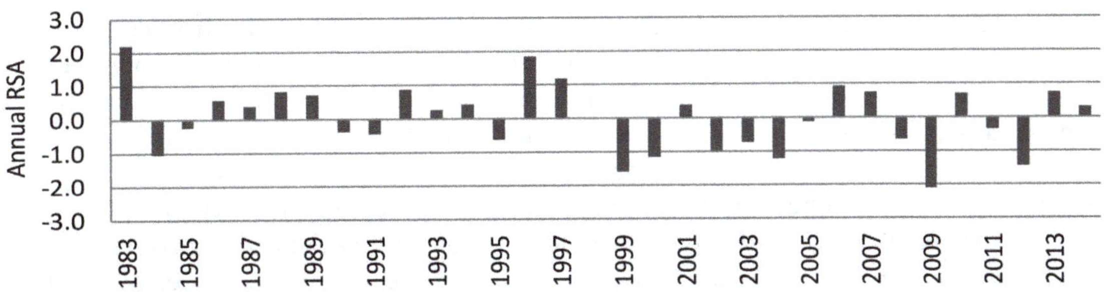

Source: Own computation from Ethiopian National Meteorological Agency data.

Figure 3. Annual RF standard anomalies in the study area (1983–2014).

The RF Standard Anomalies (RSA) helps to estimate the extent of drought based on the RF data [46] (Table 4).

Table 4. Drought severity index based on standard rainfall anomalies (SRA).

Anomaly	Temperature	RF
<−1.65	Extremely hot	Extreme drought
−1.65 to −1.28	Severe hot	Severe drought
−1.28 to −0.84	Moderate hot	Moderate drought
>−0.84	Cool	No drought

Source: Janowiak et al. 1986.

The analysis of annual RSA pointed out seven droughts incidents of different extent during the three decades under consideration. The year 2009 with annual SRA value of −2.11 was the driest year in the last three decades, which coincides with the 2009/10 extreme drought all over the country. The years 1999 and 2012 were also years of severe drought in the study area with SRA vales of −1.6 and −1.45, respectively. The other four moderate drought years were in 1984 (SRA = −1.04) 2000 (SRA = −1.16) 2002 (SRA = −0.98), and 2004 (SRA = −1.22) (Figure 3).

4.2. *Types of Climate Change-Induced Impacts*

Drought and flooding are the most common climate change variability-induced risks in Ethiopia [2,39,47]. The findings of this study nevertheless revealed other climate change-induced impacts that are recurrent in the study area. The common ones include climate-induced seasonal epidemics, drought, harvest loss, flooding, hunger, migration, and school dropout [48].

4.2.1. Climate Change and Variability Related Epidemics

Climate change-related epidemics globally claim over 150,000 lives per year [26]. Studies on the health impact of anthropogenic climate change show that climate change is affecting human health. Diseases such as cardiovascular mortality, respiratory illnesses, water-related contagious diseases, and malnutrition are directly or indirectly associated with climate change [24,26–50]. Malaria, cholera, and animal diseases were the major climate change and variability induced epidemics reported by the HHs in the study area. Of the total respondents asked about the incidence of malaria in their locality, 46% (n = 183) said that they had suffered from a malaria epidemic in their family. Malaria has been the most prevalent disease in the study area and its prevalence increases in the wet season from May to November. During wet seasons, the availability of moisture creates a conducive environment for mosquitos to breed and spread, such as those places left open to harvest rainwater.

On the other hand, thirty-five percent (n = 138) of the households reported cholera incidences. Water supply coverage in the study area has been marginal because of poor hydro-geological conditions and limited resources to improve water supply coverage. Water scarcity forces people to use untreated water for domestic consumptions. Key informants in Boricha district said that such incidences usually happen during water-scarce seasons where families resort to using unsafe water for domestic consumptions. Diarrheal diseases incidents are very common during the dry season (from December to March) affecting health and household labor availability in the area. Twenty-one percent (n = 85) of the study participants also reported the prevalence of animal diseases. The problem is more acute for agro-pastoral communities living in the western parts of the study area, especially in Loka Abaya district. Some HHS also reported the prevalence of animal diseases, such as trypanosomiasis. It affects animal farming in the study area. About 21% (n = 81) of respondents reported that they had experienced climate change-related livestock diseases. Trypanosomiasis is one of the tropical livestock diseases, which has widely spread all over the SSA [51,52]. The World Health Organization (WHO) has also pointed out that climate change induced animal diseases are affecting cattle header Maasai pastoralist communities living in Tanzania and Kenya [53].

4.2.2. Drought

Drought is a climatic condition, which occurs when there is a RF below its normal pattern for a long period of time [54]. Drought has happened in the study area every year since 2000, compelling large-scale relief intervention by government and international donors. A prolonged delay or a total absence of the rains during the two growing seasons (*belg* and *kiremt*) adversely affects crop and livestock productivity. About 84% (n = 336) of the households interviewed reported to have suffered from drought-related effects in the last ten years. During the dry season, pastoral household members travel eight to ten hours every other day to water their livestock at the Hawassa Lake or Bilate River. Such long-distance trekking undoubtedly hinders household heads or other family members from devoting enough time to other livelihoods and, therefore, incurring further economic losses. To support the farmers' drought incident experience with the RF data, the researchers carried out RSA analysis. The result pointed out decreasing annual, summer, and spring RF amounts for the years under consideration. The SRA results pointed out that seven drought events of different extent occurred in the study area of which one was extreme, two were severe, and four were moderate (Figure 3). The incidence of drought had increased in the study area because 88% of the drought happened in the last fifteen years of the three decades.

4.2.3. Harvest Loss

Climate change-induced harvest loss due to rainfall shortages in the growing season is one reason for food insecurity in developing countries [55]. There had been harvest losses because of climate change in the study area. This happened due to the shortage of RF during the growing season. Out of 401 HHs interviewed, eighty-two percent (n = 329) said that they have experienced harvest losses in the last ten years. The farmers in the study area practice a mixed livelihood: crop cultivation and animal husbandry. When drought occurs, the productivity of both suffers. The exception is for *enset* crop. Because of its biological characteristics, *enset* has strong drought resistance capacity and has a strong link with animal husbandry. The mixed livelihoods are dependent on RF since there are no small-scale irrigation projects in the study area. During the drought years when animal pasture/grass is scarce, farmers fed their cattle *enset* (false banana). This alternative is less feasible nowadays because of the scarcity of animal dung, which is a critical fertilizer to grow *enset*. As a result, as the number of cattle owned by households dwindles, *enset* production also declines. Thus, climate change is affecting the mixed (crop cultivation and animal husbandry) rural livelihood by degrading its natural connection.

4.2.4. Flooding

Climate change-induced flooding is one of the impacts of climate change, which has vast economic and social consequences. It has been occurring in Ethiopia for years inflicting heavy social and economic costs [56]. Flooding during rainy seasons is one of the primary climate-induced effects in the study area. Of the 402 households asked about their exposure to flooding, 74% (n = 297) said that flooding had affected them in the last ten years. Most parts of the study area, especially the western part, have suffered from devastating flooding due to the seasonal overflow of the Bilat River. On the other hand, since the area is located in lower elevations, there are run-offs from the surrounding highlands that destroy terraces built to control soil erosion. Also, several gullies occurred because of heavy rain that has affected the landscape as noted by the researcher during field visits. Flooding during rainy seasons causes property and harvest loss incurring a significant economic cost to the farmers. The social and economic effects mentioned by households include loss of human lives and livestock, property damages, destruction of crops, water-borne diseases, and soil erosion.

4.2.5. Climate Change-Induced Hunger

Famine is a condition where people lack food for a longer period whereas hunger is a seasonal lack of food because of natural or human-made causes. Hunger becomes famine when it persists and affects many people [57]. RF shortage is one of the main reasons for famine and hunger in SSA [4,5]. Of the 401 respondents included in this study, 77% ($n = 309$) did experience hunger in the last ten years. Data collected from the Sidama Zone Agriculture and Natural Resources Management Department show that climate change-induced hunger did happen in the three districts in the previous fifteen years. Many people received emergency food aid in 2000, 2004, 2009, and 2016. These years coincide with the years of lower SRA discussed in part 4.1. According to key informants, signs of hunger include a delayed start of rain during rainy seasons (*belg* or *kiremt*) or its absence during critical sowing seasons. Both affect crop and livestock productivity and food security.

4.3. Determinants of Climate Change-Induced Impacts on Smallholder Farmers

4.3.1. Determinants of Exposure to the Effects of Drought

The result of the logit model shows that several reasons decide the likelihood of exposure to the effects of drought in the study area. The model is statistically significant for the determinants of drought (N = 357, $Chi^2 = 59.99$, $p = 0.0000$). The pseudo R^2 and adjusted R^2 (after marginal effect) values for this category of explanatory variables are 18.56% and 88.83%, respectively (Table 5). Out of the explanatory variables considered in the model, educational status, growing enset, membership in 1 to 5 groups, and being head in 1 to 5 groups affect the exposure to the effects of drought. Besides, the use of chemical fertilizer and the participation of women in family decision-making are found to be the main determinants of exposure to the effects of climate change-induced drought. Specifically, the findings show that a one year increase in schooling decreases the likelihood of being affected by drought by 10% (at $p < 0.05$ level).

Growing enset is another determinant of climate change-induced drought. Households who grow enset are 111.3% less likely to experience the effects of drought than those without *enset* at $p < 0.05$ level (Table 5). *Enset* (false banana) is a perennial crop known for its drought-resistant nature and is wildly used as a staple food in the most southern part of Ethiopia [58,59]). *Enset* is important for the local economy because its products, locally called *waasa* and *bu'la*, are sold in markets and are a source of income. Its roots, leaves, and stem are also used to feed animals during drought years. The Peasant Association (Kebele Administration) is the lowest administrative unit in Ethiopia that is accountable to the district (district) administration. Locally, households are organized in small groups of five households known as the 1 to 5 group. The government designed it without having local farmers say in its formation. The government widely uses this group for economic, political, and social purposes. Of the total 402 households involved in this study, 60% ($n = 242$) were members in the 1 to 5 group organization and the remaining 40% ($n = 160$) were not. On the other hand, 11% ($n = 42$) were serving as group leaders. There is widespread debate and controversies on the merits and demerits of such grouping. However, the purpose here is as a social organization, to what extent being a member or a leader in 1 to 5 groups is related to the likelihood of facing climate change-induced impacts?

The model revealed that being a member or having a position in the 1 to 5 groups affects the likelihood of exposure to climate change-induced drought. Being a member of the group increases the likelihood of exposure to drought effects by 120.2% (at $p < 0.01$ level). On the other hand, having a position of leadership of a group decreases the exposure to climate change-induced drought by 149.7% at $p < 0.01$ level (Table 5). A further investigation on the socio-economic status of the 1 to 5 group heads pointed out that they are were individuals with a better resource base. Furthermore, the group heads have more decision making power on resources owned by the community. On the other hand, the members of these groups were poor farmers who joined the group to get government support. Despite the claims of the government to support the poor organized in the groups, the members were most vulnerable compared with others. Also, the model examined agricultural technology use and the likelihood of exposure to climate change-induced impacts. The results of the model show that except for improved animal fodder, other agricultural technologies did not decrease the likelihood of exposure to climate change-induced impacts. Using chemical fertilizers (Urea and NPS) increases the possibility of experiencing the effects of drought and harvest losses. It increases the likelihood of experiencing drought effects by 150.4% (at $p < 0.01$ level). This is contrary to the claims of government officials who argue using chemical fertilizers as rewarding for farmers. Interviews with farmers revealed that when RF onset is late, the application of chemical fertilizers is calamitous to farmers because the fertilizers wilt the crops in their early stages. Thus, applying chemical fertilizers during drought seasons increases their likelihood of exposure to the effects of drought. Moreover, those farmers who use chemical fertilizers are more likely to experience bankruptcy during drought years for they have to pay back the fertilizer loan, which is expensive according to key informants. Thus, the debt and risks associated with borrowing in precarious situations are found to be factors for the higher vulnerability of farmers.

Table 5. Marginal effects due to independent variables (the coefficients table).

Variables	Drought	Lost Harvest	Flooding	Hunger
Sex	−0.244	−0.370	−0.179	−0.559
	(0.360)	(0.355)	(0.278)	(0.327) *
Land certificate	0.468	0.100	−0.645	−0.273
	(0.424)	(0.410)	(0.079) *	(0.117)
Family size	0.026	0.063	0.009	0.021
	(0.080)	(0.087)	(0.063)	(0.073)
Educational status	−0.100	−0.030	0.069	−0.116
	(0.048) **	(0.051)	(0.042)	(0.045) ***
Total land size	−0.095	−0.202	−0.439	−0.571
	(0.160)	(0.172)	(0.181) **	(0.209) ***
PSNP beneficiary	0.003	−0.919	−0.923	−0.004
	(0.393)	(0.354) ***	(0.308) ***	(0.358)
Do you produce enset?	−1.113	−0.636	−0.209	0.094
	(0.497) **	(0.455)	(0.338)	(0.365)
Weather early warning information	0.176	0.436	−0.155	0.108
	(0.327)	(0.326)	(0.268)	(0.295)
Climate change adaptation training	0.202	0.127	−0.262	0.232
	(0.355)	(0.350)	(0.283)	(0.318)
Membership in 1 to 5	1.202	1.598	−0.852	1.475
	(0.380) ***	(0.387) ***	(0.301) ***	(0.343) ***
Head of 1 to 5	−1.497	−2.000	0.719	−1.427
	(0.510) ***	(0.515) ***	(0.486)	(0.482) ***
Grazing land	0.857	0.406	0.134	0.958
	(0.648)	(0.604)	(0.477)	(0.599)
Water pond	1.022	1.434	0.289	1.070
	(0.865)	(0.869) *	(0.542)	(0.870)
Improved animal fodder	−0.761	−0.816	−0.597	−0.887
	(0.521)	(0.480) *	(0.409)	(0.488) *
Fertilizer/urea/NPS	1.504	1.041	0.075	1.429
	(0.534) ***	(0.531) **	(0.506)	(0.506) ***
Drought resistant crop variety	0.437	−0.151	0.268	0.687
	(0.420)	(0.396)	(0.327)	(0.392) *
Women participation in family decision making	−1.089	−0.350	−0.020	−1.015
	(0.419) ***	(0.370)	(0.295)	(0.360) ***
Participation in adaptation decision	−0.135	−0.007	−0.529	−0.043
	(0.408)	(0.396)	(0.305) *	(0.360)
N	357	358	357	357
Chi^2	59.99	85.78	39.62	82.49
P	0.0000	0.0000	0.0024	0.0000
Pseudo R^2	0.1856	0.2444	0.0958	0.2118
Adjusted R^2 (Marginal effect after logit)	0.8883	0.8775	0.7570	0.8325

* $p < 0.1$; ** $p < 0.05$; *** $p < 0.01$; (SE) = standard error.

4.3.2. Determinants of Facing Harvest Loss

The logit model results show that several factors affect the possibility of exposure to harvest loss in the study area. The model is statistically significant for the determinants of harvest loss (N = 358, $Chi^2 = 85.78$, $p = 0.0000$). The pseudo R^2 and adjusted R^2 values are 24.44% and 87.75%, respectively (Table 5). Among the explanatory variables, membership and headship in 1 to 5 groups, PSNP, and animal fodder control the likelihood of experiencing harvest loss. Also, the use of chemical fertilizer controls the likelihood of harvest loss. The PSNP is a social protection program that government instruments in drought-prone and food insecure areas of Ethiopia. Because of the widespread food insecurity, the government has been carrying out the PSNP in the study area. It aids vulnerable farmers in cash and kind in the study area. Of the sampled households, 29% (n = 114) were PSNP beneficiaries, whereas the remaining were not. The results of the model revealed that HHs supported by the program are less likely to experience the effects of harvest loss by 91.9% (at $p < 0.01$ level) than the non-users of the program. Like the case of drought exposure, membership in the 1 to 5 group increases the possibility of facing a harvest loss by 159.8% (at $p < 0.01$ level). On the contrary, bearing the leadership position in the 1 to 5 group reduces the likelihood of exposure to harvest loss by 200% (at $p < 0.01$ level). One of the challenges for farmers during the drought season is the lack of pasture and water for their livestock. To solve the problem, the government has introduced a plant species that farmers grow to feed their livestock. Farmers call it animal fodder in the study area. The logit model analysis shows that the use of this improved animal fodder reduces the likelihood of facing harvest loss. Farmers who use improved animal fodder are 81.6% less likely to experience harvest loss than the non-users (at $p < 0.1$ level). This is because, without the fodder, farmers use crops to feed livestock. Using chemical fertilizers in crop fields increases the chance of experiencing harvest loss. Farmers who apply chemical fertilizers in their crop field are 37% more likely to experience harvest loss during a drought season because crops wilt under inadequate rain conditions (Table 5).

4.3.3. Determinants of Exposure to the Effects of Flooding

Flooding is one of the climate change-induced impacts experienced by the HHs in the study area. The model is statistically significant for the determinants of experiencing flooding (N = 357, Chi^2 = 39.62.78, $p = 0.0958$). The pseudo R^2 and adjusted R^2 values of the model are 9.58% and 75.7%, respectively (Table 5). Among the variables considered, land use right certificate, being a beneficiary of PSNP, and membership in the 1 to 5 groups are the determinants of exposure to flooding. Also, taking part in climate change adaptation decisions affects exposure to climate change-induced flooding (Table 5). Land use right certificate affects the likelihood of experiencing climate change-induced impacts. The likelihood of facing flooding decreases by 64.5% for farmers who have land ownership certificate. This result is statistically significant (at $p < 0.1$ level). This is because farmers with a land use right certificate have a stronger landownership feeling than those without and this encourages farmers to carry out different soil and water conservation practices. Households with larger landholding size are less likely to experience flooding than HHs with a smaller landholding size. A unit increase in landholding size decreases the likelihood of experiencing flooding by 43.9% (at $p < 0.05$). Being a PSNP beneficiary also decreases the likelihood of facing climate change-induced flooding by 92.3% (at $p < 0.01$ level). On the other hand, membership in the 1 to 5 groups controls the likelihood of exposure to flooding impacts. Being a member in 1 to 5 group decreases the chance of experiencing flooding impacts by 85.2% (at $p < 0.01$ level). Participation in climate change adaptation decisions at the local level is also relevant. Farmers who take part in local level decision-making are 52% less likely to experience climate change induced flooding impacts than the non-participants (Table 5). The incident of flooding is common for low-lying areas, but the model did not consider topography and is a limitation of the model.

4.3.4. Determinants of Facing Hunger

Climate change and variability-induced hunger are one of the impacts experienced by the HHs in the study area, especially during drought periods. The model is statistically significant for the determinants of experiencing hunger by the HHs (N = 357, Chi^2 = 82.49, $p = 0.0000$). The pseudo R^2 and adjusted R^2 values are 21.18% % and 83.25%, respectively (Table 5). The logit model result identified sex, educational status, farmland size, membership, and headship in the 1 to 5 groups as determinants of exposure to hunger. Besides, the use of improved animal fodder, chemical fertilizer, and improved seed variety also control the chance of exposure to hunger. Furthermore, female participation in family decision making also decides the likelihood of exposure to change-induced hunger. The sex of the household head is associated with the likelihood of facing climate-induced impacts; being a male household head decreases the chance of exposure to hunger by 55.9% (at $p < 0.1$ level). This means female-headed HHs are more vulnerable to the exposure of hunger than the male counterpart. This is due to issues related to access to and control over resources owned by the family and the community.

On the other hand, a unit increase in education decreases the likelihood of facing hunger by 11.6% (at $p < 0.01$ level). This suggests the role of education in minimizing the impacts of climate change induced hunger. Household's total farm size also lowers the chance of facing hunger. A unit increase in farm size reduces the likelihood of facing hunger by 57.1% (at $p < 0.01$ level). This means as land holding size decreases, farmers become more vulnerable to climate change-induced hunger. Like the case of exposure to drought, membership in the 1 to 5 groups increases the likelihood of exposure to hunger. The likelihood of facing hunger increases by 147.5% for members in 1 to 5 group and decreases by 142.7% for the heads of the group (at $p < 0.01$ levels). Using improved animal fodder decreases the likelihood of experiencing hunger. Users of improved animal fodder are 88.7% less likely to face climate change-induced hunger than non-users (at $p < 0.1$ level). Women's involvement in family decision-making such as selling assets and renting out lands affects the likelihood of exposure to hunger. It reduces the chance of experiencing climate change-induced hunger by 101.5% (at $p < 0.01$ level). The use of chemical fertilizers and drought-resistant varieties also affects the possibility of experiencing climate change-induced hunger. Using these inputs increases the chance of experiencing hunger by 142.9% and 69.7%, respectively (Table 5). Key informants' interviews to justify the reasons behind these findings suggested the use of chemical fertilizer and improved seed varieties is disadvantageous for farmers. This is because farmers who use agricultural inputs usually get them in the form of a loan. The farmers have to sell their assets to pay back the debt since there is no surplus produce to sell in the market during drought years. This further degrades their asset base, exposing them to food security.

5. Conclusions

The analyses of data collected from different sources have shown that RF had been scarce, erratic and declining from time to time, exposing rained smallholder farmers to various climate change-induced impacts. Annual and growing season (*belg* and *kiremt*) RF had shown a declining trend in the study area. The analysis annual RSA has shown that the study area had experienced seven droughts of different extent during the three decades (1983–2014) under consideration. The monthly minimum, maximum, and average temperature had increased over the years under consideration, all of which are statistically significant. These changes in climatic variables resulted in climate change-induced epidemics, drought, harvest losses, flooding, and hunger.

The logit model results have shown that different several socio-economic and institutional factors control the likelihood of exposure to the impacts. The likelihood of exposure to the effects of drought is controlled by the educational status of the HH head, growing enset, and membership in the 1 to 5 groups. Furthermore, leadership in 1 to 5 groups, use of chemical fertilizers, and female involvement in family decision-making are other determinants. Furthermore, membership and headship in the 1 to 5 groups, being the beneficiary of the PSNP, the use of improved animal fodder, and the use of chemical fertilizer affect the likelihood of exposure to harvest loss. On the other hand, land use right certificate, participation in the PSNP, membership in the 1 to 5 groups, and participation in climate change adaptation decisions are determinants of climate change-induced flooding. Moreover, the logit model has identified that sex, educational status, farmland size, and membership and headship in the 1 to 5 groups affect the likelihood of experiencing climate change-induced hunger. Besides, the use of improved animal fodder, chemical fertilizers, and improved seed variety also control the likelihood of exposure to climate change-induced hunger. Participation of women in family decision-making also reduces climate change-induced hunger.

Climate change and variability-induced impacts are widely available in the study area despite various adaptive strategies of the government and the households. Government policies on education, female empowerment, land use right certification, and participation of the households in adaptation decisions can help to minimize climate change-induced impacts. The empirical literature also suggests that working on human capital is key to successful livelihood diversification [60–62]. In this regard, education (both formal and informal) and skills training need to be emphasized. On the contrary, organizing farmers into 1 to 5 groups contributed positively only in minimizing the incidence of flooding. Memberships in such organization did not help to reduce the possibility of facing the effects of drought, harvest loss, and hunger. Thus, the government needs to reform such organizations in a manner that they can contribute to lessening climate change impelled impacts. Besides, the supplies of agricultural inputs (improved seed varieties and chemical fertilizers) specifically during drought years increased the likelihood of exposure to climate change-induced impacts. Therefore, there is a need to build human capital through expanding education, strengthening female participation in family decision-making, and improve public participation in climate change adaptation undertakings to manage climate change-induced impacts sustainably.

Acknowledgments: I would like to thank Tesfaye Semela and Sintayehu Hailu for commenting and shaping the statistical model used in the study. I also thank Girma Kebede for commenting the first draft of the article.

References

1. O'Brien, K.L.; Leichenko, R.M. Double exposure: Assessing the impacts of climate change within the context of economic globalization. *Glob. Environ. Chang.* **2000**, *10*, 221–232. [CrossRef]
2. Bewket, W.; Radeny, M.A.O.; Mungai, C. *Agricultural Adaptation and Institutional Responses to Climate Change Vulnerability in Ethiopia*; CGIAR Research Program on Climate Change, Agriculture and Food Security (CCAFS): Copenhagen, Denmark, 2015.
3. IPCC. *Climate Change 2014: Synthesis Report*; Contribution of Working Groups I, II and III to the Fifth Assessment Report of the Intergovernmental Panel on Climate Change; Pachauri, R.K., Meyer, L.A., Eds.; IPCC: Geneva, Switzerland, 2014; p. 151.
4. Kotir, J.H. Climate change and variability in Sub-Saharan Africa: A review of current and future trends and impacts on agriculture and food security. *Environ. Dev. Sustain.* **2011**, *13*, 587–605. [CrossRef]
5. Adeniyi, P.A. Climate change induced hunger and poverty in Africa. *J. Glob. Biosci.* **2016**, *5*, 3711–3724.
6. AGRA (Alliance for a Green Revolution in Africa Change). *Smallholder Agriculture in Sub-Saharan Africa*; AGRA: Nairobi, Kenya, 2014.
7. Adger, W.N.; Huq, S.; Brown, K.; Conway, D.; Hulme, M. Adaptation to climate change in the developing world. *Prog. Dev. Stud.* **2003**, *3*, 179–195. [CrossRef]
8. Morton, J.F. The impact of climate change on smallholder and subsistence agriculture. *Proc. Natl. Acad. Sci. USA* **2007**, *104*, 19680–19685. [CrossRef] [PubMed]
9. Schlenker, W.; Lobell, D.B. Robust negative impacts of climate change on African agriculture. *Environ. Res. Lett.* **2010**, *5*, 014010. [CrossRef]
10. FAO. Agriculture in Sub-Saharan Africa: Prospects and challenges for the next decade. In *Agricultural Outlook 2016–2025*; FAO: Rome, Italy, 2016.
11. MoA. *The Federal Democratic Republic of Ethiopia. Agriculture Sector Programme of Plan on Adaptation to Climate Change*; Technical Working Group: Addis Ababa, Ethiopia, 2011.
12. World Bank. *Measuring the Economic Impact of Climate Change on Ethiopian Agriculture: Ricardian Approach*; The World Bank: Washington, DC, USA, 2007.
13. Senbeta, A.F.; Olsson, J.A. Climate Change Impact on Livelihood, Vulnerability and Coping, Mechanisms: A Case Study of West-Arsi Zone, Ethiopia. Unpublished MA Thesis, Lund University, Lund, Sweden, 2009.
14. Tafesse, A.; Ayele, G.; Ketema, M.; Geta, E. Adaptation to climate change and variability in eastern Ethiopia. *J. Econ. Sustain. Dev.* **2013**, *4*, 91–103.
15. IPCC. *Climate Change 2007—The Physical Science Basis: Working Group I Contribution to the Fourth Assessment Report of the IPCC*; Solomon, S., Qin, D., Manning, M., Averyt, K., Marquis, M., Eds.; Cambridge University Press: Cambridge, UK, 2007.
16. IPCC. *Climate Change IPCC Third Assessment Report. Intergovernmental Panel on Climate Change*; IPCC Secretariat: Geneva, Switzerland, 2001.

17. Cooper, P.J.M.; Dimes, J.; Rao, K.P.C.; Shapiro, B.; Shiferaw, B.; Twomlow, S. Coping better with current climatic variability in the rain-fed farming systems of sub-Saharan Africa: An essential first step in adapting to future climate change? *Agric. Ecosyst. Environ.* **2008**, *126*, 24–35. [CrossRef]
18. Shiferaw, B.; Tesfaye, K.; Kassie, M.; Abate, T.; Prasanna, B.M.; Menkir, A. Managing vulnerability to drought and enhancing livelihood resilience in sub-Saharan Africa: Technological, institutional and policy options. *Weather Clim. Extrem.* **2014**, *3*, 67–79. [CrossRef]
19. Mertz, O.; Halsnæs, K.; Olesen, J.E.; Rasmussen, K. Adaptation to climate change in developing countries. *Environ. Manag.* **2009**, *43*, 743–752. [CrossRef] [PubMed]
20. Ravindranath, N.H.; Sathaye, J.A. Climate change and developing countries. In *Climate Change and Developing Countries*; Springer: Dordrecht, The Netherlands, 2002; pp. 247–265.
21. Mendelsohn, R.; Dinar, A.; Williams, L. The distributional impact of climate change on rich and poor countries. *Environ. Dev. Econ.* **2006**, *11*, 159–178. [CrossRef]
22. Carter, M.R.; Little, P.D.; Mogues, T.; Negatu, W. Poverty traps and natural disasters in Ethiopia and Honduras. *World Dev.* **2007**, *35*, 835–856. [CrossRef]
23. Ribot, J. Vulnerability does not fall from the sky: Toward multiscale, pro-poor climate policy. *Soc. Dimens. Clim. Chang. Equity Vulnerability A Warm. World* **2010**, *2*, 47–74.
24. McMichael, A.J. *Global Climate Change and Health: An Old Story Writ Large. Climate Change and Human Health: Risks and Responses*; World Health Organization: Geneva, Switzerland, 2003.
25. Kjellstrom, T. Climate change, direct heat exposure, health and well-being in low and middle-income countries. *Glob. Health Action* **2009**, 2. [CrossRef]
26. Patz, J.A.; Campbell-Lendrum, D.; Holloway, T.; Foley, J.A. Impact of regional climate change on human health. *Nature* **2005**, *438*, 310. [CrossRef]
27. Rosenzweig, C.; Iglesias, A.; Yang, X.B.; Epstein, P.R.; Chivian, E. Climate change and extreme weather events; implications for food production, plant diseases, and pests. *Glob. Chang. Hum. Health* **2001**, *2*, 90–104. [CrossRef]
28. Nordás, R.; Gleditsch, N.P. Climate change and conflict. *Political Geogr.* **2007**, *26*, 627–638. [CrossRef]
29. UN-DESA. *World Urbanization Prospects, the 2014. Population Division, Department of Economic and Social Affairs 2014*; United Nations Secretariat: New York, NY, USA, 2014.
30. Asaminew, A.E. *Climate Change, Growth and Poverty in Ethiopia. The Robert S. Strauss Center for International Security and Law*; The University of Texas at Austin: Austin, TX, USA, 2013.
31. Evangelista, P.; Young, N.; Burnett, J. How will climate change spatially affect agriculture production in Ethiopia? Case studies of important cereal crops. *Clim. Chang.* **2013**, *119*, 855–873. [CrossRef]
32. Mersha, A.A.; van Laerhoven, F. The interplay between planned and autonomous adaptation in response to climate change: Insights from rural Ethiopia. *World Dev.* **2018**, *107*, 87–97. [CrossRef]
33. Seleshi, Y.; Zanke, U. Recent changes in RF and rainy days in Ethiopia. *Int. J. Climatol.* **2004**, *24*, 973–983. [CrossRef]
34. Adem, A.; Bewket, W. *A Climate Change Country Assessment Report for Ethiopia. Submitted to Forum for Environment*; Epsilon International R&D: Addis Ababa, Ethiopia, 2011.
35. Adnew, M.; Bewket, W. Variability and trends in RF amount and extreme event indices in the Omo-Ghibe River Basin, Ethiopia. *Reg. Environ. Chang.* **2014**, *14*, 799–810.
36. Yalew, A.W.; Hirte, G.; Lotze-Campen, H.; Tscharaktschiew, S. *Economic Effects of Climate Change in Developing Countries: Economy-Wide and Regional Analysis for Ethiopia*; Center of Public and International Economics: Dresdon, Germany, 2017.
37. Mideksa, T.K. Economic and distributional impacts of climate change: The case of Ethiopia. *Glob. Environ. Chang.* **2010**, *20*, 278–286. [CrossRef]
38. Kreft, S.; Eckstein, D.; Dorsch, L.; Fischer, L. *Global Climate Risk Index 2016: Who Suffers Most from Extreme Weather Events? Weather-related Loss Events in 2014 and 1995 to 2014*; Germanwatch: Bonn, Germany, 2015.
39. Bewket, W. Climate change perceptions and adaptive responses of smallholder farmers in central highlands of Ethiopia. *Int. J. Environ. Stud.* **2012**, *69*, 507–523. [CrossRef]
40. Atinkut, B.; Mebrat, A. Determinants of farmers choice of adaptation to climate variability in Dera district, south Gondar zone, Ethiopia. *Environ. Syst. Res.* **2016**, *5*, 6. [CrossRef]
41. SNNPRS-BoFED. *Annual Abstract SNNPRS Bureau of Finance and Economic Development Annual Statistical Abstract. 2003 E.C. Data Collection—Dissemination Core Process*; SNNPRS-BoFED: Hawassa, Ethiopia, 2015.

42. SZBoFED. *Sidama Zone Socio-Economic Profile. Sidama Zone Finance and Economic Development Department*; SZBoFED: Hawassa, Ethiopia, 2007.
43. Krejcie, R.V.; Morgan, D.W. Determining sample size for research activities. *Educ. Psychol. Meas.* **1970**, *30*, 607–610. [CrossRef]
44. Pallant, J. *SPSS Survival Manual: A Step by Step Guide to Data Analysis Using SPSS for Windows (Versions 10 and 11): SPSS Student Version 11.0 for Windows*; Open University Press: Milton Keynes, UK, 2001.
45. Bewket, W.; Conway, D. A note on the temporal and spatial variability of RF in the drought-prone Amhara region of Ethiopia. *Int. J. Climatol.* **2007**, *27*, 1467–1477. [CrossRef]
46. Janowiak, J.E.; Ropelewski, C.F.; Halpert, M.S. The precipitation anomaly classification: A method for monitoring regional precipitation deficiency and excess on a global scale. *J. Clim. Appl. Meteorol.* **1986**, *25*, 565–574. [CrossRef]
47. Arragaw, A.; Bewket, W. Smallholder farmers' coping and adaptation strategies to climate change and variability in the central highlands of Ethiopia. *Local Environ.* **2017**, *22*, 825–839.
48. Tafesse, M. An Assessment of Domestic Water Supply Status of Rural Household in Selecting Kebeles of Boricha District, Sidaama Zone, Southern Ethiopia. Unpublished Master's Thesis, Addis Ababa University, Addis Ababa, Ethiopia, 2009.
49. Haines, A.; Kovats, R.S.; Campbell-Lendrum, D.; Corvalán, C. Climate change and human health: Impacts, vulnerability and public health. *Public Health* **2006**, *120*, 585–596. [CrossRef]
50. McMichael, C.; Barnett, J.; McMichael, A.J. An ill wind? Climate change, migration, and health. *Environ. Health Perspect.* **2012**, *120*, 646–654. [CrossRef] [PubMed]
51. Allsopp, R. Options for vector control against trypanosomiasis in Africa. *Trends Parasitol.* **2001**, *17*, 15–19. [CrossRef]
52. Hotez, P.J.; Kamath, A. Neglected tropical diseases in sub-Saharan Africa: Review of their prevalence, distribution, and disease burden. *PLoS Negl. Trop. Dis.* **2009**, *3*, e412. [CrossRef]
53. Kissui, B.M. Livestock predation by lions, leopards, spotted hyenas, and their vulnerability to retaliatory killing in the Maasai steppe, Tanzania. *Anim. Conserv.* **2008**, *11*, 422–432. [CrossRef]
54. Philander, S.G. *Encyclopedia of Global Warming and Climate Change: AE*; Sage Publishers: Thousand Oaks, CA, USA, 2008.
55. Wheeler, T.; Von Braun, J. Climate change impacts on global food security. *Science* **2013**, *341*, 508–513. [CrossRef]
56. MoWR and NMA. *Climate Change National Adaptation Programme of Action (NAPA) of Ethiopia. National Meteorological Services Agency, Ministry of Water Resources*; Federal Democratic Republic of Ethiopia: Addis Ababa, Ethiopia, 2007.
57. Ayalew, M. *What Is Food Security and Famine and Hunger. Using Science against Famine: Food Security, Famine Early Warning, and El Nino*; National Center for Atmospheric Research: Boulder, CO, USA, 1997; pp. 1–8.
58. Tsegaye, A.; Struik, P.C. Analysis of enset (*Ensete ventricosum*) indigenous production methods and farm-based biodiversity in major enset-growing regions of southern Ethiopia. *Exp. Agric.* **2002**, *38*, 291–315. [CrossRef]
59. Tesfaye, B.; Lüdders, P. Diversity and distribution patterns of enset landraces in Sidama, Southern Ethiopia. *Genet. Resour. Crop Evol.* **2003**, *50*, 359–371. [CrossRef]
60. Ellis, F. *Rural Livelihood Diversity in Developing Countries: Evidence and Policy Implications*; Overseas Development Institute: London, UK, 1999.
61. Eneyew, A. Determinants of livelihood diversification in pastoral societies of southern Ethiopia. *J. Agric. Biodivers. Res.* **2012**, *1*, 43–52.
62. Adepoju, A.O.; Obayelu, O.A. Livelihood diversification and welfare of rural households in Ondo State, Nigeria. *J. Dev. Agric. Econ.* **2013**, *31*, 482–489.

8

An Economic Assessment of Local Farm Multi-Purpose Surface Water Retention Systems under Future Climate Uncertainty

Pamela Berry [1,*], Fuad Yassin [1], Kenneth Belcher [2] and Karl-Erich Lindenschmidt [1]

1 School of Environment and Sustainability, University of Saskatchewan, 11 Innovation Boulevard, Saskatoon, SK S7N 3H5, Canada; fuad.yassin@usask.ca (F.Y.); karl-erich.lindenschmidt@usask.ca (K.-E.L.)

2 Department of Agricultural and Resource Economics, University of Saskatchewan, Saskatoon, SK S7N 5A8, Canada; ken.belcher@usask.ca

* Correspondence: pamela.berry@usask.ca

Academic Editor: Hossein Azadi

Abstract: Regions dependent on agricultural production are concerned about the uncertainty associated with climate change. Extreme drought and flooding events are predicted to occur with greater frequency, requiring mitigation strategies to reduce their negative impacts. Multi-purpose local farm water retention systems can reduce water stress during drought periods by supporting irrigation. The retention systems' capture of excess spring runoff and extreme rainfall events also reduces flood potential downstream. Retention systems may also be used for biomass production and nutrient retention. A sub-watershed scale retention system was analysed using a dynamic simulation model to predict the economic advantages in the future. Irrigated crops using water from the downstream reservoir at Pelly's Lake, Manitoba, Canada, experienced a net decrease in gross margin in the future due to the associated irrigation and reservoir infrastructure costs. However, the multi-purpose benefits of the retention system at Pelly's Lake of avoided flood damages, nutrient retention, carbon sequestration, and biomass production provide an economic benefit of $25,507.00/hectare of retention system/year. Multi-purpose retention systems under future climate uncertainty provide economic and environmental gains when used to avoid flood damages, for nutrient retention and carbon sequestration, and biomass production. The revenue gained from these functions can support farmers willing to invest in irrigation while providing economic and environmental benefits to the region.

Keywords: climate change; multi-purpose retention systems; agriculture; irrigation

---

1. Introduction

Across Canada, annual mean temperature has increased since the 1950s. As of 2010, Canada has experienced an annual average surface air temperature warming of 1.5 °C. Stronger warming trends have been found for the north and west of Canada, with the greatest warming occurring in winter and spring [1,2]. These climatic changes are expected to increase potential evapotranspiration and lead to moisture deficits [3]. Changes in precipitation timing are also expected, resulting in less snow-cover, shorter snow-cover duration, increased winter river flows, and decreased summer flood events [3–5]. If the timing of seasonal precipitation begins to change and temperatures continue to rise under climate change, there will be severe effects on agriculture, ecosystems, water runoff rates and quantities, as well as groundwater storage. On the Canadian Prairies, where 80 percent of Canada's farmland is situated, strategic water management solutions are needed to deal with the uncertainty associated with climate change and its impact on agricultural production [4,6,7]. Changes in temperature and precipitation expected with climate change will impact a wide range of variables, affecting the productivity of annual crops. Alterations in the growing season length, frost timing, heat waves, precipitation, and moisture availability will be witnessed with temperature and precipitation increases. This will require farmers to be ready to adapt to new climatic patterns [8]. The uncertainty associated with climate also increases risk for farmers, requiring water management solutions that will provide benefits to farmers under all conditions, while reducing agricultural risk [8,9]. Economic consequences of drought or flooding events will depend on the agricultural and water management sectors success in preparing and adapting for climatic extremes [4]. The chosen water management strategy needs to be economically viable, benefitting the farmer and the provincial economy, while reducing risk in agricultural production systems [10].

Climate change is predicted to increase the frequency of extreme drought and flooding events which will impact agricultural sectors, such as southern Manitoba, Canada [4,7,9,11]. Historically, the trend within Manitoba has been to remove excess water from agricultural land as quickly as possible in spring using a series of ditches and drains [3,12]. This strategy to deal with flooding in the province is common throughout the landscape [3,13]. While drainage systems are meant to remove excess water from inundated land quickly, they can actually increase the negative effects of floods by amplifying flood peaks, which then have greater force to cause damage [3]. Drainage also increases the amount of nutrients being removed from the landscape, subsequently impacting water quality as they flow into Manitoba's water bodies [3]. Quickly removing water from the landscape may also leave agricultural lands vulnerable to soil moisture deficits under future climate uncertainty as evapotranspiration quickly removes summer precipitation from the soil [3]. The province of Manitoba aims to increase their adaptive capacity to prepare for future climate uncertainties [9,14,15]. Increasing the adaptive capacity of southern Manitoba communities will require the development of techniques allowing farmers to drought proof their crops as well as to limit damages caused by floods in non-drought years. Strategies should also allow for sustainable water management by providing multiple benefits when possible, such as bio production and nutrient retention [14]. Strategies currently being used on the Prairies include crop insurance, soil and water conservation, improved irrigation (where applicable), exploration of groundwater supplies, as well as introduction of new infrastructure [4,7,9]. Infrastructure implementations range from new wells and pipelines to dugouts [4]. Dryland farmers look to decrease drought risk by conserving soil moisture and nutrients through crop rotation and minimizing tillage practices [9].

The creation of multi-purpose local farm water retention systems, designed to capture and store surface water, may be a viable option that would reduce water stress during droughts by providing water for irrigation [16]. Additionally, the retention systems would serve to capture excess runoff in spring and during extreme rainfall events to help mitigate flood events. The stored water can be used for biomass production and nutrient retention [14,17]. Retention systems also serve to reduce downstream peak flow and aid in retaining flood waters which reduces associated flood risks downstream [16,18]. If water is released from reservoirs, they serve to replenish groundwater stores downstream [16]. Researchers have found these systems to be effective for increasing and stabilizing crop yields via irrigation in locations such as Texas, Kansas, Kentucky, India, and Thailand [19]. Berry [20] reported an average increase in crop yield of \$13.00/hectare/year when irrigation was applied to canola, alfalfa, barley, and spring wheat using water from a local farm retention system in Manitoba. However, there is limited pre-existing irrigation infrastructure in Manitoba, so adopting irrigation as a management strategy would require the costly installation of irrigation infrastructure. This would require farmers to earn incremental gross margins of \$147.00/hectare of crop land/year to cover the cost of irrigation infrastructure and a surface water retention system as a water source [20]. While irrigation did not provide an increase in gross margins under present day conditions, future climate change scenario predictions of increased temperatures and precipitations may result in more economically advantageous conditions in the future.

Surface water retention systems additionally aid in improving water quality. Retention systems have shown success in reducing nutrient and sediment loading in various locations worldwide. Several examples from the literature provided by [21] on retention systems in America and Europe have shown effectively reduced nutrient loading. A runoff detention pond in Oklahoma, USA reduced sediment discharge downstream by 82%, total nitrogen by 56%, and total phosphorus by 60% [22]. A reduction in total nitrogen of 38%, and 56% in total phosphorus loading from a constructed wetland (a detention basin formed by berms adjacent to a stream) in Illinois, USA was reported by [23]. A shallow predam, a small reservoir aimed at improving water quality of a larger main reservoir downstream, in Luxembourg, was found to retain total phosphorus up to 60%, and a deep predam retained up to 82% [21,24]. A small dam in Spain reduced total phosphorus loads downstream by over 25% [25]. Small ponds in Finland and Sweden reduced total phosphorus loading by 17% and constructed wetlands in Norway and Finland reduced total phosphorus loading by 41% [21,26].

In Manitoba, small on-farm surface water retention systems are scattered throughout agricultural watersheds. The South Tobacco Creek Watershed, in south central Manitoba, is home to twenty-six dams providing management of almost 30% of the watersheds drainage area [18]. The watershed is now home to five dry dams, six back-flood dams, and fifteen multipurpose dams. Each dam was designed to retain 20–25 mm of runoff at full capacity from their catchment area [21]. These dams' capacity to reduce flood risk has been under study since the 1990s. The Watershed Evaluation of Beneficial Management Practices (WEBs) program, an Agriculture and Agri-Food Canada (AAFC) research program, began in 2004 to expand the research on the South Tobacco Creek Watershed dams to include sediment, nitrogen, and phosphorus loadings downstream [18]. Both the dry flood control dam and the multi-purpose dam were effective in reducing total suspended sediment (65%–85% reduction), particulate nitrogen (41%–43% reduction during snowmelt, 7%–11% reduction from summer rainfall events), and particulate phosphorus (27%–38% reduction in snowmelt runoff) [18]. The entire system of dams within the watershed provided a reduction in peak flow of 9%–19% from spring snowmelt runoff and 13%–25% from rainfall runoff [18]. Another on-farm retention pond in Saint-Samuel, Quebec was found to reduce peak flows by 38%, on average, from rainfall runoff events [27]. The pond was also effective at removing total suspended sediment, total nitrogen, and total phosphorus with mean removal efficiency ratios of 50%–56%, 42%–52%, and 48%–59%, respectively.

The ecosystem benefits of reduced nutrient and sediment loading downstream, carbon sequestration, and avoided flood damages downstream have been monetized for surface water retention systems on the Prairie landscape. Berry [20] reported retention system's ability to reduce downstream phosphorus and nutrient loading, sequester carbon, and contribute to avoided flood damages downstream can provide $17,600/hectare of retention basin/year. This estimate used a carbon credit value of $25.00/tonne of carbon dioxide equivalent, which the Manitoba Liquor and Lotteries Corporation is currently offering in Manitoba [28]. Benefit transfer methodology was used to estimate monetary values for reductions in nutrient loading and avoided flood damages downstream [20,29–33]. Cattail harvest from surface water retention systems as a biomass crop has recently been commercialized in Manitoba. Multi-purpose surface water retention systems gain additional ecosystem benefits of biomass production, increased nutrient management, and carbon dioxide emission offsets when cattail harvest is introduced [14,34]. Berry [20] reported that harvesting cattail for biomass production, nutrient removal, and carbon credits from multi-purpose retention systems can provide an additional $7657.00/hectare of retention basin/year. This estimate included monetary compensation for nitrogen and phosphorus removal from the ecosystem from cattail harvest, carbon credit production from biomass replacing coal with a value of $25.00/tonne of carbon dioxide equivalent, and a net value of dry cattail biomass of $16.59/tonne [17,28,34,35].

The purpose of this study is to determine the economic feasibility of the adoption of a multi-purpose local farm surface water retention system as a water management strategy in southern Manitoba to reduce the increased agricultural risks to farmers under future climate uncertainty [9,16,36]. Future climatic conditions will be modelled for the study area to determine the economic and environmental benefits water retention systems can offer as a water management strategy under climate change.

Study Site

The study site for this analysis is a pre-existing water retention system at Pelly's Lake in southern Manitoba (Figure 1). Prime agricultural lands surround the lake, with the lake's watershed currently used to produce a range of crops [6]. The four main crops in the watershed are canola, spring wheat, alfalfa, and barley [37]. Landowners in the area had historically attempted to drain the area of Pelly's Lake to increase hay production. However, these efforts ultimately failed due to the presence of an underground spring, poor drainage, poor water retention potential, and widespread flooding during times of excess water [6,13]. This led to an agreement with the La Salle Redboine Conservation District (LSRCD) to create a back flood system offering multiple benefits [38]. Conservation districts in Manitoba promote sustainable development in working to protect their districts natural resources [39]. The frequently saturated land at Pelly's Lake, MB now provides flood water retention, capturing the spring freshet along with rain runoff, and nutrient retention [17,40,41]. The retained water is released in mid-June to maintain baseflow downstream. Hay production can then occur on the drained land. A water storage capacity of 2,100,000 m^3 additionally provides a large water source if irrigation development in the area is pursued.

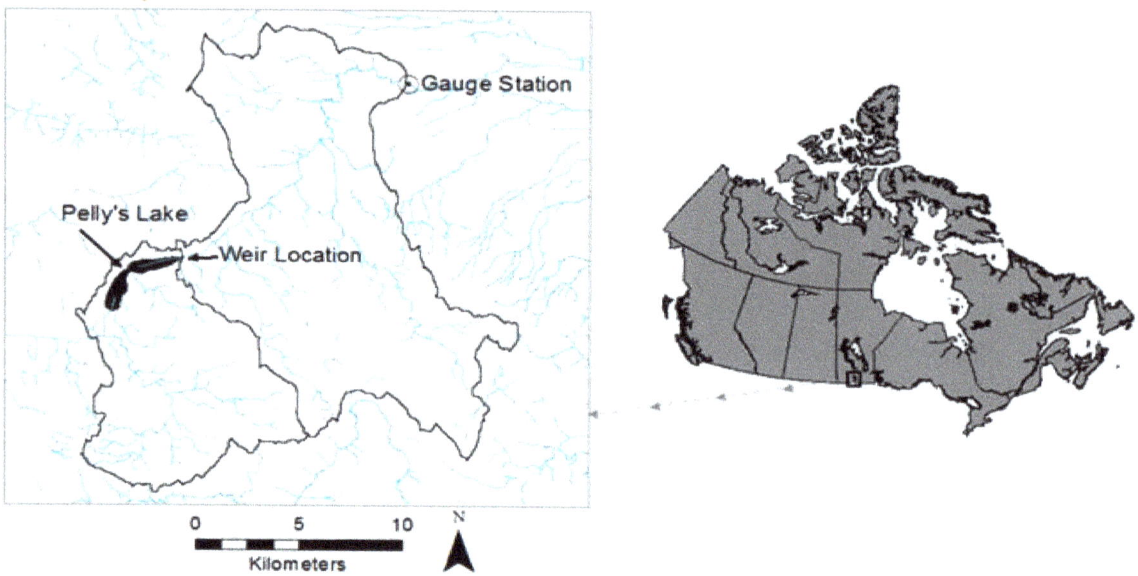

Figure 1. Pelly's Lake, Manitoba, Canada, situated within its watershed. To the right of this is the downstream gauge station and adjacent watershed boundary.

2. Materials and Methods

In order to explore the potential economic advantages of retention ponds under future climate conditions, a hydrologic model was parameterized using future precipitation and temperature data at a fine resolution for the study site. The output from the hydrologic model was input into a modeling system to provide projections of how water storage at the Pelly's Lake multi-purpose retention system would be affected by climate change. Stella modeling software, a program designed specifically for modeling complex system dynamics, allowed for an integrated hydrologic, reservoir, irrigation, plant growth, and economic model to be created [42]. The modeling system was developed based on a daily time step using a growing season simulation period running from April through September. The daily time step captured short-term components of the system while allowing for multiple years to be analyzed for a long-term analysis of the problem [10]. Spatially, study simulations were confined to the

Pelly's Lake watershed as the multi-purpose retention system would collect runoff and precipitation within this boundary (Figure 1).

2.1. Modeling System

2.1.1. Hydrologic Model and Reservoir Module

To model the hydrologic component of the target watershed, the Environment and Climate Change Canada environmental modeling system Modélisation Environnementale Communautaire—Surface and Hydrology (MESH) was chosen. MESH is a distributed land surface model commonly used in Canada for medium to large scale simulations [43,44]. Environment and Climate Change Canada uses MESH as part of an operational forecasting tool and the modeling system is currently being used within research projects such as the Drought Research Initiative (DRI) [45]. MESH requires multiple inputs to provide a complete distributed land surface model. The energy and water balance requirements for the model were determined using the Canadian Land Surface Scheme (CLASS) 1 [44] and CLASS 2 [46]. CLASS 1 is a physically based land surface model which calculates heat and moisture transfer at the surface, while CLASS 2 calculates energy and moisture fluxes at the canopy level [43,46]. Precipitation data for MESH were taken from the Canadian Precipitation Analysis (CaPA) project which produces rainfall accumulations at a six-hour time step and resolution of 15 km over North America in real-time [47]. Further required climatic data such as long wave and short wave radiation, humidity, pressure, and wind speed was acquired from the Global Environmental Multiscale (GEM) Model [43,48]. Routing of water within the study area was performed within the MESH model using a storage-routing technique which applies the continuity equation as outlined in [49]. Optimization of the MESH model was performed. Streamflow outputs from the MESH model were summed for 1 January to 14 April each year to provide an initial reservoir volume from spring freshet. Streamflow values for 15 April to 15 September each year were input into the model (Figure 2). Reservoir outflow considered the height of the emergency spillway, outflow over a rectangular weir, evaporation rates, and withdrawals taken for irrigation purposes (Figure 2). Table 1 provides detailed input parameters and equations for the modeling system.

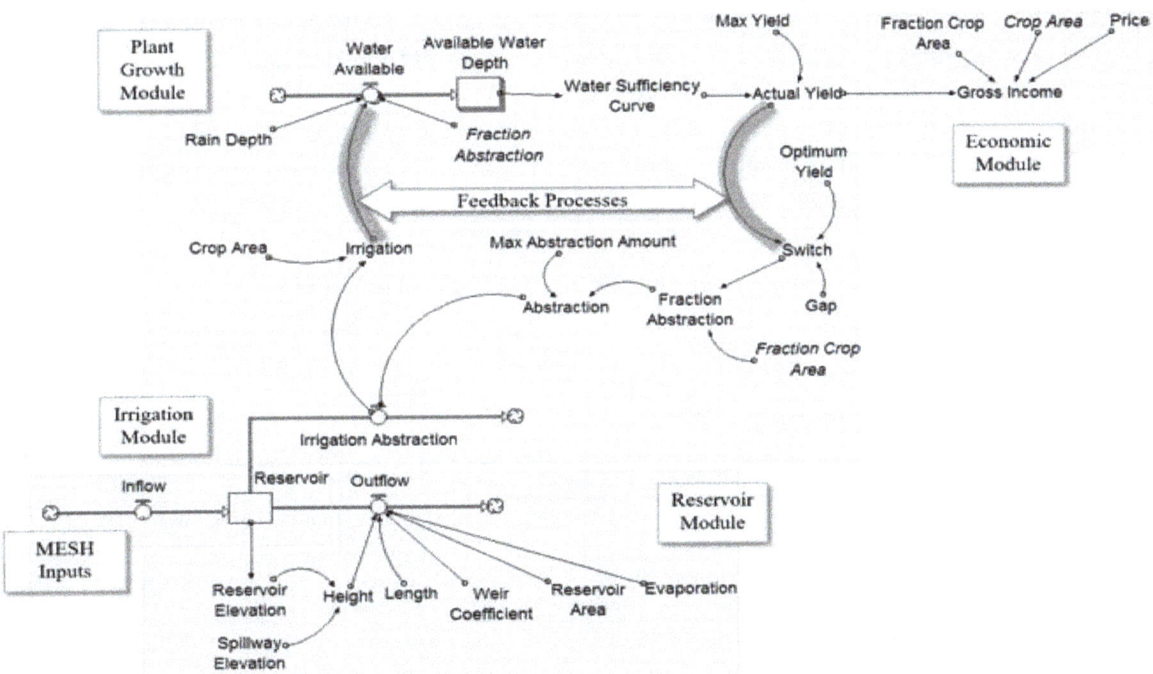

Figure 2. A stock-flow diagram of the modeling system, visually divided into its five component modules.

2.1.2. Irrigation and Plant Growth Module

The irrigation module consisted of irrigation withdrawals and precipitation during the growing season informing soil water volume available for crops (Figure 2). The four most prevalent crops in the study watershed as of 2011, canola, spring wheat, alfalfa, and barley, were modelled with a crop area of 6697 hectares [40]. On average within the study area, there is some initial spring soil moisture associated with snowmelt, however for the purposes of this model, soil moisture was recharged with precipitation and/or irrigation water. Soil moisture was recharged with precipitation and/or irrigation water. With the primary focus of the analysis on water as a crop production input and to maintain the tractability of the model, we assumed that all other production inputs, including nitrogen and phosphorus fertilizer and pesticides, were applied at rates which met crop growth requirements such that water was the only limiting factor to crop yield. This assumption was supported by our use of Government of Manitoba estimated, region specific, production costs including input and insurance cost. These costs are estimated based on the assumption that production inputs are used at rates to meet all crop growth requirements. The resultant crop yields were used in combination with crop prices to determine gross income (Table 1). Crop production costs and input costs were subtracted from gross income in Microsoft Excel to estimate gross margins under irrigation.

2.1.3. Economic Module

Crop prices, production costs, and insurance costs used in the economic module were 2015 values, provided by the Government of Manitoba, and were held constant for all future simulations (Figure 2, Table 1) [50]. Production costs refer to costs for seed and seed treatment, fertilizer, fungicide, herbicide, and insecticide application, machinery operation, fuel, leases, land taxes, and interest costs. Agricultural input costs and crop prices will fluctuate over time due to changes in food demand, changes in crop varieties and production technology, as well as changes in energy prices and climate. However, analysis of Canadian agriculture suggests that the ratios between farm expenses and receipt have been relatively stable over time [51]. As a result, we adopt the relatively strong assumption that input costs including reservoir and irrigation installation and upkeep costs as well as production costs and output prices associated with each crop type were constant for all simulations with the model. The total adjusted cost of the retention system at Pelly's Lake, which included upkeep and accrued interest for a twenty-year time horizon, was $45,167/year ($5.26/ha/year) [35,52]. A twenty-year time horizon was chosen as this represents reservoir infrastructure's typical serviceable life [52]. Centre-pivot irrigation infrastructure installation, labour, and maintenance over a twenty-year time horizon totalled $966,010 ($112.50/ha/year) [53].

Table 1. Parameter values and equation inputs for each module within the modeling system.

Parameters/Units	Inputs and Equations with Descriptions
Reservoir Module	
Inflow (m^3/day)	Input graphically using output data from the MESH hydrologic model on a daily time step.
Reservoir (m^3)	Initial reservoir volume calculated based on cumulative output from the MESH hydrologic model from 1 January–14 April.
Outflow (m^3/day)	=(3 × Weir_Coefficient × Length × Height$^{1.5}$ × 86,400) − (Evaporation × Reservoir_Area)
	The established engineering equation for discharge over a rectangular weir was multiplied by 86,400 to convert from m^3/s to m^3/day. Evaporation over the reservoir area was subtracted from the discharge equation [54].
Weir Coefficient (dimensionless)	=0.6, An established engineering value was used.
Length (m)	=12, Spillway length was taken from the engineering drawings for the Pelly's Lake weir.
Height (m)	=IF (Reservoir Elevation − Spillway Elevation) > 0 THEN (Reservoir Elevation − Spillway Elevation) ELSE 0, An established engineering equation.

Table 1. *Cont.*

Parameters/Units	Inputs and Equations with Descriptions	
Reservoir Elevation (m)	$=9 \times 10^{-7} \times$ Reservoir + 378.23	
	This equation was determined from the engineering storage rating curve for the Pelly's Lake weir.	
Spillway Elevation (m)	=379.1, This value was provided on the engineering drawings for the Pelly's Lake weir.	
Evaporation (m^3/day)	=0.00182 (April) =0.00422 (May) =0.00460 (June)	=0.00454 (July) =0.00469 (August) =0.00346 (September)
	Mean monthly evaporation values from 1981–2010 were converted to daily values at Brandon, MB [54]. These values were used due to insufficient data available to calculate evaporation at the study site.	
Reservoir Area (m^2)	=85,867,480, calculated in ArcGIS.	
Irrigation Module		
Irrigation Abstraction (m^3/day)	=Abstraction[Canola] + Abstraction[Wheat] + Abstraction[Barley] + Abstraction[Alfalfa]. This variable calculated the total water abstraction volume abstracted from the reservoir.	
Abstraction [Crop]	=Max Abstraction Amount × Fraction Abstraction This equation calculated irrigation withdrawal volumes for each crop.	
Max Abstraction Amount (m^3)	=15,000, this amount was calibrated to allow the reservoir to drain at a rate to provide sufficient water for irrigation for the entire growing season.	
Fraction Abstraction [Crop](dimensionless)	=IF ((Switch[Canola] + Switch[Wheat] + Switch[Barley] + Switch[Alfalfa]) > 0) THEN(Switch[Crop] × Fraction_Crop_Area[Crop]/(Switch[Canola] × Fraction_Crop_Area[Canola] + Switch[Wheat] × Fraction_Crop_Area[Wheat] + Switch[Barley] × Fraction_Crop_Area[Barley] + Switch[Alfalfa] × Fraction_Crop_Area[Alfalfa])) ELSE (0)	
	When water requirements were not being met by a specific crop, this algorithm directed water withdrawals to the crop requiring irrigation. It also ensured water was not applied unless required to optimize crop growth.	
Fraction Crop Area	=0.46 (Canola) =0.38 (Wheat)	=0.11 (Alfalfa) =0.05 (Barley)
	Historical patterns of crop production in Manitoba were used to determine the fraction of total crop area each crop comprised [37,55].	
Switch[Crop]	=IF (Actual Yield < Gap × Optimum Yield AND TIME > 30 THEN 1 ELSE 0	
	Irrigation was triggered for a specific crop if the crop's actual yield fell below 80% of optimum yield on day 30 (15 May).	
Gap	=0.8, Irrigation application occurred when available water only allowed for 80% or less of optimum yield growth. Yield reductions due to water stress occur when available water falls below 60% of optimum plant requirements. The threshold of 80% ensured there was always sufficient water available to the plant [53].	
Irrigation (mm/day)	=(Irrigation Abstraction/Crop Area) × 1000	
	This equation served to convert irrigation from a volume to a depth.	
Crop Area (m^2)	=66,976,634, This was calculated in ArcGIS using data on land use downloaded from the Manitoba Land Initiative [55].	
Plant Growth Module		
Rain Depth (mm/day)	Input graphically using values from Environment and Climate Change Canada for Holland, MB [56].	
Water Available [Crop] (mm/day)	=Rain Depth + Irrigation × Fraction Abstraction	
	Daily water available for crop growth [57].	
Available Water Depth [Crop] (mm)	Initial value set at 0. Snowmelt would contribute to initial spring soil moisture, however for the purposes of the model soil moisture was assumed to be recharged solely from precipitation and/or irrigation water.	
Water Sufficiency Curve [Crop]	These curves were input graphically and represented the unique optimal water requirements of each crop. They allowed for crop yield to be calculated based on water availability [57].	
Max Yield [Crop](tonnes/m^2)	= 0.000224124 (Canola) = 0.000336063 (Wheat)	=0.000376588 (Barley) =0.000672126 (Alfalfa)
	Max yield values were held constant for all simulations [50].	

Table 1. *Cont.*

Parameters/Units	Inputs and Equations with Descriptions	
Actual Yield [Crop] (tonnes/hectare)	=Max Yield × Water Sufficiency Curve × 10,000	
	Crop yield was calculated based on water availability. Each crop's water sufficiency curve provided a proportion of maximum growth based on water availability. This proportion was multiplied by max yield and 10,000 to convert from m^2 to hectares.	
Optimum Yield [Crop]	For each crop, values were input graphically. The variable represented the maximum yield of each crop over time when its water requirements were being met. Values were constant for all simulations.	
Economic Module		
Gross Income [Crop] ($)	=Actual Yield × Price × Crop Area × Fraction Crop Area/10,000 Calculated landscape level gross income.	
Price [Crop] ($/tonne)	=418.87 (Canola) =238.83 (Wheat)	=173.23 (Barley) =132.28 (Alfalfa)
	Crop prices were not available for years before 2015. Thus, 2015 crop prices were used in all simulations [50].	

2.2. Climate Change

Statistically downscaled climate data for the study area were acquired from the Pacific Climate Impacts Consortium (PCIC) for the present study [58,59]. Climate scenarios are available across Canada from PCIC. Data are produced at a gridded resolution of roughly 10 km or 300 arc-seconds for 1950–2100. Three output variables on a daily time step are available from PCIC: precipitation, minimum temperature, and maximum temperature [59]. Scenarios for all four representative concentration pathways (RCPs) (Table 2) are available and multi-model ensemble tables are provided to aid the researcher in climate model selection with the widest breadth of future climate simulations. Due to constraints on time and resources, the model ensemble list was narrowed down to contain only four models for this study (Table 3). Historical daily gridded climate data for Canada were used in combination with General Circulation Model (GCM) projections from the Coupled Model Intercomparison Project Phase 5 (CMIP5) [59]. The chosen GCMs have been studied and shown to model climate change most effectively for regional applications over North America [58,60,61]. Of the two downscaling methods provided by PCIC [59], Bias Corrected Spatial Disaggregation (BSCD) was chosen for this study due to its extensive application in previous hydrologic modeling research across North America, its ease of use, daily time series output of gridded temperature and precipitation, and its ability to capture emission scenario transience effectively [58,61–64]. The BSCD method bias-corrects monthly GCM data against GCM gridded observed data. BSCD also downscales monthly GCM data to allow for regional analysis at a daily time step [58,62,63]. Further description of the application of BSCD to PCIC scenarios can be found in Werner [58]. The second downscaling option, Bias Correction/Constructed Analogues with Quantile mapping reordering (BCCAQ), is a recently developed method. As such, at the time of this study its application and accuracy in modeling climate change effects had not been extensively tested and was thus not suitable for this application [63].

Table 2. Representative concentration pathways (RCPs) overview.

RCP	Description
RCP2.6	Radiative forcing will peak at approximately 3 W/m^2 before 2100 and then levels will decline.
RCP4.5	Radiative forcing will stabilize at 4.5 W/m^2 after 2100.
RCP6	Radiative forcing will stabilize at 6 W/m^2 after 2100.
RCP8.5	Radiative forcing will rise resulting in 8.5 W/m^2 in 2100.

Table 3. Selected models used in multi-model ensemble of future climate scenarios.

Modeling Center	Institute ID	Model Name
Canadian Centre for Climate Modeling and Analysis	CCCMA	CanESM2
Meteorological Office Hadley Centre	MOHC	HadGEM2-ES
Max Planck Institute for Meteorology	MPI-M	MPI-ESM-LR
NOAA Geophysical Fluid Dynamics Laboratory	NOAA GFDL	GFDL-ESM2G

Emission scenarios RCP2.6, RCP4.5, and RCP8.5 were used for this study. The emission scenario RCP6 was excluded due to time constraints. The RCP4.5 emission scenario represented a median radiative forcing scenario in place of simulating RCP4.5 and RCP6 as median scenarios. Outputs from the four chosen climate models were downloaded for the study area. As the downscaled models have a grid resolution of approximately 10 km, outputs were spatially constrained over the study area between latitudes 49° N and 50° N and longitudes 98° W and 97° W.

The multi-model ensemble mean historical climate data were validated against observed climate data for the 2005–2014 (present day) study period to confirm its representation of observed values. As shown in Figure 3 there is some discrepancy between modelled and observed precipitation. This is due to the downscaling method application to the GCMs introducing bias. However, it does appear that the multi-model ensemble mean is capturing the annual and monthly cycle of precipitation. Figure 4 illustrated the multi-model ensemble mean's capacity to simulate observed temperature at the study location. These figures provided validation that the GCMs are suitable for simulating future climate change at this location.

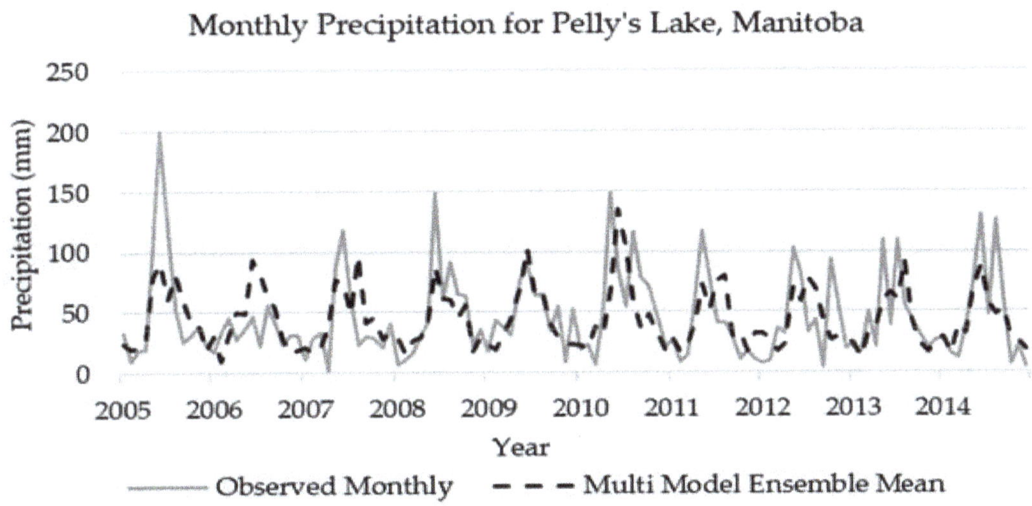

Figure 3. Comparison of observed and modelled monthly precipitation at Pelly's Lake, Manitoba for 2005–2014.

Climate trends at Pelly's Lake for the median emission scenario, RCP4.5, along with the two extreme emission scenarios, RCP2.6 and RCP8.5, were graphed to provide a comparison (Figures 5 and 6). The data were divided into summer and winter to determine multi-model ensemble mean increment changes in precipitation and temperature for the two seasons. These plots indicate a consensus in future climate precipitation trends between the four different models, increasing confidence that the climate models are performing as desired. Confidence in the model's abilities to simulate future climate conditions was further increased by the clear trends between models for each RCP. Outputs from the PCIC downscaled models support the findings published by Warren and Lemmon [1] that precipitation is projected to increase for all seasons across Canada in the future.

Future precipitation simulation outputs in the same report also indicated precipitation increases will be greater in the winter than the summer [1].

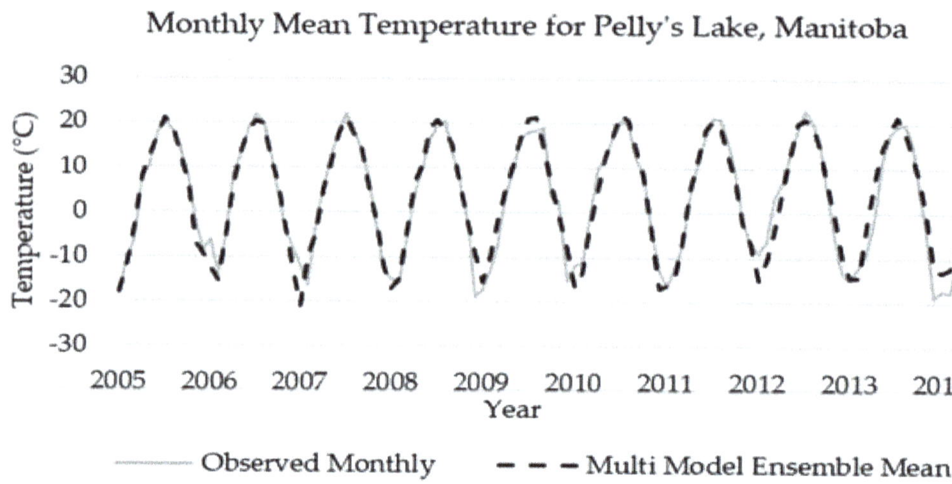

Figure 4. Comparison of observed and modelled monthly mean temperature at Pelly's Lake, Manitoba for 2005–2014.

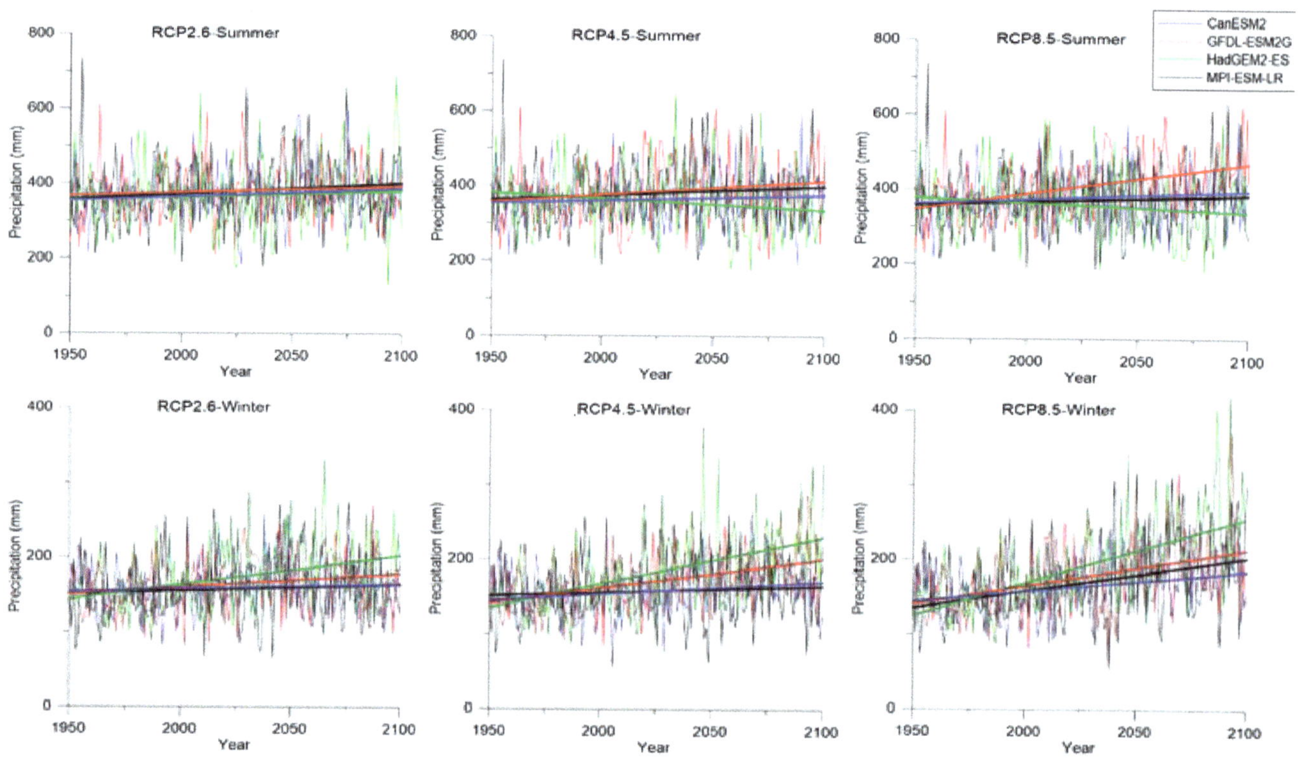

Figure 5. Multi-model ensembles for each representative concentration pathway (RCP) showing summer and winter precipitation with lines representing mean precipitation for each climate model. The spread between model simulations illustrates uncertainty.

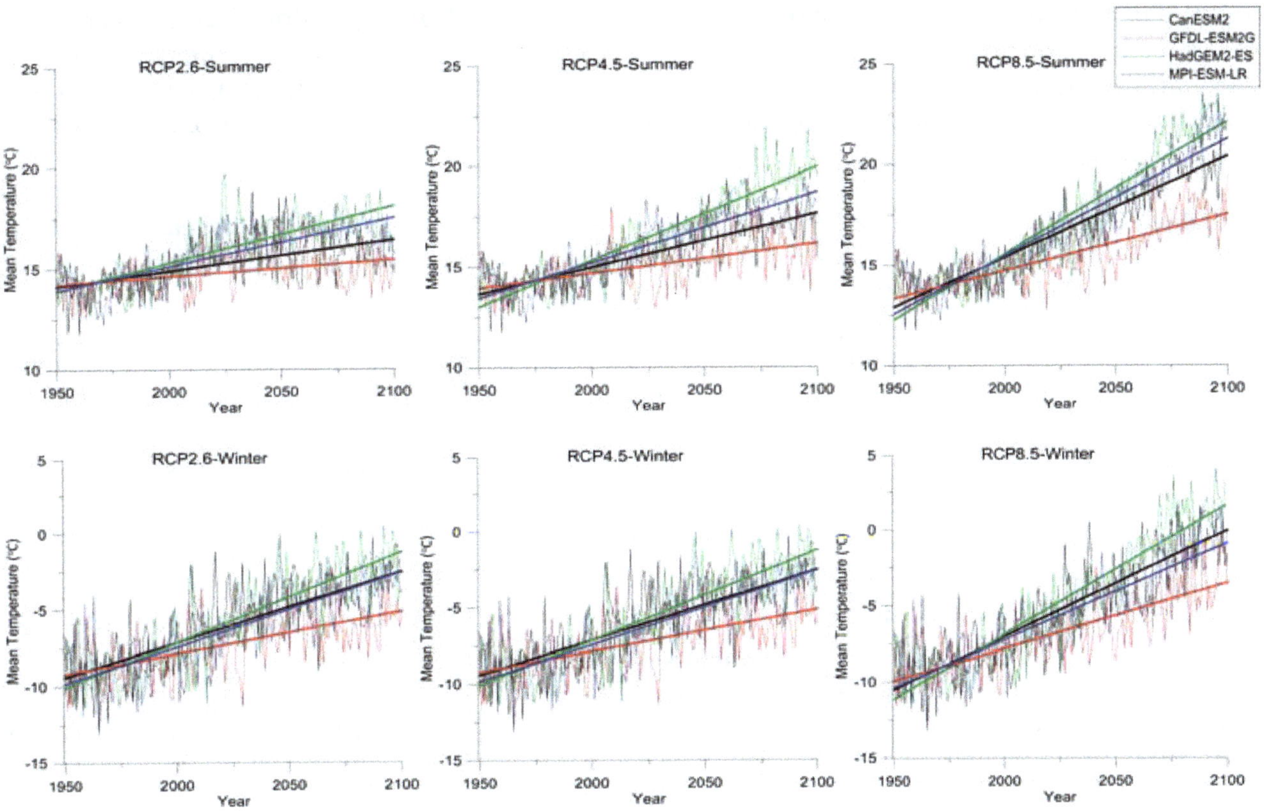

Figure 6. Multi-model ensembles for each representative concentration pathway (RCP) showing summer and winter temperature with lines representing mean temperature for each climate model. The spread between model simulations illustrates uncertainty.

2.2.1. Delta Method

As this study deals with the benefits of retention basins under future climate uncertainty, climate data were required to provide long term trends in future climate change that will impact the retention system's water volume. For this reason, a simple delta method application was chosen to simulate the two future climate periods. This method has been extensively used for studies that aim to provide a sensitivity analysis of future climate uncertainty [52,65–68]. The delta method applies changes in climate variables extracted from GCMs or Regional Climate Models (RCMs) for the study area to a baseline observed climatology [66,68]. Precipitation is adjusted multiplicatively and temperature is adjusted additively [62,66,68]. Spatial variability of the climate variables within the observed time series are preserved in any future climate simulations [62,66,68]. This assumption is a key limitation of the delta method [62,68]. However, for this study, it is more important to have accurate representation of future spring runoff volumes and precipitation volumes for the growing season [68].

Using the downscaled GCMs, climate outputs taken from PCIC future climate conditions were extracted for two ten-year time periods, 2050–2059 and 2090–2099. This allowed for representation of the middle (2050–2059) and end of the century (2090–2099). Multi-model ensemble means were used to calculate incremental changes in temperature and precipitation for the two future study periods based on the present-day time period (2005–2014). An ensemble mean value was used for this calculation as research has determined that ensemble means reduce bias within individual models, providing a more robust output than any individual GCM output [60]. Incremental changes in precipitation and temperature from the three RCP4.5 multi-model ensemble means were applied to 2005–2014 present day climate data within the hydrologic model (Table 4). Precipitation changes were applied to the baseline time period multiplicatively while temperature changes were applied additively. All other parameters within the modeling system remained the same as for the 2005–2014 time period. Streamflow outputs from the hydrologic model provided the 2050s and 2090s hydrologic input for the modeling system.

Table 4. Mean season precipitation totals, mean season temperature increases and increases between study periods based on the multi-model ensemble means for representative concentration pathways (RCP)2.6, RCP4.5, and RCP8.5.

Decade	Mean Precipitation Total (mm)			Incremental Increase %			Mean Temperature (°C)			Temperature Increase (°C)		
	RCP			RCP			RCP			RCP		
	2.6	4.5	8.5	2.6	4.5	8.5	2.6	4.5	8.5	2.6	4.5	8.5
Summer (May–October)												
2005–2014	371.9	371.9	371.9	-	-	-	15.29	15.44	15.25	-	-	-
2050–2059	413.3	376.8	377.7	11.1	1.3	1.56	16.17	16.86	17.56	0.9	1.4	2.3
2090–2099	382.2	381.2	408.9	2.77	2.5	9.95	16.00	17.82	20.51	0.7	2.4	5.3
Winter (November–April)												
2005–2014	162.7	162.7	162.7	-	-	-	−6.867	−6.525	−6.857	-	-	-
2050–2059	174.5	176.9	199.3	7.25	8.8	22.5	−5.786	−4.489	−4.241	1.1	2.0	2.6
2090–2099	159.6	189.6	218.7	−1.91	16.6	34.4	−6.158	−3.178	−0.430	0.7	3.4	6.4

2.2.2. Uncertainty

It is important to note uncertainties associated with climate projections from GCMs as well as downscaled methods [69,70]. Future anthropogenic emission levels involve uncertainty. Models used for simulating future climate scenarios have uncertainties linked to imperfect representation of climate processes. Current understanding of climate conditions is imperfect leading to imperfect knowledge being fed into projections [69,70]. Finally, variability at the interannual and decadal level is difficult to represent accurately in long-term projections. However, this does not mean future climate projections are false as uncertainty can be quantified [69]. Each GCM and downscaling method has a unique set of parameters as initial conditions within the model. By using future climate scenario results for as many models as possible and producing a multi-model ensemble mean or median, a more probable future climate scenario can be determined. The spread in results between models illustrates the level of uncertainty in the obtained multi-model ensemble results [58,69,71].

3. Results

Annual gross margin was estimated from model simulations with and without irrigation for the middle of the century, 2050 to 2059, and the end of the century, 2090 to 2099 for three radiative forcing scenarios. The simulation results indicated that irrigated crops using water abstractions from the reservoir experienced a decrease in gross margin when compared to gross margins without irrigation and associated infrastructure for both study simulation periods. Results for the three radiative forcing scenarios, RCP2.6, RCP4.5, and RCP8.5 using incremental percentage increases to the 2005–2014 MESH climate data for the 2050s and 2090s are provided in Tables 5 and 6, respectively. The differences in gross margin when irrigation and associated infrastructure costs were considered are provided in the second column of Tables 5 and 6. The estimated yearly cost of the retention pond and irrigation infrastructure was $160.00/hectare. Any gross margin value above −$160.00 reflects that the increased crop yield enabled by irrigation water offset the yearly cost of the retention pond and irrigation infrastructure. A value above zero would indicate all retention pond and irrigation infrastructure costs are being offset by increased crop yield under irrigation application.

Table 5. Difference in crop gross margins without irrigation and crop gross margins with irrigation and associated variable and infrastructure costs, and increase in crop gross income under irrigation for the 2050s using three different radiative forcing scenarios.

Year	Difference in Crop Gross Margins without Irrigation and Crop Gross Margins with Irrigation and Associated Variable and Infrastructure Costs ($/hectare)			Increase in Crop Gross Margins under Irrigation ($/hectare)		
	RCP2.6	RCP4.5	RCP8.5	RCP2.6	RCP4.5	RCP8.5
2050	−174.00	−172.00	−173.00	−13.60	−11.80	−12.30
2051	−145.00	−151.00	−150.00	15.10	9.42	10.60
2052	−155.00	−153.00	−153.00	4.79	7.31	6.94
2053	−150.00	−146.00	−147.00	9.99	14.50	12.80
2054	−146.00	−143.00	−144.00	14.01	17.40	16.50
2055	−159.00	−159.00	−159.00	0.83	1.53	1.53
2056	−157.00	−146.00	−147.00	3.60	14.00	13.50
2057	−143.00	−138.00	−142.00	17.40	24.50	17.90
2058	−125.00	−127.00	−127.00	35.40	33.10	33.70
2059	−126.00	−126.00	−126.00	33.90	33.90	33.90
Average	−148.00	−146.00	−147.00	12.10	14.40	13.50

Note: Difference = Gross margin with retention pond used for irrigation - Gross margin without retention pond installation and associated irrigation.

Table 6. Difference in crop gross margins without irrigation and crop gross margins with irrigation and associated variable and infrastructure costs, and increase in crop gross income under irrigation for the 2090s using three different radiative forcing scenarios.

Year	Difference in Crop Gross Margins without Irrigation and Crop Gross Margins with Irrigation and Associated Variable and Infrastructure Costs ($/hectare)			Increase in Crop Gross Margins under Irrigation ($/hectare)		
	RCP2.6	RCP4.5	RCP8.5	RCP2.6	RCP4.5	RCP8.5
2090	−173.00	−173.00	−174.00	−13.10	−13.20	−13.50
2091	−151.00	−150.00	−142.00	9.42	10.10	18.00
2092	−154.00	−154.00	−155.00	6.39	6.35	4.86
2093	−146.00	−147.00	−153.00	14.30	12.80	7.15
2094	−143.00	−144.00	−146.00	17.00	16.30	14.00
2095	−159.00	−159.00	−159.00	1.53	1.53	0.82
2096	−147.00	−147.00	−157.00	13.30	13.00	3.60
2097	−135.00	−142.00	−144.00	25.20	17.90	16.30
2098	−125.00	−125.00	−125.00	35.30	35.30	35.40
2099	−126.00	−126.00	−129.00	34.10	33.90	30.90
Average	−146.00	−147.00	−148.00	14.30	13.40	11.80

Note: Difference = Gross margin with retention pond used for irrigation - Gross margin without retention pond installation and associated irrigation.

All years, with the exceptions of 2050 and 2090, under all radiative forcing scenarios, within the two study time periods would experience increased crop yields from irrigation water (Tables 5 and 6). The years 2050 and 2090 experienced the highest growing season precipitation amounts of the simulation periods (497–546 mm and 503–541 mm, respectively). A rainfall event on 14 July overwhelmed canola and wheat crops, reducing yields and subsequently triggering more irrigation water to be applied until the end of the growing season. This additional irrigation application after the crops had already received excess water was detrimental to crop yields. It is important to note that the model does not account for the dynamic decision making of farmers regarding irrigation application. In a high precipitation year, the farmer would recognize crop yield reductions were due to an excess of water and would abstain from irrigation application.

The average impact of irrigation application for the 2050s simulations was an increase in annual average gross income of \$12.10 to \$14.40/hectare, depending on the radiative forcing scenario (Tables 5 and 6). However, due to the cost of irrigation and reservoir installation this left the farmer with an average gross margin of \$146.00 to \$148.00/hectare, dependent on radiative forcing scenario, each year to cover the reservoir and irrigation infrastructure and operation costs. For the 2090s simulations, the average impact of irrigation application was an increase in annual average gross income of \$11.80 to \$14.40/hectare, decreasing as radiative forcing increased. This left the farmer with an average net cost of \$146.00 to \$148.00/hectare each year, increasing as radiative forcing increased, to cover the reservoir and irrigation infrastructure and operation costs. Therefore, although the availability of irrigation water did increase crop production, the increased gross income was insufficient to offset the costs of the irrigation water.

Yearly gross margins with and without irrigation, yearly precipitation amounts, and reservoir volumes are provided in Figures 7 and 8 for the 2050s and 2090s, respectively. There was one year under all radiative forcing scenarios in each simulation period, 2057 and 2097, which required the total reservoir storage volume for irrigation application. In addition, 2091 under RCP2.6 required the total reservoir storage volume for irrigation application. This year experienced the lowest precipitation levels of all years and radiative forcing simulations, which required substantial irrigation application to optimize crop yield. While initial reservoir volumes and irrigation withdrawal volumes varied between radiative forcing scenarios, during the course of the growing seasons the reservoir always filled to capacity, allowing for irrigation withdrawals. As the MESH simulations used climate data from 2005 to 2014, variability in precipitation and temperature reflected that time period.

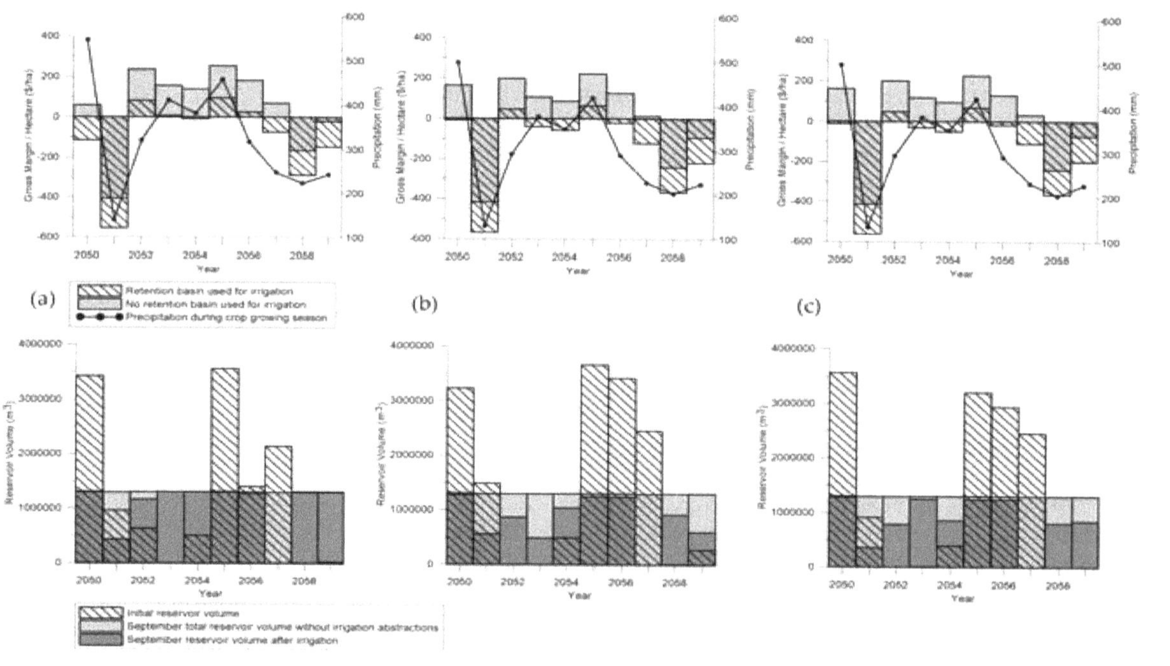

Figure 7. Yearly 2050–2059 crop gross margins with and without irrigation application and yearly water availability using incremental precipitation and temperature increases for: (**a**) Representative concentration pathway (RCP)2.6; (**b**) RCP4.5; and (**c**) RCP8.5. Reservoir levels for: (**a**) RCP2.6; (**b**) RCP4.5; and (**c**) RCP8.5 are also provided.

Results of this study can be explained by the precipitation and temperature increases experienced in the 2050s and 2090s. Incremental increases in temperature and precipitation during the 2090s increased from RCP2.6 to RCP8.5. These changes to temperature and precipitation caused crop gross income under irrigation to decrease from RCP2.6 to RCP8.5 and excess water overwhelmed the water capacity of the crops. Precipitation changes during the 2050s were less consistent with increases in radiative forcing. The 2050s experienced the largest increase in summer precipitation and smallest increase to winter precipitation under RCP2.6. Subsequently, the pattern of crop gross income increases does not align with increases in radiative forcing scenarios. Instead, RCP4.5 provided the highest gross income under irrigation and RCP2.6 provided the smallest increase to crop gross income under irrigation. Winter precipitation, which impacts initial reservoir volumes, increased for all scenarios except RCP2.6 for the 2090s. This however, did not provide enough water to fill the reservoir in all years. The increased summer precipitation levels did ensure the reservoir always filled to capacity by the end of the growing season. The variation in temperature and precipitation under different radiative forcing scenarios and time periods were not significant enough to have a significant impact on gross margins. Based on these simulation results, irrigation and reservoir installation at Pelly's Lake are not economically viable and do not generate positive crop gross margins in the middle or end of the century.

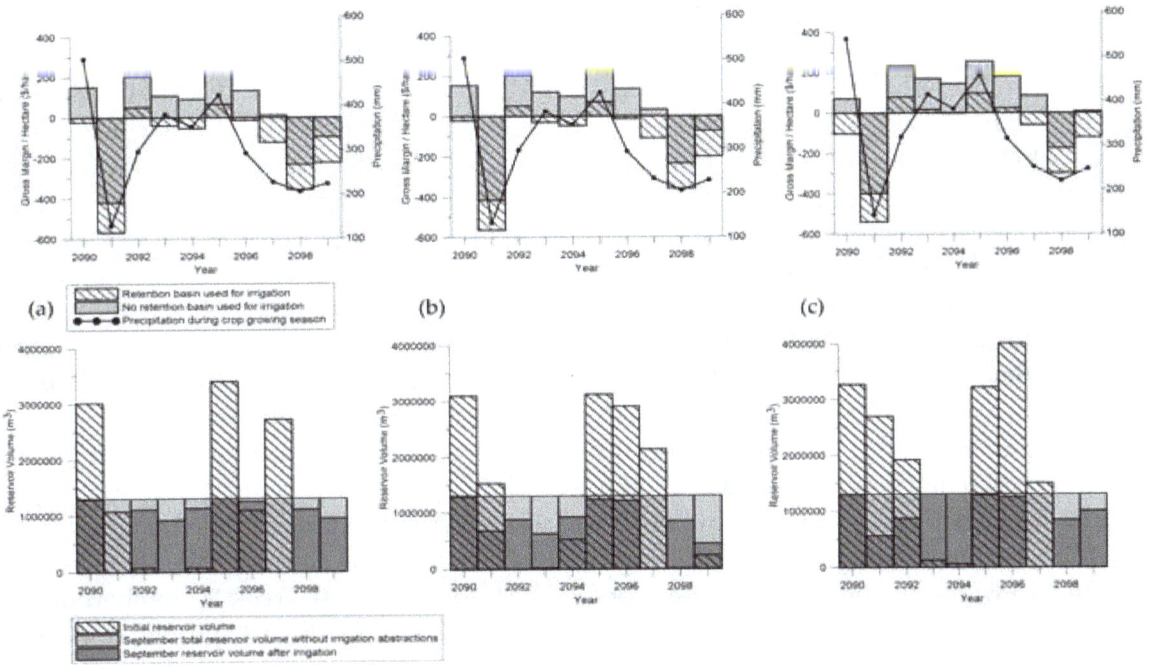

Figure 8. Yearly 2090–2099 crop gross margins with and without irrigation application and yearly water availability using incremental precipitation and temperature increases for: (**a**) Representative concentration pathway (RCP)2.6; (**b**) RCP4.5; and (**c**) RCP8.5. Reservoir levels for: (**a**) RCP2.6; (**b**) RCP4.5; and (**c**) RCP8.5 are also provided.

4. Discussion

The economic advantages of multi-purpose local farm retention pond systems and their use for irrigation under future climatic conditions were investigated using a dynamic modeling system. The middle of the century, 2050–2059, and the end of the century, 2090–2099, were simulated under three radiative forcing scenarios, RCP2.6, RCP4.5, and RCP8.5. The multi model ensemble future climate scenarios indicated precipitation increases will occur in the 2050s and 2090s under each RCP scenario, with the exception of winter precipitation under RCP2.6 for the 2090s, which decreased. Precipitation increases were higher for winter than summer. These results are supported by several studies which have predicted a general trend of increasing precipitation over Canada with higher projected increases to precipitation for winter months [1,3–5,72,73].

Using the delta method, incremental changes to temperature and precipitation from the present day period (2005–2014) were used to simulate future climate conditions. This method allowed for comparison of how changes in water volumes will impact irrigation water availability and subsequent crop gross margin. Due to the high costs of irrigation infrastructure and maintenance, irrigated cropping systems utilizing water abstractions from the reservoir experienced a decrease in gross margin when compared to gross margins without irrigation for both simulation periods under all RCP scenarios. To cover the costs of the irrigation and reservoir infrastructure, the farmer would have to earn incremental gross margins of $146.00 to $148.00/hectare of crop land/year in the 2050s and 2090s. Compared to present day simulations, the cost of irrigation and reservoir infrastructure to the farmer did not change under predicted future climate conditions [20]. However, increases in crop yield under irrigation were more consistent and increases in spring runoff and precipitation provided more stable reservoir water volumes which increased irrigation water availability.

The predicted increases in reservoir water volumes indicates that flooding may become more severe in the future, increasing multi-purpose surface water retention systems importance and value for flood reduction. However, this increase in reservoir water volumes also points to an opportunity for water storage for irrigation application. As future climate change is predicted to increase the severity and duration of floods and droughts [4,7,9,11], the ability to capture surface runoff in times of flood also provides water stores to draw on during times of drought. While irrigation installation remains costly, it is still an important adaptation strategy to reduce agricultural risk during times of drought. Policy providing irrigation subsidization may be implemented in the future to increase its adoption, in which case knowledge of strategies that provide sufficient water sources to utilize for irrigation application will be important.

Based on the findings of this research, multi-purpose retention basins are not economically justified when considering only the irrigation benefits provided to the participating farmers. However, a range of other benefits may be provided by the water retention system. As illustrated by Berry [20], Grosshans [17], and Dion and McCandless [35], multi-purpose retention systems provide significant economic benefits when you consider the ecosystem services they provide. Multi-purpose retention systems provide avoided flood damages downstream, sequester carbon, and reduce downstream nutrient and sediment runoff. There was substantial flow produced from spring melt that exceeded the capacity of the modelled reservoir under climate simulations for the middle and end of this century (Figure 7). A network of several multi-purpose retention systems, similar to the installed network in the South Tobacco Creek Watershed, may be required to deal with the future increases in spring runoff volumes. As part of the retention system network in the South Tobacco Creek Watershed in Manitoba, a multi-purpose dam reduced peak flow caused by spring snowmelt by an average of 72% per year, with a range of 38% to 100% peak flow reduction/year. Summer rainfall generated peak flow was reduced an average of 48% per year by the same multi-purpose dam. There is discussion regarding the construction of a second upstream reservoir at the Pelly's Lake retention site. This would increase the systems storage capacity by 1,600,000 m^3. As a result, the retention system would retain the predicted future volumes of spring melt runoff (Figure 9).

In additional to the economic benefits associated with avoided flood damages downstream, multi-purpose surface water retention systems can provide a biomass source. Harvesting cattail for biomass production and nutrient removal from the flooded area of multi-purpose retention systems is being commercialized in Manitoba and can provide monetary benefits from the production of carbon credits and bio products [17,28,34]. These ecosystem benefits of multi-purpose retention systems can provide additional economic revenue to compensate for the cost of the reservoir and irrigation infrastructure.

Using multi-purpose retention basins for avoided flood damages, nutrient retention, and biomass production also allows the public to benefit from these systems. The province of Manitoba has expressed interest in retention basins as a nutrient abatement option as part of their commitment to reducing downstream nutrient loading. The Manitoba Surface Water Management Strategy [14] states that water storage and associated release strategies should optimize production and harvest of biomass resources to remove phosphorus from the aquatic environment. As the South Tobacco Creek Watershed in Manitoba has illustrated, a series of retention systems on the Manitoba landscape has the potential to reduce downstream loading of phosphorus and nitrogen. Over a nine year period from 1999–2007, the retention system network decreased downstream nutrient loading above the Manitoba government's targets of 10% and 13% for phosphorus and nitrogen, respectively [21]. As the average phosphorus and nitrogen concentrations in the watershed were still in excess of recommended levels in the Canadian Prairies, Tiessen et al. [21] suggested using the reservoirs for local benefits, such as irrigation, to reduce downstream nutrient loading further. With the addition of cattail harvest, downstream loading of phosphorus and nitrogen would be further reduced [14,34,74,75]. Additionally, the removal of phosphorus during cattail harvest increases the wetlands ability to store more phosphorus, reducing downstream loading. This is essential for combating algal blooms and increasing water quality in aquatic environments such as Lake Winnipeg, Manitoba [34].

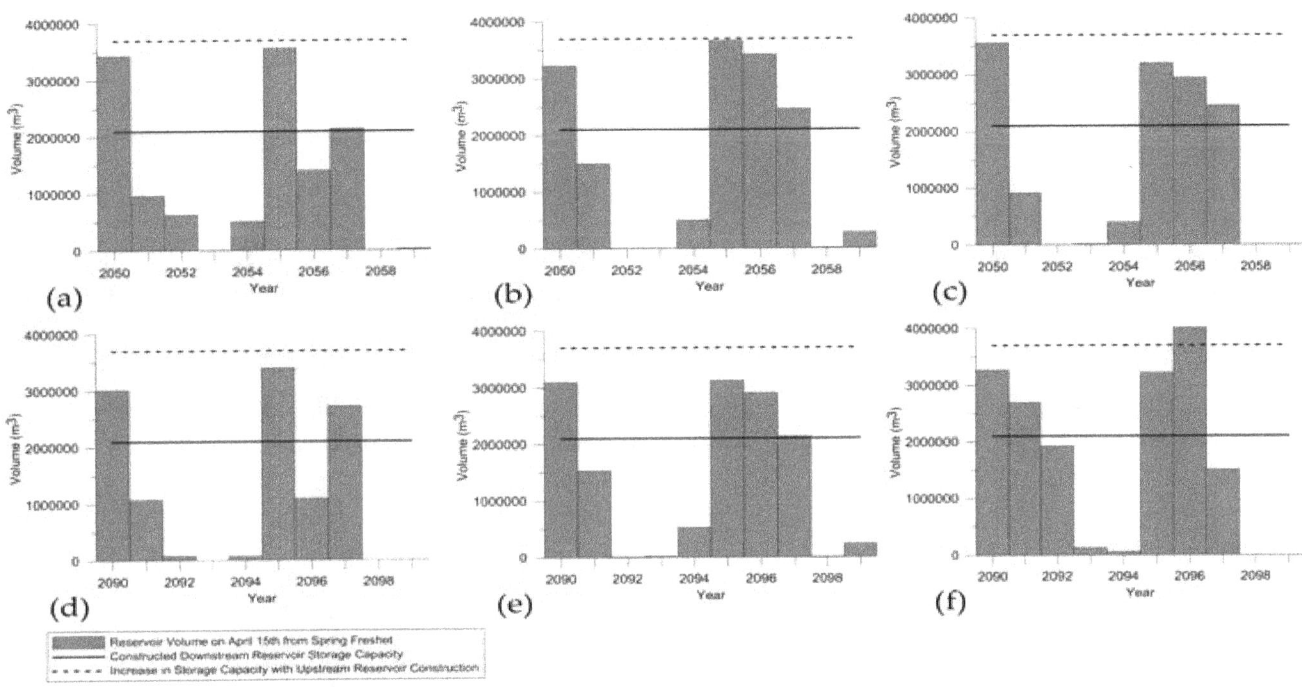

Figure 9. Reservoir capacities and initial reservoir volumes for future radiative forcing scenarios for the 2050s time period: (**a**) Representative concentration pathway (RCP)2.6; (**b**) RCP4.5; (**c**) RCP8.5; and the 2090s time period: (**d**) RCP2.6; (**e**) RCP4.5; (**f**) RCP8.5.

The current strategy for quickly removing water from the Manitoba landscape, via a series of ditches and drains, increases downstream flood peaks and decreases water quality. This method is only sustainable when there is adequate access to water and land use practices do not create nutrient pollution issues [3]. This quick drainage is already proving problematic for downstream nutrient loading into Lake Winnipeg in Manitoba. Future predictions of increased spring runoff volumes indicate increased issues with this strategy due to increased downstream flood peaks and increased nutrient loading. Moving forward, investing in multi-purpose local farm retention systems decreases flood peaks, increases water quality, while also providing water security during times of drought, as well as opportunities for biomass production and irrigation development. The reductions in phosphorus and nitrogen multi-purpose local farm retention systems can provide aid in Manitoba's goal of reducing nitrogen and phosphorus concentrations by 50% to Lake Winnipeg [14]. Rural municipalities and landowners benefit from the savings associated with avoided flooding damages while the province of Manitoba and its population benefit from the reduction to downstream nutrient loading and carbon storage providing climate regulation. Due to the economic and environmental gains multi-purpose retention systems provide to the province, subsidies could also be provided to incentivize widespread adoption.

Future Directions

The modelling system developed for this research could easily be adapted to include additional reservoirs within the catchment area, enabling an analysis of the regional impacts. As the current study is localized, it is difficult to state how well retention systems would work throughout the Red River Valley landscape. Regionalization of the study would also allow for the calculation of flow reductions over a larger area due to the installation of multiple retention systems. Comparisons could then be drawn between the effectiveness of water retention systems vs. current drainage systems on the Red River Valley landscape. The modelling system could also be easily expanded to include additional modules of interest to the researcher. Water samples are being collected for Pelly's Lake, upstream and downstream of the reservoir. Inclusion of a module on sediment and nutrient levels would allow for a more accurate economic assessment based on the amount of phosphorus and nitrogen loading the retention system at Pelly's Lake is reducing.

5. Conclusions

This paper aimed to estimate the monetary benefits of multi-purpose local farm retention systems under future climate scenarios. When there was insufficient precipitation to allow for maximum crop growth during the study time periods, irrigation using water stored from spring runoff and rainfall events in the water retention system provided increased annual crop gross income. However, a loss in gross margin was estimated due to the costs of developing the retention and irrigation systems. Retention basins' additional capacity for avoided flood damages, nutrient retention, biomass production, and carbon sequestration provide substantial economic and environmental gains. The addition of a second reservoir basin to the Pelly's Lake retention system would accommodate the predicted increases to future spring runoff, reducing downstream flood damage. The value multi-purpose retention basins can provide provincially supports the adoption of multi-purpose local farm surface water retention systems as an effective water management strategy. The recommended use of multi-purpose local farm retention systems is for avoided flood damages, nutrient retention, and biomass production to support farmers wanting to invest in irrigation. This use will subsequently provide economic and environmental benefits for the government of Manitoba.

Acknowledgments: This work was supported by Environment Canada's Lake Winnipeg Basin Stewardship Fund, Growing Forward 2, and SSHRC which is gratefully acknowledged.

Author Contributions: Pamela Berry, Ken Belcher, and Karl-Erich Lindenschmidt conceived and designed the study. Pamela Berry performed the experiment. Fuad Yassin performed the MESH hydrologic modelling. Pamela Berry wrote the paper.

References

1. Warren, F.J.; Lemmon, D.S. (Eds.) *Synthesis; in Canada in a Changing Climate: Sector Perspectives on Impacts and Adaptation*; Government of Canada: Ottawa, ON, Canada, 2014. Available online: http://www.nrcan.gc.ca/environment/resources/publications/impacts-adaptation/reports/assessments/2014/16309 (accessed on 3 January 2016).
2. Vincent, L.A.; Wang, X.L.; Milewska, E.J.; Wan, H.; Yang, F.; Swail, V. A second generation of homogenized Canadian monthly surface air temperature for climate trend analysis. *J. Geophys. Res. Atmos.* **2012**, *117*, 1–13. [CrossRef]
3. Venema, H.D.; Oborne, B.; Neudoerffer, C. *The Manitoba Challenge: Linking Water and Land Management for Climate Adaptation*; International Institute for Sustainable Development: Winnipeg, MB, Canada, 2010.
4. Bonsal, B.R.; Wheaton, E.E.; Chipanshi, A.C.; Lin, C.; Sauchyn, D.J.; Wen, L. Drought research in Canada: A review. *Atmos. Ocean* **2011**, *49*, 303–319. [CrossRef]
5. Intergovernmental Panel on Climate Change (IPCC). *Climate Change 2007: Synthesis Report. Contribution of Working Groups I, II, and III to the Fourth Assessment Report of the Intergovernmental Panel on Climate Change*; Core Writing Team, Pachauri, R., Reisinger, A., Eds.; IPCC: Geneva, Switzerland, 2007.

6. Hearne, R.R. Evolving water management institutions in the Red River Basin. *Environ. Manag.* **2007**, *40*, 842–852. [CrossRef] [PubMed]
7. Wheaton, E.; Kulshreshtha, S.; Wittrock, V.; Koshida, G. Dry times: Hard lessons from the Canadian drought of 2001 and 2002. *Can. Geogr.* **2008**, *52*, 241–262. [CrossRef]
8. Wall, E.; Smit, B. Climate Change Adaptation in Light of Sustainable Agriculture. *J. Sustain. Agric.* **2005**, *27*, 113–123. [CrossRef]
9. Pittman, J.; Wittrock, V.; Kulshreshtha, S.; Wheaton, E. Vulnerability to climate change in rural Saskatchewan: Case study of the Rural Municipality of Rudy No. 284. *J. Rural Stud.* **2011**, *27*, 83–94. [CrossRef]
10. Belcher, K. Agroecosystem Sustainability: An Integrated Modeling Approach. Ph.D. Thesis, University of Saskatchewan, Saskatoon, SK, Canada, 1999.
11. Samarawickrema, A.; Kulshreshtha, S. Value of irrigation water for drought proofing in the South Saskatchewan River Basin (Alberta). *Can. Water Resour. J.* **2008**, *33*, 273–282. [CrossRef]
12. Bower, S.S. Watersheds: Conceptualizing Manitoba's Drained Landscape, 1985–1950. *Environ. Hist. Durh. N. C.* **2007**, *12*, 796–819. [CrossRef]
13. La Salle Redboine Conservation District. La Salle River Watershed: State of the Watershed Report, 2007. pp. 1–296. Available online: https://www.gov.mb.ca/waterstewardship/iwmp/whitemud/documentation/summary_lasalle.pdf (accessed on 23 November 2015).
14. Government of Manitoba. *Surface Water Management Strategy*; Government of Manitoba: Winnipeg, MB, Canada, 2014. Available online: http://gov.mb.ca/waterstewardship/questionnaires/surface_water_management/pdf/surface_water_strategy_final.pdf (accessed on 1 October 2015).
15. Manitoba Government. Conservation and Water Stewardship. Available online: http://mli2.gov.mb.ca/ (accessed on 1 September 2014).
16. Pavelic, P.; Srisuk, K.; Saraphirom, P.; Nadee, S.; Pholkern, K.; Chusanathas, S.; Munyou, S.; Tangsutthinon, T.; Intarasut, T.; Smakhtin, V. Balancing-out floods and droughts: Opportunities to utilize floodwater harvesting and groundwater storage for agricultural development in Thailand. *J. Hydrol.* **2012**, *470–471*, 55–64. [CrossRef]
17. Grosshans, R.E.; Gass, P.; Dohan, R.; Roy, D.; Venema, H.D.; McCandless, M. *Cattail Harvesting for Carbon Offsets and Nutrient Capture: A Lake Friendly Greenhouse Gas Project*; International Institute for Sustaianble Development: Winnipeg, MB, Canada, 2012; Available online: http://www.iisd.org/library/cattails-harvesting-carbon-offsets-and-nutrient-capture-lake-friendly-greenhouse-gas-project (accessed on 3 November 2015).
18. Government of Canada. *Agriculture and Agri-Food Canada Positive Effects of Small Dams and Reservoirs: Water Quality and Quantity Findings from a Prairie Watershed*; WEBS Fact Sheet #7; Government of Canada: Ottawa, ON, Canada, 2012. Available online: http://www.agr.gc.ca/eng/?id=1351881784186 (accessed on 24 April 2016).
19. Arnold, J.G.; Stockle, C.O. Simulation of Supplemental Irrigation from On-Farm Ponds. *J. Irrig. Drain. Eng.* **1991**, *117*, 408–424. [CrossRef]
20. Berry, P. An economic Assessment of On-Farm Surface Water Retention Systems. Ph.D. Thesis, University of Saskatchewan, Saskatoon, SK, Canada, 2016.
21. Tiessen, K.H.D.; Elliott, J.A.; Stainton, M.; Yarotski, J.; Flaten, D.N.; Lobb, D.A. The effectiveness of small-scale headwater storage dams and reservoirs on stream water quality and quantity in the Canadian Prairies. *J. Soil Water Conserv.* **2011**, *66*, 158–171. [CrossRef]
22. Sharpley, A.; Smith, S.J.; Zollweg, J.A.; Coleman, G.A. Gully treatment and water quality in the Southern Plains. *J. Soil Water Conserv.* **1996**, *51*, 498–503.
23. Kovacic, D.A.; Twait, R.M.; Wallace, M.P.; Bowling, J.M. Use of created wetlands to improve water quality in the Midwest-Lake Bloomington case study. *Ecol. Eng.* **2006**, *28*, 258–270. [CrossRef]
24. Salvia-Castellvi, M.; Dohet, A.; Vander Borght, P.; Hoffmann, L. Control of the eutrophication of the reservoir of Esch-sur-Sûre (Luxembourg): Evaluation of the phosphorus removal by predams. *Hydrobiologia* **2001**, *459*, 61–71. [CrossRef]
25. Avilés, A.; Niell, F.X. The control of a small dam in nutrient inputs to a hypertrophic estuary in a Mediterranean climate. *Water Air Soil Pollut.* **2007**, *180*, 97–108. [CrossRef]
26. Uusi-Kämppä, J.; Braskerud, B.; Jansson, H.; Syversen, N.; Uusitalo, R. Buffer zones and constructed wetlands as filters for agricultural phosphorus. *J. Environ. Qual.* **2000**, *29*, 151–158. [CrossRef]

27. Chrétien, F.; Gagnon, P.; Thériault, G.; Guillou, M. Performance analysis of a wet-retention pond in a small agricultural catchment. *J. Environ. Eng.* **2016**, *142*, 4016005. [CrossRef]
28. Manitoba Liquor and Lotteries Corporation. Environmental Innovation. Available online: http://www.mbll.ca/content/environmental-innovation (accessed on 12 December 2016).
29. Brander, L.; Brouwer, R.; Wagtendonk, A. Economic valuation of regulating services provided by wetlands in agricultural landscapes: A meta-analysis. *Ecol. Eng.* **2013**, *56*, 89–96. [CrossRef]
30. Olewiler, N. *The Value of Natural Capital in Settled Areas of Canada*; Ducks Unlmited and the Nature Conservancy of Canada: Stonewall, MB, Canada, 2004; Available online: http://www.cmnbc.ca/sites/default/files/natural%2520capital_0.pdf (accessed on 5 May 2016).
31. Wilson, S. *Ontario's Wealth, Canada's Future: Appreciating the Value of the Greenbelt's Eco-Services*; David Suzuki Foundation: Vancouver, BC, Canada, 2008. Available online: http://www.davidsuzuki.org/publications/downloads/2008/DSF-Greenbelt-web.pdf (accessed on 5 May 2016).
32. Sohngen, B.; King, K.W.; Howard, G.; Newton, J.; Forster, D.L. Nutrient prices and concentrations in Midwestern agricultural watersheds. *Ecol. Econ.* **2015**, *112*, 141–149. [CrossRef]
33. Collins, A.R.; Gillies, N. Constructed Wetland Treatment of Nitrates: Removal Effectiveness and Cost Efficiency. *J. Am. Water Resour. Assoc.* **2014**, *50*, 898–908. [CrossRef]
34. Grosshans, R.; Grieger, L.; Ackerman, J.; Gauthier, S.; Swystun, K.; Gass, P.; Roy, D. *Cattail Biomass in a Watershed-Based Bioeconomy: Commercial-Scale Harvesting and Processing for Nutrient Capture, Biocarbon and High-Value Bioproducts*; International Institute for Sustainable Development: Winnipeg, MB, Canada, 2014. Available online: http://www.iisd.org/sites/default/files/publications/cattail-biomass-watershed-based-bioeconomy-commerical-scale-harvesting.pdf (accessed on 17 January 2015).
35. Dion, J.; McCandless, M. *Cost-Benefit Analysis of Three Proposed Distributed Water Storage Options for Manitoba*; International Institute for Sustainable Development: Winnipeg, MB, Canada, 2013. Available online: http://www.iisd.org/pdf/2014/mb_water_storage_options.pdf (accessed 23 December 2014).
36. McMinn, W.R.; Yang, Q.; Scholz, M. Classification and assessment of water bodies as adaptive structural measures for flood risk management planning. *J. Environ. Manag.* **2010**, *91*, 1855–1863. [CrossRef] [PubMed]
37. Government of Canada. *Government of Canada*; Statistics Canada; Government of Canada: Ottawa, ON, Canada, 2015; pp. 2–6. Available online: www.statcan.gc.ca/eng/start (accessed on 15 September 2015).
38. La Salle Redboine Conservation District. *Pelly's Lake Backflood Project*; La Salle Redboine Conservation District: Holland, MB, Canada, 2013; pp. 1–2. Available online: www.lasalleredboine.com/pellys-lake-backflood-project (accessed on 28 August 2015).
39. La Salle Redboine Conservation District. About. Available online: http://www.lasalleredboine.com/ (accessed on 24 November 2015).
40. Armstrong, R.N.; Pomeroy, J.W.; Martz, L.W. Estimating evaporation in a prairie landscape under drought conditions. *Can. Water Resour. J.* **2010**, *35*, 173–186. [CrossRef]
41. Pomeroy, J.; Fang, X.; Williams, B. *Modelling Snow Water Conservation on the Canadian Prairies*; University of Saskatchewan: Saskatoon, SK, Canada, 2011. Available online: http://www.usask.ca/hydrology/reports/CHRpt11_Prairie-Shelterbelt-Study_Apr11.pdf (accessed on 5 February 2015).
42. Costanza, R.; Duplisea, D.; Kautsky, U. Ecological Modelling on modelling ecological and economic systems with STELLA. *Ecol. Model.* **1998**, *110*, 1–4.
43. Pietroniro, A.; Fortin, V.; Kouwen, N.; Neal, C.; Turcotte, R.; Davison, B.; Verseghy, D.; Soulis, E.D.; Caldwell, R.; Evora, N.; et al. Development of the MESH modelling system for hydrological ensemble forecasting of the Laurentian Great Lakes at the regional scale. *Hydrol. Earth Syst. Sci.* **2007**, *11*, 1279–1294. [CrossRef]
44. Verseghy, D.L. CLASS—A Canadian Land Surface Scheme for GCMS. I. Soil Model. *Int. J. Climatol.* **1991**, *11*, 111–133. [CrossRef]
45. University of Saskatchewan. MEC—Surface and Hydrology (MESH). Available online: http://www.usask.ca/ip3/models1/mesh.htm (accessed on 15 October 2015).
46. Verseghy, D.L.; McFarlane, N.A.; Lazare, M. CLASS—A Canadian Land Surface Scheme for GCMS. II. Vegetation Model and Coupled Runs. *Int. J. Climatol.* **1993**, *13*, 347–370. [CrossRef]
47. Mahfouf, J.-F.; Brasnett, B.; Gagnon, S. A Canadian Precipitation Analysis (CaPA) project: Description and preliminary results. *Atmos. Ocean* **2007**, *45*, 1–17. [CrossRef]

48. Côté, J.; Desmarais, J.-G.; Gravel, S.; Methot, A.; Patoine, A.; Roch, M.; Staniforth, A. The operational CMC—MRB Global Environmental Multiscale (GEM) Model. Part II: Results. *Mon. Weather Rev.* **1998**, *126*, 1397–1418. [CrossRef]
49. Kouwen, N.; ASCE, M.; Soulis, E.D.; Pietroniro, A.; Donald, J.; Harrington, R.A. Grouped Response Units for Distributed Hydrologic Modeling. *J. Water Resour. Plan. Manag.* **1993**, *119*, 289–305. [CrossRef]
50. Government of Manitoba. *Guidelines for Estimating Crop Production Costs 2015*; Government of Manitoba: Winnipeg, MB, Canada, 2015. Available online: https://www.gov.mb.ca/agriculture/business-and-economics/financial-management/cost-of-production.html (accessed on 15 January 2015).
51. Statistics Canada. The Financial Picture of Farms in Canada. Available online: http://www.statcan.gc.ca/ca-ra2006/articles/finpicture-portrait-eng.htm (accessed on 25 January 2017).
52. Waelti, C.; Spuhler, D. Retention Basin. Available online: http://www.sswm.info/content/retention-basin (accessed on 14 August 2014).
53. Grinder, B. *Alberta's Irrigation Infrastructure*; Government of Alberta: Lethbridge, AB, Canada, 2000. Available online: http://www1.agric.gov.ab.ca/\protect\T1\textdollardepartment/deptdocs.nsf/all/irr7197 (accessed on 3 March 2015).
54. Government of Manitoba. Personal communication, 2015.
55. Government of Manitoba. *Manitoba Land Initiative: Core Maps—Data Warehouse*; Government of Manitoba: Winnipeg, MB, Canada. Available online: http://mli2.gov.mb.ca/mli_data/index.html (accessed on 23 September 2014).
56. Government of Canada. Historical Climate Data. Available online: http://climate.weather.gc.ca/ (accessed on 15 February 2015).
57. Belcher, K.W.; Boehm, M.M.; Fulton, M.E. Agroecosystem sustainability: A system simulation model approach. *Agric. Syst.* **2004**, *79*, 225–241. [CrossRef]
58. Werner, A.T. *BCSD Downscaled Transient Climate Projections for Eight Select GCMs over British Columbia, Canada*; Pacific Climate Impacts Consortium, University of Victoria: Victoria, BC, Canada, 2011. Available online: https://pdfs.semanticscholar.org/5dc0/a283b244d177e30c12e1d48d8a227f3720f9.pdf (accessed on 4 January 2016).
59. Pacific Climate Impacts Consortium Statistically Downscaled Climate Scenarios. Available online: https://www.pacificclimate.org/data/statistically-downscaled-climate-scenarios (accessed on 1 July 2016).
60. Randall, D.A.; Wood, R.A.; Bony, S.; Colman, R.; Fichefet, T.; Fyfe, J.; Kattsov, V.; Pitman, A.; Shukla, J.; Srinivasan, J.; et al. Climate Models and Their Evaluation. In *Climate Change 2007: The Physical Science Basis. Contribution of Working Group 1 to the Fourth Assessment Report of the Intergovernmental Panel on Climate Change*; Soloman, S., Qin, D., Manning, M., Chen, Z., Marquis, M., Averyt, K.B., Tignor, M., Miller, H.L., Eds.; Cambridge University Press: Cambridge, UK; New York, NY, USA, 2007; p. 996.
61. Schnorbus, M.; Werner, A.; Bennett, K. Impacts of climate change in three hydrologic regimes in British Columbia, Canada. *Hydrol. Process.* **2014**, *28*, 1170–1189. [CrossRef]
62. Hamlet, A.F.; Salathé, E.P.; Carrasco, P. *Statistical Downscaling Techniques for Global Climate Model Simulations of Temperature and Precipitation with Application to Water Resources Planning Studies*; Chapter 4 in Final Report for the Columbia Basin Climate Change Scenarios Project; Climate Impacts Group, Center for Science in the Earth System, Joint Institute for the Study of the Atmosphere and Ocean, University of Washington: WA, USA, 2010. Available online: https://cig.uw.edu/publications/ statistical-downscaling-techniques-for-global-climate-model-simulations-of-temperature-and-precipitation-with-application-to-water-resources-planning-studies/ (accessed on 27 December 2015).
63. Werner, A.T.; Cannon, A.J. Hydrologic extremes—An intercomparison of multiple gridded statistical downscaling methods. *Hydrol. Earth Syst. Sci.* **2016**, *20*, 1483–1508. [CrossRef]
64. Shrestha, R.; Schnorbus, M.A.; Werner, A.T.; Zwiers, F.W. Evaluating Hydroclimatic Change Signals from Statistically and Dynamically Downscaled GCMs and Hydrologic Models. *J. Hydrometeorol.* **2014**, *15*, 844–860. [CrossRef]
65. Barrow, E. Climate Change Scenarios. In *The Availability, Characteristics and Use of Climate Change Scenarios*; Prairie Adaptation Research Collaborative: Regina, SK, Canada, 2001. Available online: http://www.parc.ca/pdf/conference_proceedings/jan_01_barrow1.pdf (accessed on 1 February 2017).
66. Diaz-Nieto, J.; Wilby, R.L. A comparison of statistical downscaling and climate change factor methods: Impacts on low flows in the River Thames, United Kingdom. *Clim. Chang.* **2005**, *69*, 245–268. [CrossRef]

67. Graham, L.P.; Andreáasson, J.; Carlsson, B. Assessing climate change impacts on hydrology from an ensemble of regional climate models, model scales and linking methods—A case study on the Lule River basin. *Clim. Chang.* **2007**, *81*, 293–307. [CrossRef]
68. Chen, J.; Brissette, F.P.; Leconte, R. Uncertainty of downscaling method in quantifying the impact of climate change on hydrology. *J. Hydrol.* **2011**, *401*, 190–202. [CrossRef]
69. Trzaska, S.; Schnarr, E. *A Review of Downscaling Methods for Climate Change Projections*; Tetra Tech ARD: Burlington, VT, USA, 2014.
70. Intergovernmental Panel on Climate Change—Task Group on Data and Scenario Support for Impact and Climate Analysis (IPCC-TGICA). *General Guidelines on the Use of Scenario Data for Climate Impact and Adaptation Assessment*; Intergovernmental Panel on Climate Change: Geneva, Switzerland, 2007; Volume 2. Available online: http://www.ipcc-data.org/guidelines/TGICA_guidance_sdciaa_v2_final.pdf (accessed on 16 January 2016).
71. Charron, I. *A Guidebook on Climate Scenarios: Using Climate Information to Guide Adaptation Research and Decisions*; Ouranos: Montreal, QC, Canada, 2014. Available online: https://www.ouranos.ca/publication-scientifique/GuideCharron2014_EN.pdf (accessed on 5 January 2016).
72. Nyirfa, W.; Harron, B. *Assessment of Climate Change on the Agricultural Resources of the Canadian Prairies*; Prairie Adaptation Research Collaborative: Regina, SK, Canada, 2001. Available online: http://www.parc.ca/pdf/research_publications/agriculture4.pdf (accessed on 13 February 2016).
73. Sauchyn, D.J.; Barrow, E.M.; Hopkinson, R.F.; Leavitt, P.R. Aridity on the Canadian Plains. *Géogr. Phys. Quat.* **2002**, *56*, 247–259. [CrossRef]
74. Bourne, A.; Armstrong, N.; Jones, G. *A Preliminary Estimate of Total Nitrogen and Total Phosphorus Loading to Streams in Manitoba, Canada*; Government of Manitoba: Winnipeg, MB, Canada, 2002. Available online: http://www.gov.mb.ca/sd/eal/registries/4864wpgww/mc_nitrophosload.pdf (accessed 10 March 2016).
75. Lake Winnipeg Stewardship Board. *Reducing Nutrient Loading to Lake Winnipeg and Its Watershed: Our Collective Responsibility and Commitment to Action*; Government of Manitoba: Winnipeg, MB, Canada, 2006. Available online: https://www.gov.mb.ca/waterstewardship/water_quality/lake_winnipeg/lwsb2007-12_final_rpt.pdf (accessed on 5 April 2016).

9

Geographic Information and Communication Technologies for Supporting Smallholder Agriculture and Climate Resilience

Billy Tusker Haworth [1,*], Eloise Biggs [2], John Duncan [2], Nathan Wales [3], Bryan Boruff [2] and Eleanor Bruce [4]

1 Humanitarian and Conflict Response Institute, University of Manchester, Ellen Wilkinson Building, Oxford Road, Manchester M13 9PL, UK
2 UWA School of Agriculture and Environment, University of Western Australia, Perth, WA 6009, Australia; eloise.biggs@uwa.edu.au (E.B.); john.duncan@uwa.edu.au (J.D.); bryan.boruff@uwa.edu.au (B.B.)
3 School of Geography, Earth Science & Environment, University of the South Pacific, Suva, Fiji; wales_n@usp.ac.fj
4 School of Geosciences, University of Sydney, Sydney, NSW 2006, Australia; eleanor.bruce@sydney.edu.au
* Correspondence: billy.haworth@manchester.ac.uk

Abstract: Multiple factors constrain smallholder agriculture and farmers' adaptive capacities under changing climates, including access to information to support context appropriate farm decision-making. Current approaches to geographic information dissemination to smallholders, such as the rural extension model, are limited, yet advancements in internet and communication technologies (ICTs) could help augment these processes through the provision of agricultural geographic information (AGI) directly to farmers. We analysed recent ICT initiatives for communicating climate and agriculture-related information to smallholders for improved livelihoods and climate change adaptation. Through the critical analysis of initiatives, we identified opportunities for the success of future AGI developments. We systematically examined 27 AGI initiatives reported in academic and grey literature (e.g., organisational databases). Important factors identified for the success of initiatives include affordability, language(s), community partnerships, user collaboration, high quality and locally-relevant information through low-tech platforms, organisational trust, clear business models, and adaptability. We propose initiatives should be better-targeted to deliver AGI to regions in most need of climate adaptation assistance, including SE Asia, the Pacific, and the Caribbean. Further assessment of the most effective technological approaches is needed. Initiatives should be independently assessed for evaluation of their uptake and success, and local communities should be better-incorporated into the development of AGI initiatives.

Keywords: climate change adaptation; livelihoods; geographic information; agriculture; resilience

---

1. Introduction

The agricultural industry is supported by 500 million smallholder farms, responsible for approximately 56% of global agricultural production [1,2]. Smallholder farmers are increasingly resource-poor and confronted by challenges associated with climate change, natural disasters, resource availability and access, and food insecurity [1,3]. Global climatic changes are influencing crop growth and yield, water balances, input availability, and agricultural system management components [4], with ensuing impacts on farming practices [5–7]. Smallholders are faced with both long-term climate stressors and short-term shocks [8]. Geographic variability in climate impacts coupled with low levels of coping and adaptive capacity results in high levels of vulnerability for marginalised farmers [9–11].

Vulnerability varies geographically (often at very local levels). This arises from the complexity of smallholder livelihoods, with multiple on-farm/off-farm activities [12], variation in asset levels and market orientation [13], local (within-farm) variability in productivity [14], gendered roles and access to resources [15], and differential capacity to manage risk [16], affecting smallholder capacity to respond and adapt to climatic challenges.

Incorporating geographic components (i.e., locational properties) into information for climate adaptation is valuable for enhancing environmental decision-making in high risk sectors, such as agriculture. Rapid advancements in geographic information technologies (e.g., geographic information systems (GIS)) and the availability of geospatial data allow for sophisticated capture, analysis, storage, dissemination and access of information across space and time. Concurrently, advancements in information communication technologies (ICTs) (e.g., short message service (SMS); smartphones; Web 2.0), have further increased the usability of geographic information derived from a diversity of sources [17].

Note, while popularity in use of the term geospatial has grown (e.g., geospatial web [18]; geospatial semantics [19]), ambiguity remains over the difference between geospatial and geographic information. Geographic describes information with a reference to Earth's surface and near-surface [20], and geospatial data has been defined as location properties (any descriptive information about the location or area of, and relationships among geographic features) related to any terrestrial feature/ phenomena [21]. We adopt the term geographic information/data, despite much of the material reviewed employing the term geospatial. We consider geographic information to be any information to which location on the Earth is a relevant feature, including both explicit and implicit [22] locational data.

Geographic information used within the agriculture sector—here termed agricultural geographic information (AGI)—is increasingly available to smallholders, yet uptake is limited. Despite a range of geographic information types, such as remote sensing, household surveys, or climate/market reports, accessibility and/or availability is often not in useful/usable formats. Traditionally, information provision to smallholders in developing countries is provided via agricultural extension organisations through farmer field schools, innovation networks and farming associations [23]. However, resource constraints and the diverse needs of smallholders limit the flow of top-down information [24]. For example, resource constraints of agricultural extension staff have been identified as a challenge under climate change in the South Pacific [25] and the lack of transparency and connectivity a constraint to information delivery in India [26].

To this end, we suggest a different or complementary model to supply smallholders with information is necessary, whereby smallholders can harness AGI to make better-informed and cost-saving decisions [27]. Using ICTs to communicate with farmers directly offers a potential for AGI to enhance sustainable agriculture [28], particularly through resources provision for increasing climate resilience at multiple landscape scales [29]. For example, access to geographic information regarding which drought-resistant crops to plant, including when and how, may increase smallholders' capacities to prepare for and withstand such long term climate stresses. Or, localised and context-specific weather forecasts delivered directly to farmers' mobile phones may allow timely decisions and mitigating actions to be taken that reduce the impacts of storms on farming livelihoods. The World Bank, African Development Bank, and African union claim that the greatest opportunities for economic growth and poverty alleviation (in Africa) are provided by ICTs in the agriculture industry [30]. Yet, the evidence base for ICT and use of AGI to support adaptive capacity of smallholders is poorly documented [31]. Baumüller [32] argues that the potential use of ICTs, such as mobile services for smallholder agriculture remains largely unfulfilled. Consequently, here we review recent trends and approaches to utilising geographic information and ICTs for agriculture, and in particular, initiatives for communicating climate and other agriculture-related information to smallholder farmers for improved livelihood security, climate change adaptation and landscape resilience. Our aim is not only to contribute to rectifying the dearth of systematically documented and analysed uses of ICTs in smallholder agriculture, but also to uncover valuable lessons for the design and application of future

AGI initiatives. We achieve this through a systematic review of multi-source literature to address the following research questions:

i. What are the key challenges that AGI initiatives aim to address?
ii. What technological approaches have been adopted to provide AGI to smallholder farmers?
iii. Who are the target users of AGI initiatives and how have initiatives been adopted?
iv. What are the factors promoting or limiting the success of AGI initiatives?

We acknowledge that earlier review works exist on related topics with similar aims and methods to those we present here. The Food and Agriculture Organisation of the United Nations (FAO) [33] reviewed a decade of ICT advancements with applications to agriculture and rural development presenting important findings, such as the significant influence of elements like quality partnerships and the digital divide on project success. But this report was largely descriptive and based on a narrow selection of projects and therefore lacks the analytical depth and rigour associated with our systematic review of AGI initiatives. The World Bank [34] also produced a report on ICT in agriculture, but a similar critique to above could be applied. Baumüller [32] systematically analysed the impact of various mobile services for smallholder agriculture, offering useful lessons for future service developments and an assessment of current shortcomings, including a lack of useful empirical evidence and limitations to current methodologies for evaluating project impact. Our work differs in that it is not constrained to examining only mobile services, but includes a broader range of ICTs used in AGI initiatives, and specifically considers delivery of information of a geographic nature. Duncombe [35] also analysed mobile phone use for agriculture in developing countries, and again, our work examines a more technologically-diverse breadth of AGI initiatives. Further, our work includes the review of AGI initiatives found and described in multiple sources, as opposed to reviews based on only practice-based literature (e.g., [34]) or academic research articles (e.g., [35]).

We first provide a brief background to geographic information and farmer information needs in agriculture, followed by a detailed methodology, presentation of results and discussion in relation to the stated research questions, with particular emphasis on lessons learned from examining a broad range of AGI initiatives. We conclude by identifying critical knowledge gaps and future opportunities.

2. Geographic Information in Agriculture

AGI encompasses a wide range of information types and can be provided through a similarly wide range of technologies. This includes any agricultural information provided through ICTs that has a geographic component, such as location-specific information delivered via SMS, telephone or the Internet, as well as geographic information produced through more sophisticated technological approaches, such as GIS mapping and spatial modelling. GIS technologies provide flexible spatially-explicit tools that support decision making for environmental and natural resource management [36]. Combined with remote sensing technologies, mapping, modelling and monitoring environmental change aids climate change adaptation and mitigation initiatives across the agriculture sector [37,38]. These technologies have contributed to advances in precision agriculture and improved crop management in commercial broad acre agriculture [39–41], yet AGI utilisation by smallholders remains limited. Reflecting on successes from other sectors, geographic information has been used to respond to natural disasters and increase community resilience across a range of environments [42,43], and resilience building in the agricultural sector, particularly in smallholder communities, has similar use potential. Such an aspiration aligns well with the concept of climate smart agriculture (CSA)—to increase food and livelihood security, and farming and landscape resilience [8,44,45]—but explicitly identifies smallholders' needs for improved information access to enable better decision making for sustainable agriculture.

2.1. *Information Needs of Smallholders*

Smallholder farmers require diverse information to support their livelihoods, with development in the agriculture sector dependent on success in generating, sharing, and applying knowledge [1,46]. Information can be obtained from scientists, educators, advisors, policy makers, and informal networks and smallholders themselves [31]. Information needs differ between farmers based on multiple factors, including socio-economic circumstance, literacy levels, access to resources, size of landholding, and agroclimatic conditions [28]. These factors, in conjunction with a range of socio-political conditions, such as governance structures, cultural norms and gender roles, influence how different individuals obtain and seek (applicable) information (e.g., [47]).

2.1.1. Information Availability

Availability of appropriate climate change adaptation information for smallholders often varies by geography and culture. For example, public media and personal experience form dominant information sources amongst Vietnamese farmers [48]. Conversely, in India, farmers rely on external experts such as non-governmental agricultural research for advice, despite their long histories of traditional knowledge [49]. Less formal agricultural knowledge transfer takes place through face-to-face interactions and verbal communication via mobile phones in rural communities [49]. Television, radio, agriculture offices/departments, neighbours and progressive farmers provide the most useful information sources, at least in part due to exposure and availability [50]. Further, the availability of precise and timely weather-based agro-advisory messages are useful in making informed and cost-saving decisions regarding cultivation conditions [27].

2.1.2. Information Accessibility

Information is commonly delivered to farmers through agriculture extension and advisory services [23]. Primarily top-down approaches, these transfer technologies, skills and knowledge to rural farmers and families to enhance crop/livestock production systems, household food security, and livelihoods, through increasing incomes, nutrition, education, and strengthening natural resource management [3]. However, several deficiencies of extension systems restrict their effectiveness, including limited staff, rigid organisation, poor capacity, a top-down linear culture, weak links to the research sector, and limited reach to farmers [28]. In India, for example, there are many [often duplicate] extension systems, yet the majority of farmers still suffer from inadequate information access [28]. Compounding these issues, women in rural communities bear considerable proportions of farming workloads, but have limited roles in receiving information and making decisions (see [27]). Women are often poorer with less land ownership and have difficulty accessing agricultural information from sources aside from other farmers [51]. Munyna [52] argues that women being ill-informed about technologies, markets, and other agriculture information is detrimental to agricultural development.

2.1.3. Information Applicability

Scale of agricultural systems can influence who has access to [relevant] information. For example, national information produced at the government level may not be effective for improving farming practices at more localised scales. At the local scale, farmer field schools are a variation of extension services. Small groups of farmers routinely gather to observe and evaluate potential suitability of agricultural interventions for their farms [53]. This approach also builds social capital, but often exhibits fiscal limitations [54]. Researchers have argued for an increased emphasis on local rather than global initiatives in developing countries with improved relevance and applicability of information (see [55]). This includes the exchange of knowledge in appropriate formats that respect the oral traditions of many indigenous cultures [56].

3. Methodology

To identify AGI initiatives for analysis, literature was assessed from (i) peer-reviewed academic journals, and (ii) projects listed elsewhere or in grey literature, such as through government/non-government organisation, and other key development organisations and/or private sector agency databases. Assessing academic literature involved multiple keyword searches of the Web of Science Core Collection database, which focused on the topic areas of information, climate, and agriculture practices (in that hierarchical order) (Figure 1). Articles were constrained to include only current or recent literature (published after the year 2000; the time period considered to represent the growth of relevant geographic information, the internet, and other ICTs; when mobile technology penetration rates began to expand in developing countries [32]), those published in English language, and only items with full-text versions available. We acknowledge relevant literature will also exist in other languages, such as French, Spanish, Mandarin, or Hindi, among others, and hence incapacity to analyse non-English sources is a limitation of this study [57]. Articles which met all criteria (n = 156) were read and either entered into a spreadsheet for summarisation and analysis, or discarded if deemed not relevant. Assessment of relevance was made in relation to the research questions presented in Section 1. An article may have met all search criteria by using geographic information technologies to examine some aspect of improving agricultural practices in the context of climate change, but if the article did not describe initiatives specifically for communicating such information with farmers it was deemed not applicable to our research questions and thus was excluded. This process was performed initially by one author, and afterwards verified by another. Articles were also discarded if they only provided duplication (e.g., multiple articles describing the same initiative).

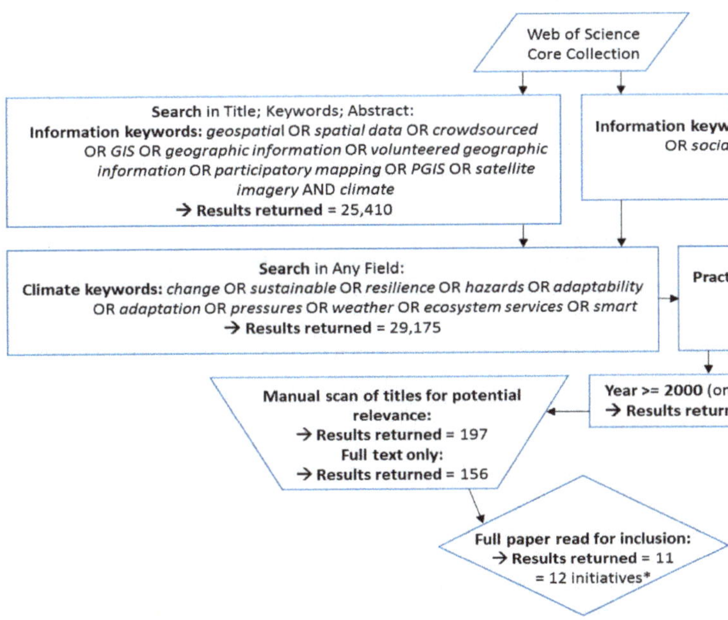

Figure 1. Flow diagram for the academic literature search resulting in 11 relevant papers (12 agricultural geographic information (AGI) initiatives) for analysis (see Table 1 for sources). * One paper described multiple initiatives.

Assessing grey literature involved identifying databases, sources, agencies, and other websites that may contain information on relevant community, agriculture and climate-related AGI initiatives. Where a database had a large number of initiatives, filtering based on keywords (in line with those presented in Figure 1) produced a subset which was manually reviewed for inclusion in a database used for summarisation and analysis. Grey literature assessment was inherently less systematic and could not be automated, and we note the limitation to findings for inclusion in this paper, as once a perceived cross-section of different types of initiatives was obtained the search was ceased. This resulted in

15 AGI initiatives identified. In total, 27 individual AGI initiatives were identified through the above scholarly and grey literature search methods (See Table 1). All initiatives were summarised and analysed in a spreadsheet according to key information relevant for answering the predefined research questions. This included descriptive information (such as initiative name, source, and year), target location and users, initiative aims and approach to achieving aims, climate-related challenges being addressed (short and long term), geographic technologies adopted, the participatory nature of each initiative, adoption and usage information, and details of if/how the initiative was evaluated and by whom.

Table 1. All AGI initiatives identified for this review, including description, target locations and source.

Initiative	Description	Targeted Country or Region	Source
Agriculture Monitoring System	Agriculture monitoring system and technologies for collecting, analysing, and disseminating information. Includes satellite remote sensing, GIS, and mobile GPS. Provides a knowledge base for government, NGOs, rural communities and other stakeholders that will aid sustainable land use and agriculture.	Afghanistan	[58]
Airtel Kilimo	Mobile phone and SMS advisory service. Dissemination of information related to crops, weather and market prices for improved farmer livelihood security.	Kenya	[59]
Avaaj Otalo	Top-down mobile phone advisory service. Delivery of weather, crop, fertiliser and other agriculture information to farmers. Addresses shortcomings of the extension system.	India	[60]
Climate Wizard Tool	Web-based system for climate change data analysis and mapping. Provides practical information for local and regional agriculture managers. Facilitates advanced statistical analyses for more technical users.	Global	[61]
CROPROTECT	Internet and smartphone application utilising GIS and Google Earth. Knowledge exchange system for farmers to acquire and share information relating to pest, weed and disease management.	United Kingdom	[62]
Digital Green	Participatory videos (local languages) used to involve local communities in sharing scientific agriculture information and local knowledge to improve livelihoods through better and more adaptive farming practices.	India, Ghana, Ethiopia	[53,63]
Farmer Decision Support System (FDSS)	Advisory information for registered farmers via SMS to assist farming decisions e.g., when and how to plant, harvest, fertilise and manage crops. 7-day weather forecasts also provided.	Philippines	[64]
Farmforce	SMS and smartphone application to link farmers with other actors in the agro-value chain to reduce transaction costs, aid compliance with food standards, and increase information exchange.	Asia, Africa, Latin America	[65]
Geospatial Information for Rice Crop Monitoring (GIRCM)	Agriculture information derived from image classification and rice crop area estimation to enhance food security. Still in proposal stage.	Afghanistan	[66]
Indian Farmers Fertiliser Cooperative (IFFCO) Kisan Agriculture App	Smartphone application to provide crop information in various formats for enhanced decision making. Aimed at farmers who are receptive to new technologies and business approaches.	India	[67]
Information Technology and Indigenous Knowledge with Intelligence (ITIKI)	Early warning system that integrates information from sensor networks and local knowledge on droughts. Communication using SMS, mobile phone calls, website posts, digital billboards and radio broadcasts to disseminate forecast information to farmers.	Kenya, Sub-Saharan Africa	[68]
iska	GPS-located weather forecasts (various time intervals) distributed via SMS to farmers to improve decision making and reduce weather-related crop losses.	West Africa	[69,70]

Table 1. *Cont.*

Initiative	Description	Targeted Country or Region	Source
Jayalaxmi Agro Tech	Crop-specific smartphone applications for access to agriculture, horticulture and animal husbandry information (English and regional languages).	India	[71]
LandCaRe DSS	Spatial simulation modelling to produce information for stakeholders and farmers involved in decision making related to land management and long-term impacts of climate change at regional and farm scales.	Germany	[72]
Mobile geospatial information for African farmers (MGIAF)	Mobile phone alerts regarding purchasing of drought-tolerant crops for farmers in remote regions. GIS maps for extension officers and community development workers for information dissemination to farmers.	Kenya	[73]
Mobile market information service (MMIS)	SMS request service for rural farmers to receive information on market information (e.g., product prices) to improve selling practices and decision making.	Papua New Guinea	[74]
Mobile soil information for African farmers (MSIAF)	Web-mapping platform for providing soil information to farmers and government workers. Accessed via the internet or mobile phone.	Kenya	[73]
(M)obile Solutions	Mobile phone voice and SMS messages (Hindi or a local language) sent to farmers. Contain information relating to weather, pests, seed varieties, climate change and climate-smart technologies. Provides recommended actions. Option for farmers to provide feedback to inform future messaging.	India	[27]
Participatory Mapping Disaster Risk Reduction Local Knowledge (PMDRRLK)	Participatory approaches and co-produced mapping to improve local resilience to climate change related hazards and increase the use of local environmental knowledge.	Switzerland	[75]
Plantwise Knowledge Bank	Online and smartphone-based knowledge bank with pest identification tools and factsheets on plant health to aid community farming.	Global	[62]
Radio Monsoon	National meteorological information and local knowledge for weather forecasts disseminated to fishermen via social media and the internet, landline and mobile phones, and loudspeakers positioned in fishing communities.	India	[76]
SmartScape	Internet and GIS tool to allow users to experiment with policy options, predict cropping system changes, and compare cropping scenarios. Produces information to be shared with stakeholders, such as policymakers, community agriculture groups, or non-government organisations.	United States of America	[77]
Sowing Application	Smartphone application and SMS used to advise registered farmers best times for sowing seeds based on soil health indicators and rainfall and weather information. Alerts issued for extreme weather conditions that may damage crops or impact farmers.	India	[78]
Tigo Kilimo	Mobile phone dissemination of information on weather, crops and markets for enhanced decision making to improve food security, livelihoods and household income for farmers.	Tanzania	[51]
Watershed Management Information System (WATMIS)	Web-based information and decision support system integrating soil, vegetation, climate and other environment information to assist agriculturalists, resource managers and the rural extension community in managing water scarcity.	India	[79]
World AgroMeteorological Information Service (WAMIS)	Web-server for disseminating agrometeorological products and information bulletins. Provides knowledge and training to large numbers of agriculture stakeholders cost effectively via the internet.	Global	[80]
Wireless Sensor Network—Decision Support System (WSN-DSS)	Wireless sensor network and web-based decision support system for irrigation scheduling. Supports farmers in restructuring agricultural land to address issues of food security and inefficient farming.	Tunisia	[81]

4. Results

4.1. AGI Initiatives

Target users of the AGI initiatives and the key challenges they seek to address are reflected in the distribution of where implementation occurred (see Table 1 for name and summary description of each initiative). Initiatives were concentrated in the global south, particularly south Asia, and east/west Africa. India and Kenya were highlighted as individual countries with the highest numbers of initiatives reviewed. Initiatives largely targeted smallholder farmers and rural communities (n = 18). Some AGI initiatives specifically targeted women farmers (Tigo Kilimo), farmers with low education levels (Tigo Kilimo), fishing households (Radio Monsoon), and progressive farmers more receptive to new technologies and practices (IFFCO Kisan Agriculture App). These target user groups are synonymous with those of more traditional approaches to agricultural extension and advisory services [3]. Other target users included scientists (e.g., PMDRRLK), governments (Smartscape), the agriculture extension community (WATMIS), NGOs and conservation organisations (Agriculture Monitoring System; Smartscape; LandCaRe DSS), risk management agencies (PMDRRLK), and the private sector (Agriculture Monitoring System).

Almost all initiatives adopted a top-down approach (n = 23), with only a few employing bottom-up practices (Digital Green, PMDRRLK and CROPROTECT). Greater emphasis was on communicating AGI to farmers, or providing a service that farmers can receive information from, rather than working with farmers to utilise AGI to support livelihoods. Of the initiatives adopting a bottom-up approach, Digital Green identified 'champions' from a local community to film and edit videos on new farming practices and topics, such as health (outputs were in local languages and topic selections were informed by scientists). Videos were then screened regularly in the community to share learnings. The localised participatory nature of Digital Green was important for people to relate to AGI information and increased adoption of sustainable livelihood practices throughout the community. IFFCO directly targeted progressive farmers, or those more likely to trial and adopt new practices based on capacity, circumstance, and interest. This assumed that farmers who receive AGI through the app, and adopt new practices, will then influence others in the community, either directly through sharing learnings or indirectly through demonstrated success.

4.1.1. Agro-Climatic Challenges Being Addressed

Many initiatives addressed climate adaptation of farmers through increasing livelihood security (n = 19), with some initiatives specifically aiming to increase household income or food security (n = 15). Several initiatives focus on addressing both long-term and short-term climate change to combat adverse impacts on livelihoods [53] and agricultural productivity [60]. In Kenya, where rainfed agriculture supports the majority of subsistence livelihoods, ITIKI sought to address the challenge of limited rainfall monitoring through the development of an integrated communication framework for indigenous knowledge and scientific drought forecast information. In Tunisia, issues of agricultural water wastage and mal-management of resources were being addressed by WSN-DSS, supporting farmers with weather information, improved irrigation scheduling and water management. In rural Africa, MSIAF aimed to mitigate the long-term stress of drought by alerting farmers to market locations to purchase drought-tolerant beans. Initiatives addressing short-term climate shocks were largely related to weather variability, including increased frequency and intensity of meteorological natural disasters (PMDRRLK; WAMIS; iska; Digital Green), extreme conditions like hailstorms and unseasonal rains (Sowing Application), and erratic weather (Radio Monsoon; (M)obile Solutions).

4.1.2. Technological Provisioning to Smallholder Farmers

Various technologies were utilised in the AGI initiatives (Figure 2). MMIS, Tigo Kilimo, Airtel Kilimo, (M)obile Solutions, and FDSS provided simple weather, crop or market information to farmers via low-tech tools, such as SMS and mobile phones, whereby farmers could either receive automatic updates (push notifications) or request information through SMS request or calling a helpline. Varying degrees of complexity were built into these basic mobile phone-based solutions. The inclusion of multiple languages and a peer-to-peer chat function were provided in the Airtel Kilimo mobile service. iska harnessed GPS technology to provide location-specific weather information via SMS. Other AGI initiatives employed internet capabilities to develop custom platforms and smartphone applications, expanding the possible information services offered in terms of both content and format, including support of images, video, animation, interactive content and maps, and hyperlinks to additional online resources. Jayalaxmi Agro Tech offered a range of crop-specific smartphone applications that aimed to enhance food and livelihood security by providing text, audio and visual content on crop information, pricing analytics, and on-demand weather to farmers in English and local languages. Similarly, IFFCO Kisan Agriculture App and Sowing Application aided farmer decision making through the provision of crop or weather information through text, voice, photo and video content. Plantwise Knowledge Bank used smartphones to augment their community-based information exchange activities by pooling information into a central resource for farmers and stakeholders to access; this is particularly useful for remote access by individuals. While GPS was explicitly stated for few AGI initiatives (WATMIS; iska; Agriculture Monitoring System), other initiatives using smart devices likely exploited this technology to provide their locational services.

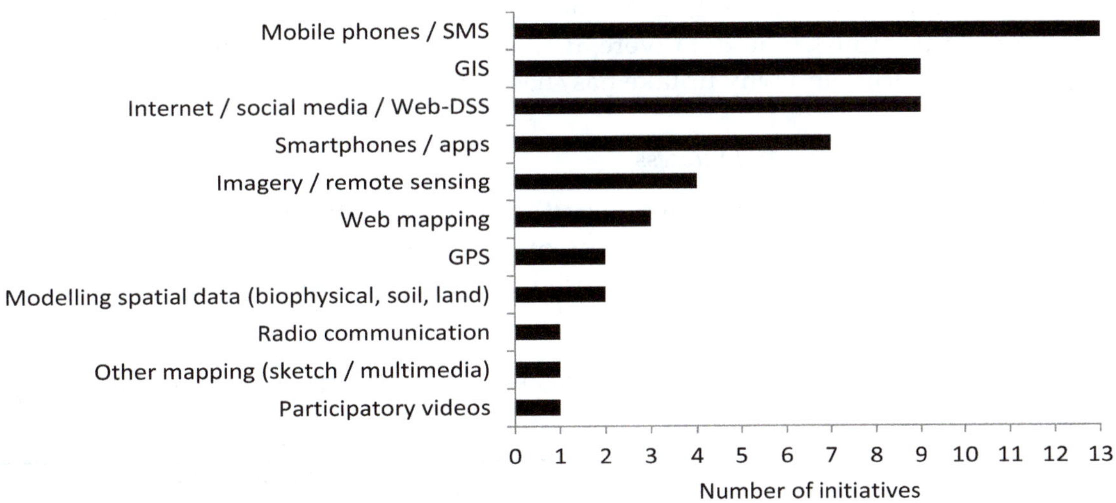

Figure 2. Technologies featured in reviewed AGI initiatives.

Some web-based platform initiatives included and disseminated more data-rich geographic information, such as fine resolution satellite imagery e.g., WATMIS and Agriculture Monitoring System. Satellite imagery and other forms of remote sensing are valuable for detailed depictions of landscape environments and remote capture of data [82]. In Afghanistan, experimentation with methods of classifying satellite imagery was undertaken to strengthen national capacity on rice crop monitoring for sustainable development and food security (GIRCM). WATMIS incorporated GIS and remote sensing data for viable and cost-effective integrated watershed and natural resource planning and management, used by agriculturalists, rural communities and extension services, and land managers. Many AGI initiatives used GIS in combination with ICTs to increase landscape resilience. For example, environmental mapping of drought extent, soils and crops were disseminated to extension workers and farmers, through mobile phones (MSIAF). Online capabilities of technologies have allowed user feedback and sharing of local knowledge for a range of applications, in particular, through social media and crowdsourcing platforms [17]. Radio Monsoon included social media through multiple AGI dissemination methods, and participatory mapping activities that harness local knowledge were used in PMDRRLK. Aside from these two initiatives, social media was absent in all other initiatives. More traditional and primitive forms of information communication, such as radio, loudspeakers and billboards in communities were utilised in some initiatives (e.g., Radio Monsoon; ITIKI).

4.1.3. Adoption of Initiatives

Our review identified limited details of AGI initiative adoption details, with a number being at the proposal, pilot or development stage ($n = 10$). For those with uptake statistics, assessments of adoption were complex. While the number of users or downloads (e.g., of a smartphone application) of an initiative seemed a standard measure of uptake, more nuanced patterns and differences between numbers of downloads, active users, repeat users, and those who implemented changes to their livelihood practices were also observed. Tigo Kilimo reported 400,000 registered users in two years. Of these, 61% were repeat users, with many trialling the service once but not returning. 30% of users reported continued use with concurrent use of new agricultural practices or growing new crops more likely. 39% were more likely to experience increased income than those not engaging with the service. An analogous service, Airtel Kilimo, reported similar adoption patterns, observing 6432 of their total 22,438 registered users (December 2014) as active, with approximately 50% of users implementing farming changes. IFFCO Kisan Agriculture App reported 170,000 users (October 2016), of which 10–20% were estimated to be active. Iska self-reported to have reached more than 80,000 farmers [70] and sent more than 8.5 million weather forecasts [71]. However, no data were provided on how farmers benefited from this service, or how weather forecasts improved livelihoods and were received/read. Digital Green claimed to have reached one million individuals across 13,592 villages through their participatory video approach, with 574,222 farmers adopting at least one of the best-practice video promotions. Yet, similar to iska, no data were available regarding individuals that have/have not implemented new practices, and why uptake has/has not occurred.

4.2. Factors Promoting or Limiting AGI Success

Given the results of the initiatives reviewed, four cross-cutting themes emerged which are important for promoting or limiting the success of AGI initiatives for climate change adaptation: Farmer capacity, delivery approach, technology used, and the organisation delivering the information (summarised in Table 2).

Table 2. Summary of factors promoting and limiting the success of AGI initiatives for addressing key agro-climatic challenges.

	Factors Promoting Success	Factors Limiting Success
Farmer capacity	Affordability to farmers	Participation capacity (exclusion through gender, costs, digital divide)
	Available languages	Limited languages
		Information alone often not enough for meaningful change
Approach	Partnerships with existing community groups	Methods for incorporating community knowledge into GIS
	User collaboration/sharing	Purely top-down approach—lack of interactivity
	Farmers involved in design	User registration required
Technological	High quality, locally-relevant information	Acquisition and sourcing of suitable and quality information/data
	Low tech and user friendly—ease of use	Availability and capacity of telecommunications infrastructure
	Allows participant feedback—interactivity functionality	Personal and community information security
Organisational	Organisational trust	Low user retention
	Potential for expansion—agile service	(In)ability to reach target users
	Marketing and endorsements	
	Clear business model, including funding	Funding of initiatives

4.2.1. Farmer Capacity

The most sophisticated AGI initiatives may be ineffective if target users are unable to access or utilise the information. Various socio-economic factors potentially limit accessibility for smallholder farmers, e.g., the level of disposable income required to acquire and/or access technologies like the internet, computers, smartphones, or televisions. Even relatively low-cost technologies like mobile phones may be inaccessible for many individuals, particularly in developing nations [83]. Consequently, poorer farmers are disadvantaged with increased difficulty in accessing AGI, despite often being the most in need. With reference to increasing participation in CROPROTECT, Bruce [62] described a lessening of digital divides in recent years, but poorer minorities still may lack access to ICTs. Communication technologies for enhancing knowledge access are often most beneficial for younger and more highly educated individuals [49]. Conversely, Bojovic et al. [84] demonstrated a weakening of digital divides for online participation in climate adaptation with groups that are typically excluded appearing as active participants (e.g., older or uneducated individuals). The contrasting ability of geographic information and ICTs to disproportionately benefit those who have access could be exacerbated if existing socioeconomic divisions within and across communities become greater [85].

One measure to increase farmer capacity is to incorporate local and additional languages in AGI initiatives, to ensure the usefulness of information and geographic information reach to maximise farmers benefitted. Information services provided only in English, for example, reduce the capacity of farmers who have first/only language to access the information. Producing and providing content in local languages facilitates comprehension and immediate connection with the local community (see Digital Green; [63]). However, using a local language alone reduces opportunities to expand platform use into other populations/geographical areas. Provision of information in both local/regional and national/international languages increases the probability of meeting a target user's preference [59]. Projects incorporating detailed information in multiple languages relevant to the scale of operation, including regional and local dialects (e.g., Airtel Kilimo, Jayalaxmi Agro Tech, and Digital Green) are likely to exhibit improved information dissemination and utilisation.

4.2.2. Approach

Approaches with participatory elements offer multiple potential benefits over purely top-down approaches. Where individuals can share their own information with others and/or feedback with AGI initiative developers they may feel their input is more valued and subsequently more interconnected to build community resilience [86]. Partnering with existing community groups can be a useful approach to increasing community participation. Digital Green leveraged community groups, such as women's self-help groups or farmers' groups by actively partnering with government, non-government, and private agencies with strong integration and relationships with communities, and cites these partnerships as critical to their success. Whilst having users involved in initiative development is beneficial, requiring registration for participation is seen as a limiting factor. Registering and then subscribing to content causes confusion with some users and has deterred people from using AGI services ([59]; e.g., CROPROTECT).

4.2.3. Technological

A major consideration for the successful implementation of any AGI initiative is the availability and capacity of the information and telecommunications infrastructure. This includes infrastructure for capturing and disseminating information, and for farmers to receive and use it. For example, if an initiative requires high-speed internet access to deliver high-resolution images/videos, then internet coverage is essential, as is the accessibility of affordable internet-enabled devices and data plans. Similarly, if AGI initiatives are designed to include mechanisms for user participation and feedback, then necessary ICT functionalities are required to facilitate interactivity. Many of the reviewed initiatives emphasised the importance of low-tech, user-friendly technological platforms, especially for those with low digital literacy. Additionally, the information itself is important in AGI initiatives, particularly in relation to content, quality and scale. High quality and trustworthy, locally-relevant information is most useful; sourcing and compiling such data can be technologically-challenging for the success of AGI projects [59]. Jayalaxmi Agro Tech attempted to ensure information was relevant to users by developing multiple smartphone applications specific to individual crops and livestock, whereby farmers can select an app to receive only relevant advice to their own farming practices. Attention also needs to be paid to ensuring the security and privacy of users and the data they might supply to the system, particularly in approaches that encourage public participation.

4.2.4. Organisational

Organisational factors include the organisation responsible for developing and implementing the AGI initiative and the kind of support an initiative receives. Initial funding and ongoing financial capital for maintenance, management, and information sourcing are vital for AGI initiatives. Monetary uncertainty may result in premature cessation of an initiative. Funded by a university competition prize, Radio Monsoon was received very positively by village fisherman and the local forecasters. However, the initiative ceased after two years of operation, as funding was no longer available [76]. Many of the reviewed initiatives were developed by universities and funded by external grants/agencies which resulted in uncertain or short-term initiative lifespans (<5 years) and funding unpredictability. This is problematic for climate change adaptation as climate impacts and building livelihood resilience occur over longer timeframes and multiple generations. Programs that are supported financially and in-kind by multiple sources congruently, including through local and international partnerships with the private sector, government agencies, non-government bodies, and the research sector, such as FDSS, and with a clear business model to manage these funds, appear to have greater success and longevity through decreased pressures of financial insecurity.

Reaching and maintaining users is essential for the success of any AGI initiative. Product marketing is imperative to reach users of relevance, and to raise awareness of initiative existence and accessibility. IFFCO Kisan Agriculture App utilised an existing mobile phone service with relevant potential users to target uptake. Search engine optimisation and social media sites can also provide effective and affordable marketing tools [67], but accessibility to these technologies and services is reflective of farmer socio-economic development levels. The IFFCO Kisan Agriculture App social media marketing strategy was augmented by the addition of local celebrity endorsements. GSMA [59] describe marketing and user retention challenges linked to brand identity and loyalty. Airtel Kilimo is provided to farmers through Kenyan mobile network provider Airtel, and multiple ownership, name and brand changes of Airtel have negatively impacted customer loyalty, and thus initiative uptake. Conversely, good reputation and high organisational trust can foster the success of AGI initiatives through user loyalty, sharing of positive experiences and promotion to other farmers (e.g., Tigo Kilimo).

5. Future Potential of AGI

We reflect upon the results and cross-cutting themes discussed above to recommend future avenues for ensuring successful adoption of AGI initiatives by smallholders for climate change adaptation and mitigation.

5.1. Geographical Targeting

Observational factors (Table 2) suggest that both demand- (by the need for climate adaptation solutions) and opportunity- (by the growth of populations with functional access to required ICTs) driven AGI initiatives are important. Geographical targeting of regions currently not utilising AGI initiatives could substantially benefit smallholder farmers in areas highly impacted by changing climates. Regarding regions of high climate change vulnerability and areas predicted for severe climate impacts on agriculture, various reports identify South and Southeast Asia, Africa, Caribbean nations, and small island developing states (SIDS), such as Vanuatu, Samoa and Tonga (see [4,87–89]. Nations in some of these regions already have targeted AGI initiatives (e.g., India, Afghanistan, and parts of Africa), but many other global priority areas remain untargeted. Further research is needed to expound the reasons for these geographical gaps, and for smallholders in these countries to develop appropriate AGI strategies utilising either existing or new infrastructure, technologies, or platforms that will be most effective for the populations of those regions. Vulnerable climate regions generally coincide with areas of increasing access to ICTs, with fast-growing global internet penetration rates observed in Africa, the Middle East, Latin America, and Asia (2000–2017; [90]).

5.2. Types of Information and Information Technologies

Better understanding of the types of information and technologies that are most useful is needed to target users more effectively. A detailed SWOT (strengths, weaknesses, opportunities, threats) analysis of technologies would be valuable, specifically to determine which technological approaches would most effectively deliver AGI to smallholders impacted by digital divides, for example, impoverished and uneducated farmers, women, and those in regions where access to ICT is limited. Mobile phones and SMS can be especially useful technologies for communicating AGI to smallholder farmers as necessary infrastructure is often already present, and data requirements/costs are comparatively low; in many rural regions, mobile phones are often accessible for farmers where other technologies are limited [1,59]. However, credit costs and access to electricity for charging phones can prohibit farmers' use of mobile technologies [83]. Additionally, the information disseminated via mobile phone may be limited by the text- or voice-only format. Technological, resource (cost), and skill components required to access and use AGI will present barriers for some farmers, which also impacts the inclusion of farmer feedback and local knowledge in initiatives. If technologies can be harnessed effectively, then community information sharing could promote greater peer learning and social connectedness, and contribute to increased community resilience [86].

5.3. Independent Assessment of Initiatives

Existing initiatives and future AGI projects should be independently assessed to provide robust success evaluations of their approaches. This is essential as current non-standardised, self-evaluative techniques provide no meaningful and comparable measures of AGI initiative effectiveness, and self-published usage statistics are often more aligned with marketing. The observed asymmetrical pattern of registered and active users is not unique to AGI initiatives, and transferability of assessment approaches by other online geographic information services could be investigated, e.g., OpenStreetMap has 0.5 million registered users (2011) with 38% having undertaken some mapping, and 5% classed as active contributors [91]. There is also a need to examine impacts for users with different characteristics (considering factors, such as gender, age, income, ethnicity, social status, religion and others), as usage and impacts will not be homogeneous among heterogeneous populations [32]. Furthermore, how project success is reported and marketed may have important implications for future funding and resource allocations, agriculture and climate policies, research and development directions, and the livelihoods of farmers. Thus, independent standardised approaches to evaluating AGI initiatives with an emphasis on more nuanced measures of success beyond simple user statistics are recommended. Moreover, the trust and collaboration often needed for farmers to adopt new practices and alternative ways of thinking takes time, and processes of social change can occur over generations [44], thus longitudinal assessments are also advised over raw user statistics.

5.4. Inclusivity for Multi-Level Stakeholder Communication

Ballantyne [31] argues the need for inclusive, participatory approaches to knowledge sharing, and to successfully use ICT to support farmers and rural communities, farming communities must be empowered to define their own needs. Public participation in GIS (e.g., participatory mapping by communities) to contribute their own unique spatial knowledge, often with support from government, nongovernmental, university and other organisations engaged in development and land-related planning [92], can develop community cohesion [93] and facilitate greater local engagement in land-related decision making [94]. Combining local knowledge on coping mechanisms with top-down strategies has enhanced the capacity of rural indigenous communities in SIDS to mitigate and withstand environmental pressures [95]. Additionally, enhancing smallholder social capital can provide opportunities for more effective articulation of individual and community goals/needs to policy makers, researchers and extension providers [3]. Challenges to inclusive AGI participation (e.g., education levels, household resources, local agro-ecological conditions, market access, availability of local producer organisations, and ability/willingness to collaborate and take risks) need careful consideration, particularly regarding equality for women [3]. Baumüller [32] reports for mobile services that study of behavioural factors impacting farmers' capacity and willingness to participate and/or take risks is a significant research shortfall. Technologies that are adapted to smallholders' capacity to take risks and integrated with relevant support services [28], especially to reach marginal farmers where traditional extension activities [3] or locations where reliability of traditional farming approaches [70] fall short, may prove useful in uptake of AGI to overcome cultural and socio-economic obstacles.

Underpinning each area of potential are important considerations and limitations to AGI that warrant further understanding. Adoption of AGI and any outcomes for smallholders are limited by the capacity to act on the knowledge or information gained. For example, a farmer may receive information of a locally-relevant drought-resistant crop, but may not have the financial means to acquire it. Capacity for decision making will also influence the success of AGI initiatives, and information provision alone may not result in meaningful change. Information accessibility is just one factor among many that significantly affect adaptation [96]. Improved comprehension is needed regarding how significant livelihood change occurs when farmers adopt AGI. This requires localised studies at the level of those users most affected (smallholder farmer communities). Further, as livelihood change is not a short-term process and may vary geographically, studies should be longitudinal and undertaken in a variety of climate-impacted regions. Significantly, the potential ability for AGI provision and adoption to address long-term systemic vulnerabilities requires further research attention.

6. Summary

Learning from past experiences and innovations to promote a successful climate adaptation and development research agenda for the future is crucial [97]. Under increasing livelihood pressures associated with short term, and long term, climate stressors, we advocate that smallholder farmers require diverse and locally-relevant geographic information to aid adaptation for increased food and livelihood security. As we identify, only a small percentage of targeted users of AGI initiatives we reviewed are using and acting on the information provided, which raises questions of the appropriateness of such approaches for addressing key agro-climatic challenges. Addressing these shortcomings is important for supporting smallholders to overcome global risks of extreme weather events, natural disasters, and failures of climate change mitigation and adaptation [98]. Our analysis has identified key recommendations that will serve as a valuable guide for the success of future AGI developments whereby knowledge gaps and implementation challenges should be addressed, particularly to align with the geographically varying needs of smallholder farmers (e.g., [99,100]. Use of AGI initiatives could greatly aid smallholders to move towards climate-smart agriculture [101] for sustainably increasing productivity [44], improving environmental livelihood security [102], and enhancing landscape resilience under a changing climate [103].

Author Contributions: Conceptualisation, B.T.H., J.D., E.B. (Eloise Biggs), B.B., E.B. (Eleanor Bruce) and N.W.; Methodology, B.T.H., J.D. and E.B. (Eloise Biggs); Validation, B.T.H. and E.B. (Eloise Biggs); Formal Analysis, B.T.H.; Investigation, B.T.H.; Resources, E.B. (Eloise Biggs) and B.B.; Writing—Original Draft Preparation, B.T.H.; Writing—Review and Editing, B.T.H., E.B. (Eloise Biggs), J.D., N.W., B.B. and E.B. (Eleanor Bruce); Visualisation, B.T.H. and E.B. (Eloise Biggs); Supervision, E.B. (Eloise Biggs), B.B. and E.B. (Eleanor Bruce); Project Administration, E.B. (Eloise Biggs), B.B. and E.B. (Eleanor Bruce); Funding Acquisition, E.B. (Eloise Biggs), B.B. E.B. (Eleanor Bruce) and J.D.

References

1. Sylvester, G. *Success Stories on Information and Communication Technologies for Agriculture and Rural Development*, 2nd ed.; Food and Agriculture Organization of the United Nations: Rome, Italy, 2015; Available online: http://www.fao.org/3/a-i4622e.pdf (accessed on 8 May 2017).
2. Samberg, L.H.; Gerber, J.S.; Ramankutty, N.; Herrero, M.; West, P.C. Subnational distribution of average farm size and smallholder contributions to global food production. *Environ. Res. Lett.* **2016**, *11*, 124010. [CrossRef]
3. Swanson, B. *Global Review of Good Agricultural Extension and Advisory Service Practice*; Food and Agriculture Organization of the United Nations: Rome, Italy, 2008.
4. Knox, J.; Hess, T.; Daccache, A.; Wheeler, T. Climate change impacts on crop productivity in Africa and South Asia. *Environ. Res. Lett.* **2012**, *7*, 034032. [CrossRef]
5. Coumou, D.; Rahmstorf, S. A decade of weather extremes. *Nat. Clim. Chang.* **2012**, *2*, 491–496. [CrossRef]
6. Trenberth, K.E. Changes in precipitation with climate change. *Clim. Res.* **2011**, *47*, 123–138. [CrossRef]
7. Lobell, D.; Schlenker, W.; Costa-Roberts, J. Climate trends and global crop production since 1980. *Science* **2011**, *333*, 616–620. [CrossRef] [PubMed]
8. Hochman, Z.; Horan, H.; Reddy, D.R.; Sreenivas, G.; Tallapragada, C.; Adusumilli, R.; Gaydon, D.S.; Laing, A.; Kokic, P.; Singh, K.K.; et al. Smallholder farmers managing climate risk in India: 2. Is it climate-smart? *Agric. Syst.* **2017**, *151*, 61–72. [CrossRef]
9. Thornton, P.K.; Lipper, L. *How Does Climate Change Alter Agricultural Strategies to Support Food Security?* IFPRI Discussion Paper 01340; IFPRI: Washington, DC, USA, 2013.
10. Dow, K.; Berkhout, F.; Preston, B.L.; Klein, R.J.; Midgley, G.; Shaw, M.R. Limits to adaptation. *Nat. Clim. Chang.* **2013**, *3*, 305–307. [CrossRef]
11. Jayaraman, T.; Murari, K. Climate change and agriculture: Future trends and implications for India. *Rev. Agrar. Stud.* **2014**, *4*, 1–49.
12. Barrett, C.B.; Reardon, T.; Webb, P. Nonfarm income diversification and household livelihood strategies in rural Africa: Concepts, dynamics, and policy implications. *Food Policy* **2001**, *26*, 315–331. [CrossRef]
13. Jayne, T.S.; Mather, D.; Mghenyi, E. Principal challenges confronting smallholder agriculture in Sub-Saharan Africa. *World Dev.* **2010**, *38*, 1384–1398. [CrossRef]
14. Tittonell, P.; Giller, K.E. When yield gaps are poverty traps: The paradigm of ecological intensification in African smallholder agriculture. *Field Crops Res.* **2013**, *143*, 76–90. [CrossRef]
15. Fisher, M.; Carr, E.R. The influence of gendered roles and responsibilities on the adoption of technologies that mitigate drought risk: The case of drought-tolerant maize seed in eastern Uganda. *Glob. Environ. Chang.* **2015**, *35*, 82–92. [CrossRef]
16. Carter, M.R.; Little, P.D.; Mogues, T.; Negatu, W. Poverty traps and natural disasters in Ethiopia and Honduras. *World Dev.* **2007**, *35*, 835–856. [CrossRef]
17. Goodchild, M.F. Citizens as sensors: The world of volunteered geography. *GeoJournal* **2007**, *69*, 211–221. [CrossRef]
18. Elwood, S. Geographic information science: Emerging research on the societal implications of the geospatial web. *Prog. Hum. Geogr.* **2010**, *34*, 349–357. [CrossRef]
19. Kuhn, W. Geospatial Semantics: Why, of What, and How. In *Journal on Data Semantics III. Lecture Notes in Computer Science*; Spaccapietra, S., Zimányi, E., Eds.; Springer: Berlin, Germany, 2005; Volume 3534.
20. Longley, P.A.; Goodchild, M.F.; Maguire, D.J.; Rhind, D.W. *Geographic Information Systems and Science*, 2nd ed.; John Wiley and Sons Ltd.: Chichester, UK, 2006.
21. Open Geospatial Consortium (OGC). Glossary of Terms—G. 2017. Available online: http://www.opengeospatial.org/ogc/glossary/g (accessed on 5 May 2017).
22. Craglia, M.; Ostermann, F.; Spinsanti, L. Digital Earth from vision to practice: Making sense of citizen-generated content. *Int. J. Digit. Earth* **2012**, *5*, 398–416. [CrossRef]
23. Benson, A.; Jafry, T. The state of agricultural extension: An overview and new caveats for the future. *J. Agric. Educ. Ext.* **2013**, *19*, 381–393. [CrossRef]
24. Anderson, J.R.; Feder, G. Agricultural extension: Good intentions and hard realities. *World Bank Res. Obs.* **2004**, *19*, 41–60. [CrossRef]
25. Taylor, M.; McGregor, A.; Dawson, B. *Vulnerability of Pacific Island Agriculture and Forestry to Climate Change*; Pacific Community: New Caledonia, 2016; Available online: https://spccfpstore1.blob.core.windows.net/digitallibrary-docs/files/6f/6fdef19c8085874a0406d7e1f64897bd.pdf?sv=2015-12-11andsr=

bandsig=qUFufXOhh7WrD9g5SQuNcg89y1d4QUHTmUiUPqK0hEo%3Dandse=2018-05-20T04%3A14%3A47Zandsp=randrscc=public%2C%20max-age%3D864000%2C%20max-stale%3D86400andrsct=application%2Fpdfandrscd=inline%3B%20filename%3D%22Vulnerability_Pacific_agriculture_climate_change.pdf%22 (accessed on 12 December 2017).
26. Duncan, J.M.; Tompkins, E.L.; Dash, J.; Tripathy, B. Resilience to hazards: Rice farmers in the Mahanadi Delta, India. *Ecol. Soc.* **2017**, *22*, 3. [CrossRef]
27. Mittal, S. Role of Mobile Phone-enabled Climate Information Services in Gender-inclusive Agriculture. *Gend. Technol. Dev.* **2016**, *20*, 200–217. [CrossRef]
28. Glendenning, C.J.; Babu, S.; Asenso-Okyere, K. Review of Agricultural Extension in India. IFPRI Discussion Paper 01048. International Food Policy Research Institute, 2010. Available online: http://cdm15738.contentdm.oclc.org/utils/getfile/collection/p15738coll2/id/7280/filename/7281.pdf (accessed on 16 June 2017).
29. Balaghi, R.; Badjeck, M.C.; Bakari, D.; De Pauw, E.; De Wit, A.; Defourny, P.; Donato, S.; Gommes, R.; Jlibene, M.; Ravelo, A.C.; et al. Managing Climatic Risks for Enhanced Food Security: Key Information Capabilities. *Procedia Environ. Sci.* **2010**, *1*, 313–323. [CrossRef]
30. Yonazi, E.; Kelly, T.; Halewood, N.; Blackman, C. *eTransform Africa 2012: The Transformational Use of Information and Communication Technologies in Africa*; World Bank, African Development Bank and African Union: Washington, DC, USA; Tunis, Tunisia, 2012.
31. Ballantyne, P. Accessing, sharing and communicating agricultural information for development: Emerging trends and issues. *Inf. Dev.* **2009**, *25*, 260–271. [CrossRef]
32. Baumüller, H. The little we know: An exploratory literature review on the utility of mobile phone-enabled services for smallholder farmers. *J. Int. Dev.* **2018**, *30*, 134–154. [CrossRef]
33. Food and Agriculture Organization of the United Nations (FAO). *e-Agriculture 10 Year Review Report*; Food and Agriculture Organization of the United Nations: Rome, Italy, 2015.
34. World Bank. *ICT in Agriculture: Connecting Smallholders to Knowledge, Networks, and Institutions*; World Bank: Washington, DC, USA, 2011.
35. Duncombe, R. Mobile phones for agricultural and rural development: A literature review and suggestions for future research. *Eur. J. Dev. Res.* **2016**, *28*, 213–235. [CrossRef]
36. Wang, J.; Chen, J.; Ju, W.; Li, M. IA-SDSS: A GIS-based land use decision support system with consideration of carbon sequestration. *Environ. Model. Softw.* **2010**, *25*, 539–553. [CrossRef]
37. Adenle, A.A.; Azadi, H.; Arbiol, J. Global assessment of technological innovation for climate change adaptation and mitigation in developing world. *J. Environ. Manag.* **2015**, *161*, 261–275. [CrossRef]
38. Kroschel, J.; Sporleder, M.; Tonnang, H.E.Z.; Juarez, H.; Carhuapoma, P.; Gonzales, J.C.; Simon, R. Predicting climate-change-caused changes in global temperature on potato tuber moth Phthorimaea operculella (Zeller) distribution and abundance using phenology modeling and GIS mapping. *Agric. For. Meteorol.* **2013**, *170*, 228–241. [CrossRef]
39. Aubert, B.A.; Schroeder, A.; Grimaudo, J. IT as enabler of sustainable farming: An empirical analysis of farmers' adoption decision of precision agriculture technology. *Decis. Support Syst.* **2012**, *54*, 510–520. [CrossRef]
40. Bongiovanni, R.; Lowenberg-DeBoer, J. Precision agriculture and sustainability. *Precis. Agric.* **2004**, *5*, 359–387. [CrossRef]
41. Zhang, N.; Runquist, E.; Schrock, M.; Havlin, J.; Kluitenburg, G.; Redulla, C. Making GIS a versatile analytical tool for research in precision farming. *Comput. Electron. Agric.* **1999**, *22*, 221–223. [CrossRef]
42. Meier, P. Crisis mapping in action: How open source software and global volunteer networks are changing the world, one map at a time. *J. Map Geogr. Libr.* **2012**, *8*, 89–100. [CrossRef]
43. Haworth, B.T.; Bruce, E.; Whittaker, J.; Read, R. The good, the bad, and the uncertain: Contributions of volunteered geographic information to community disaster resilience. *Front. Earth Sci.* **2018**, *6*, 183. [CrossRef]
44. FAO. *Climate-Smart Agriculture Sourcebook*; FAO: Rome, Italy, 2013; Available online: http://www.fao.org/docrep/018/i3325e/i3325e00.htm (accessed on 4 December 2018).
45. Thornton, P.K.; Aggarwal, P.; Parsons, D. Editorial: Prioritising climate-smart agricultural interventions at different scales. *Agric. Syst.* **2017**, *151*, 149–152. [CrossRef]
46. World Bank. *Enhancing Agricultural Innovation: How to Go beyond the Strengthening of Research Systems*;

World Bank: Washington, DC, USA, 2007; Available online: http://siteresources.worldbank.org/INTARD/Resources/EnhancingAgInnovationebook.pdf (accessed on 9 May 2017).
47. Obayelu, A.; Ogunlade, I. Analysis of the uses of information communication technology (ICT) for gender empowerment and sustainable poverty alleviation in Nigeria. *Int. J. Educ. Dev. Using ICT* **2006**, *2*, 45–69.
48. Dang, H.L.; Li, E.; Bruwer, J.; Nuberg, I. Farmers' perceptions of climate variability and barriers to adaptation: Lessons learned from an exploratory study in Vietnam. *Mitig. Adapt. Strateg. Glob. Chang.* **2014**, *19*, 531–548. [CrossRef]
49. Hudson, S.; Krogman, N.; Beckie, M. Social practices of knowledge mobilization for sustainable food production: Nutrition gardening and fish farming in the Kolli hills of India. *Food Secur.* **2016**, *8*, 523–533. [CrossRef]
50. Kumar, K.R.; Nain, M.S.; Singh, R.; Bana, R.S. Analysis of farmers' communication network and factors of knowledge regarding agro meteorological parameters. *Indian J. Agric. Sci.* **2015**, *85*, 1592–1596.
51. Palmer, T.; Pshenichnaya, N. *Tigo Kilimo Impact Evaluation*; GSMA: London, UK, 2015; Available online: http://www.gsma.com/mobilefordevelopment/programme/magri/assessing-the-impact-of-tigo-kilimo (accessed on 15 August 2017).
52. Munyna, H. Application of ICTs in Africa's agricultural sector: A gender perspective. In *Gender and the information revolution in Africa*; Rathgeber, E.M., Adera, E.O., Eds.; International Development Research Centre: Ottawa, ON, Canada, 2000; pp. 85–124.
53. Gandhi, R.; Veeraraghavan, R.; Toyama, K.; Ramprasad, V. Digital Green: Participatory video for agricultural extension. In Proceedings of the 2007 International Conference on Information and Communication Technologies and Development, Bangalore, India, 15–16 December 2007.
54. Rola, A.C.; Quizon, J.B.; Jamias, S.B. Do Farmer Field School Graduates Retain and Share What They Learn? An Investigation in Iloilo, Philippines. *J. Int. Agric. Ext. Educ.* **2002**, *5*, 65–75. [CrossRef]
55. Walsham, G.; Sahay, S. Research on information systems in developing countries: Current landscape and future prospects. *Inf. Technol. Dev.* **2006**, *12*, 7–24. [CrossRef]
56. Metcalfe, M.; Joham, C. The "ear" and "eye" digital divide. In *Organizational Information Systems in the Context of Globalization*; Korpela, M., Montealegre, R., Poulymenakou, A., Eds.; Springer: Boston, MA, USA, 2003; pp. 419–434.
57. Livoreil, B.; Glanville, J.; Haddaway, N.R.; Bayliss, H.; Bethel, A.; de la Chapelle, F.F.; Robalino, S.; Savilaakso, S.; Zhou, W.; Petrokofsky, G.; et al. Systematic searching for environmental evidence using multiple tools and sources. *Environ. Evid.* **2017**, *6*, 23. [CrossRef]
58. FAO. Enhancing Agriculture Monitoring System Based on Geospatial Technology in Afghanistan. 2016. Available online: http://www.fao.org/3/a-i5569e.pdf (accessed on 15 August 2017).
59. Groupe Spéciale Mobile Association (GSMA). *Case Study: Airtel Kilimo, Kenya*; GSMA: London, UK, 2015; Available online: http://www.gsma.com/mobilefordevelopment/wp-content/uploads/2015/02/GSMA_Case_Airtel_FinalProof02.pdf (accessed on 15 August 2017).
60. Cole, S.; Fernando, A.N. Mobile'izing Agricultural Advice: Technology Adoption, Diffusion and Sustainability. Harvard Business School Working Paper. Cambridge, UK, April 2016. Available online: https://papers.ssrn.com/sol3/papers.cfm?abstract_id=2179008anddownload=yes (accessed on 10 October 2017).
61. Girvetz, E.H.; Zganjar, C.; Raber, G.T.; Maurer, E.P.; Kareiva, P.; Lawler, J.L. Applied Climate-Change Analysis: The Climate Wizard Tool. *PLoS ONE* **2009**, *4*, e8320. [CrossRef] [PubMed]
62. Bruce, T.J.A. The CROPROTECT project and wider opportunities to improve farm productivity through web- based knowledge exchange. *Food Energy Secur.* **2016**, *5*, 89–96. [CrossRef] [PubMed]
63. Gandhi, R. Case Study 1: Digital Green: Leveraging social networks for agricultural extension. In *Success Stories on Information and Communication Technologies for Agriculture and Rural Development*, 2nd ed.; Sylvester, G., Ed.; Food and Agriculture Organization of the United Nations: Rome, Italy, 2015; Available online: http://www.fao.org/3/a-i4622e.pdf (accessed on 8 May 2017).
64. Trogo, R.; Ebardaloza, J.B.; Sabido, D.J.; Bagtasa, G.; Tongson, E.; Balderama, O. SMS-based Smarter Agriculture Decision Support System for Yellow Corn Farmers in Isabela. In Proceedings of the IEEE Canada International Humanitarian Technology Conference (Ihtc2015), Ottawa, ON, Canada, 31 May–4 June 2015.
65. GSMA. *Farmforce*; GSMA: London, UK, 2013; Available online: http://www.gsma.com/mobilefordevelopment/programme/magri/farmforce (accessed on 16 October 2017).

66. FAO. Monitoring of Rice Crop Using Satellite Remote Sensing and GIS Technologies in Northern and Eastern Afghanistan. 2016. Available online: http://www.fao.org/3/a-i6146e.pdf (accessed on 22 August 2017).
67. Darabian, N. Case Study IFFCO Kisan Agriculture App: Evolution to Data Driven Services in Agriculture. GSMA, London. 2016. Available online: http://www.gsma.com/mobilefordevelopment/programme/magri/iffco-kisan-agricultural-app-evolution-to-data-driven-services-in-agriculture (accessed on 15 August 2017).
68. Masinde, M.; Bagula, A.; Muthama, N. Implementation Roadmap for Downscaling Drought Forecasts in Mbeere Using Itiki. In Proceedings of the 2013 Itu Kaleidoscope Academic Conference: Building Sustainable Communities (K-2013), Kyoto, Japan, 22–24 April 2013; pp. 63–70.
69. Ignitia. Ignitia, Tropical Weather Forecasting. 2017. Available online: www.ignitia.se (accessed on 16 June 2017).
70. United Nations Development Program (UNDP). Using SMS Texts to Provide Weather Forecasts for Small Farmers in West Africa. December 2015. Available online: http://www.undp.org/content/undp/en/home/presscenter/pressreleases/2015/12/22/using-sms-texts-to-provide-weather-forecasts-for-small-farmers-in-west-africa.html (accessed on 16 June 2017).
71. Jayalaxmi Agro Tech. Welcome to Jayalaxmi Agro Tech. 2014. Available online: http://www.jayalaxmiagrotech.com/ (accessed on 21 August 2017).
72. Wenkel, K.O.; Berg, M.; Mirschel, W.; Wieland, R.; Nendel, C.; Kostner, B. LandCaRe DSS—An interactive decision support system for climate change impact assessment and the analysis of potential agricultural land use adaptation strategies. *J. Environ. Manag.* **2013**, *127*, S168–S183. [CrossRef] [PubMed]
73. Ogodo, O. African Farmers Get Geospatial Info on Their Phones. SciDevNet, 24 April 2009. Available online: http://www.scidev.net/global/farming/news/african-farmers-get-geospatial-info-on-their-phone.html (accessed on 19 August 2017).
74. Laraki, J. Case Study 6: Mobile market information service: A pilot project of ICT use for smallholder farmers in Papua New Guinea. In *Success Stories on Information and Communication Technologies for Agriculture and Rural Development*, 2nd ed.; Sylvester, G., Ed.; Food and Agriculture Organization of the United Nations: Rome, Italy, 2015; Available online: http://www.fao.org/3/a-i4622e.pdf (accessed on 8 May 2017).
75. Reichel, C.; Frömming, U.U. Participatory mapping of local disaster risk reduction knowledge: An example from Switzerland. *Int. J. Disaster Risk Sci.* **2014**, *5*, 41–54. [CrossRef]
76. Slawson, N. Radio Monsoon attempts to ensure safety reigns among fisherman in south India. The Guardian, 24 April 2017. Available online: https://www.theguardian.com/global-development/2017/apr/24/radio-monsoon-safety-fishermen-south-india-kerala (accessed on 15 August 2017).
77. Tayyebi, A.; Meehan, T.D.; Dischler, J.; Radloff, G.; Ferris, M.; Gratton, C. SmartScape™: A web-based decision support system for assessing the tradeoffs among multiple ecosystem services under crop-change scenarios. *Comput. Electron. Agric.* **2016**, *121*, 108–121. [CrossRef]
78. Agarwal, V. New App Promises to Tell Indian Farmers When to Sow Crops. Available online: https://blogs.wsj.com/indiarealtime/2016/06/17/new-app-promises-to-tell-indian-farmers-when-to-sow-crops/ (accessed on 17 June 2016).
79. Aher, P.D.; Adinarayana, J.; Gorantiwar, S.D.; Sawant, S.A. Information System for Integrated Watershed Management Using Remote Sensing and GIS. *Remote Sens. Appl. Environ. Res.* **2014**, 17–34. [CrossRef]
80. Aggarwal, P.K.; Baethegan, W.E.; Cooper, P.; Gommes, R.; Lee, B.; Meinkef, H.; Rathoreg, L.S.; Sivakumarh, M.V.K. Managing Climatic Risks to Combat Land Degradation and Enhance Food security: Key Information Needs. *Procedia Environ. Sci.* **2010**, *1*, 305–312. [CrossRef]
81. Fourati, M.A.; Chebbi, W.; Kamoun, A. Development of a Web-based weather station for irrigation scheduling. In Proceedings of the 2014 Third IEEE Colloquium in Information Science and Technology, Tetouan, Morocco, 20–22 October 2014; pp. 37–41. [CrossRef]
82. Lillesand, T.; Kiefer, R.W.; Chipman, J. *Remote Sensing and Image Interpretation*; John Wiley and Sons: New York, NY, USA, 2014.
83. Duncan, J.M.; Haworth, B.; Biggs, E.; Boruff, B.; Wales, N.; Bruce, E. Managing multifunctional landscapes: Local insights from a Pacific Island country context. *J. Environ. Manag.* in review.
84. Bojovic, D.; Bonzanigo, L.; Giupponi, C.; Maziotis, A. Online participation in climate change adaptation: A case study of agricultural adaptation measures in Northern Italy. *J. Environ. Manag.* **2015**, *157*, 8–19. [CrossRef]
85. Haworth, B.T. Implications of Volunteered Geographic Information for Disaster Management and GIScience:

A More Complex World of Volunteered Geography. *Ann. Am. Assoc. Geogr.* **2018**, *108*, 226–240. [CrossRef]
86. Haworth, B.; Whittaker, J.; Bruce, E. Assessing the application and value of participatory mapping for community bushfire preparation. *Appl. Geogr.* **2016**, *76*, 115–127. [CrossRef]
87. Wheeler, D. *Quantifying Vulnerability to Climate Change: Implications for Adaptation Assistance*; Center for Global Development Working Paper No. 240; Center for Global Development: Washington, DC, USA, 2011.
88. Kreft, S.; Eckstein, D.; Dorsch, L.; Fischer, L. Global Climate Risk Index 2016: Who Suffers Most from Extreme Weather Events? Weather-Related Loss Events in 2014 and 1995 to 2014. Germanwatch Nord-Süd Initiative eV. 2015. Available online: https://germanwatch.org/fr/download/13503.pdf (accessed on 15 August 2017).
89. Notre Dame Global Adaptation Initiative (NDGAI). Notre Dame Global Adaptation Index. 2017. Available online: http://index.gain.org (accessed on 14 August 2017).
90. Internet Usage Statistics. Internet World Stats. N.p., 9 June 2017. Available online: http://www.internetworldstats.com/stats.htm (accessed on 12 June 2017).
91. Neis, P.; Zipf, A. Analyzing the contributor activity of a volunteered geographic information project—The case of OpenStreetMap. *Int. J. Geo-Inf.* **2012**, *1*, 146–165. [CrossRef]
92. IFAD. *Good Practices in Participatory Mapping*; Prepared by Corbett, J.M.; The International Fund for Agricultural Development: Rome, Italy, 2009.
93. Corbett, J.M.; Keller, C.P. An analytical framework to examine empowerment associated with participatory geographic information systems (PGIS). *Cartogr. Int. J. Geogr. Inf. Geovis.* **2005**, *40*, 91–102. [CrossRef]
94. Corbett, J. "I Don't Come from Anywhere": Exploring the Role of the Geoweb and Volunteered Geographic Information in Rediscovering a Sense of Place in a Dispersed Aboriginal Community. In *Crowdsourcing Geographic Knowledge: Volunteered Geographic Information (VGI) in Theory and Practice*; Sui, D.Z., Elwood, S., Goodchild, M.F., Eds.; Springer: Berlin, Germany, 2013; pp. 223–241.
95. Mercer, J.; Dominey-Howes, D.; Kelman, I.; Lloyd, K. The potential for combining indigenous and western knowledge in reducing vulnerability to environmental hazards in small island developing states. *Environ. Hazards* **2007**, *7*, 245–256. [CrossRef]
96. Menike, L.; Arachchi, K.K. Adaptation to climate change by smallholder farmers in rural communities: Evidence from Sri Lanka. *Procedia Food Sci.* **2016**, *6*, 282–292. [CrossRef]
97. Jones, L.; Harvey, B.; Cochrane, L.; Cantin, B.; Conway, D.; Cornforth, R.J.; De Souza, K.; Kirbyshire, A. Designing the next generation of climate adaptation research or development. *Reg. Environ. Chang.* **2018**, *18*, 297–304. [CrossRef]
98. World Economic Forum (WEF). *The Global Risks Report 2017*, 12th ed.; World Economic Forum: Geneva, Switzerland, 2017; Available online: http://reports.weforum.org/global-risks-2017/ (accessed on 14 August 2017).
99. Harvey, C.A.; Saborio-Rodriguez, M.; Martinez-Rodriguez, M.R.; Viguera, B.; Chain-Guadarrama, A.; Vignola, R.; Alpizar, F. Climate change impacts and adaptation among smallholder farmers in Central America. *Agric. Food Secur.* **2018**, *7*, 57. [CrossRef]
100. Abdul-Razak, M.; Kruse, S. The adaptive capacity of smallholder farmers to climate change in Northern Region of Ghana. *Clim. Risk Manag.* **2017**, *17*, 104–122. [CrossRef]
101. Martinez-Baron, D.; Orjuela, G.; Renzoni, G.; Loboguerrero Rodriguez, A.M.; Prager, S.D. Small-scale farmers in a 1.5 °C future: The importance of local social dynamics as an enabling factor for implementation and scaling of climate-smart agriculture. *Curr. Opin. Environ. Sustain.* **2018**, *31*, 112–119. [CrossRef]
102. Biggs, E.M.; Bruce, E.; Boruff, B.; Duncan, J.M.A.; Horsley, J.; Pauli, N.; McNeil, K.; Neef, A.; Van Ogtrop, F.; Curnow, J.; et al. Sustainable development and the water-energy-food nexus: A perspective on livelihoods. *Environ. Sci. Policy* **2015**, *54*, 389–397. [CrossRef]
103. Kremen, C.; Merenlender, A.M. Landscapes that work for biodiversity and people. *Science* **2018**, *362*, eaau6020. [CrossRef] [PubMed]

10

The Value of Tactical Adaptation to El Niño–Southern Oscillation for East Australian Wheat

Bangyou Zheng [1], Scott Chapman [1,2] and Karine Chenu [3,*]

1 CSIRO Agriculture and Food, Queensland Bioscience Precinct, 306 Carmody Road, St. Lucia, QLD 4067, Australia; bangyou.zheng@csiro.au (B.Z.); scott.chapman@csiro.au (S.C.)
2 School of Agriculture and Food Sciences, The University of Queensland, Building 8117A NRSM, Gatton, QLD 4343, Australia
3 The University of Queensland, Queensland Alliance for Agriculture and Food Innovation (QAAFI), 203 Tor Street, Toowoomba, QLD 4350, Australia
* Correspondence: Karine.Chenu@uq.edu.au

Abstract: El Niño–Southern Oscillation strongly influences rainfall and temperature patterns in Eastern Australia, with major impacts on frost, heat, and drought stresses, and potential consequences for wheat production. Wheat phenology is a key factor to adapt to the risk of frost, heat, and drought stresses in the Australian wheatbelt. This study explores broad and specific options to adapt wheat cropping systems to El Niño–Southern Oscillation, and more specifically, to the Southern Oscillation Index (SOI) phases ahead of the season (i.e., April forecast) in Eastern Australia, when wheat producers make their most crucial management decisions. Crop model simulations were performed for commercially-grown wheat varieties, as well as for virtual genotypes representing possible combinations of phenology alleles that are currently present in the Australian wheat germplasm pool. Different adaptation strategies were tested at the site level, across Eastern Australia, for a wide range of sowing dates and nitrogen applications over long-term historical weather records (1900–2016). The results highlight that a fixed adaptation system, with genotype maturities, sowing time, and nitrogen application adapted to each location would greatly increase wheat productivity compared to sowing a mid-maturity genotype, mid-season, using current practices for nitrogen applications. Tactical adaptation of both genotype and management to the different SOI phases and to different levels of initial Plant Available Water ('PAW & SOI adaptation') resulted in further yield improvement. Site long-term increases in yield and gross margin were up to 1.15 $t \cdot ha^{-1}$ and AU$ 223.0 ha^{-1} for fixed adaptation (0.78 $t \cdot ha^{-1}$ and AU$ 153 ha^{-1} on average across the whole region), and up to an extra 0.26 $t \cdot ha^{-1}$ and AU$ 63.9 ha^{-1} for tactical adaptation. For the whole eastern region, these results correspond to an annual AU$ 440 M increase for the fixed adaptation, and an extra AU$ 188 M for the PAW & SOI tactical adaptation. The benefits of PAW & SOI tactical adaptation could be useful for growers to adjust farm management practices according to pre-sowing seasonal conditions and the seasonal climate forecast.

Keywords: ENSO; Southern Oscillation Index; SOI; El Niño; La Niña; soil water; environment type; climate adaptation; management practices; crop model; APSIM

---

1. Introduction

In Australia, the wheat industry is challenged by complex genotype x environment x management (GxExM) interactions [1,2], due in part to the high spatial and temporal variability of the Australian climate (e.g., [3]). In the eastern part of the continent, annual variations in temperature and rainfall that are influenced by El Niño–Southern Oscillation (ENSO) [1,4,5] affect frost, heat, and drought stress patterns [5–7], and ultimately, wheat production [1,3,5]. Drought and warmer temperatures, but also greater frost risk due to the clear night sky, are generally associated with the onset of El Niño episodes [5,8,9], and limit grain yield [10–13]. Stronger ENSO climate oscillations are expected in the near future, as climate forecasts project more frequent extreme El-Niño and La-Niña conditions [14,15].

As the major driver of inter-annual climate variability in Eastern Australian [4,5,16], ENSO is a quasiperiodic climate pattern that occurs across the tropical Pacific Ocean every 3–8 years. It is caused by variations in the surface temperature of the tropical eastern Pacific Ocean, and the air surface pressure in the tropical western Pacific [17]. The Southern Oscillation Index (SOI), as measured by surface pressure anomaly difference between Tahiti and Darwin, has been used to investigate ENSO effects on crops. Five SOI phases have been defined through grouping all sequential two-month pairs of the SOI into five clusters, using principal component analysis and a cluster analysis [18]. Hammer et al. [1] found that using the 5-phase SOI classification (based on SOI values for the current and previous month) could significantly increase wheat profits (up to 20%) and decrease failure risk (up to 35% less risk) in Goondiwindi, South-Eastern Queensland, Australia, through adapting wheat cultivars and nitrogen fertiliser.

Strategies for yield improvement include breeding new cultivars and adapting management practices to the target population of environments [19]. Climate forecasting offers new opportunities in terms of agricultural planning and operation [4]. In the Australian broad-acre dryland wheat production area, most major decisions occur prior to sowing. Producers can potentially react to early indicators of upcoming rainfall and temperature. Early estimation of SOI phases can thus help farmers adjust management practices such as which cultivar to sow, when to sow, and what nitrogen fertilisation to apply [5,20,21].

In Eastern Australia, wheat crops rely heavily on soil-stored plant available water (PAW) [6,22]. An appropriate combination of sowing data, variety maturity, and pre-sowing PAW is crucial to allow flowering and grain filling to occur with minimal stress, in particular frost, heat, and drought stress, and thus, to maximise yield potential [6,7,23–25]. In this context, crop modelling can assist farmers to adapt their practices to specific SOI phases through adequate choice of maturity type and sowing date, in order to get extra benefit and increased profit [26].

The aims of this paper were to determine the values of (i) fixed adaptation (no distinction between the years) and (ii) adaptations to specific pre-sowing plant available water (PAW) and/or SOI phase. In this study, adaptation strategies were defined in terms of sowing, maturity type, and nitrogen fertilisation, to target the greatest long-term productivity at each site. The APSIM crop model [27], together with a phenology model [28], frost impact module [12] and heat impact module [10], were used to predict flowering time and yield of wheat, and search for the best long-term adaptation strategies.

2. Materials and Methods

2.1. Climatic Data

Fifteen weather stations representing local pedo-climatic conditions from the East Australian wheatbelt [7,22] were selected (see Chenu et al., 2013 [22] for more details) to compare adaptation options to Southern Oscillation Index (SOI) phases (Figure 1, Table 1).

The SOI, which corresponds to differences in sea level pressure between Tahiti and Darwin, has been classified in five phases [18]: 'consistently negative' (I), 'consistently positive' (II), 'rapidly falling' (III), 'rapidly rising' (IV) and 'consistently near zero' (V). These phases were grouped into three classes: consistently negative and rapidly falling (SOI phases I & III), consistently positive and rapidly rising (SOI phases II & IV), and consistently near zero (SOI phase V), as suggested by Potgieter et al. [3]. Weather records from 1900 to 2016 (117 years) were extracted from the Australian weather database (SILO Patched Point Dataset [29]; http://www.longpaddock.qld.gov.au/silo/). The SOI phase classification was sourced from Seasonal Climate Outlook in the Long Paddock (http://www.longpaddock.qld.gov.au/). In this study, the SOI phases were classified using SOI values from March–April to look at the effects for a pre-season indicator. From 1900 to 2016, 36 years had been classified as SOI phases I & III, 45 years as SOI phases II & IV and 34 years as SOI phase IV.

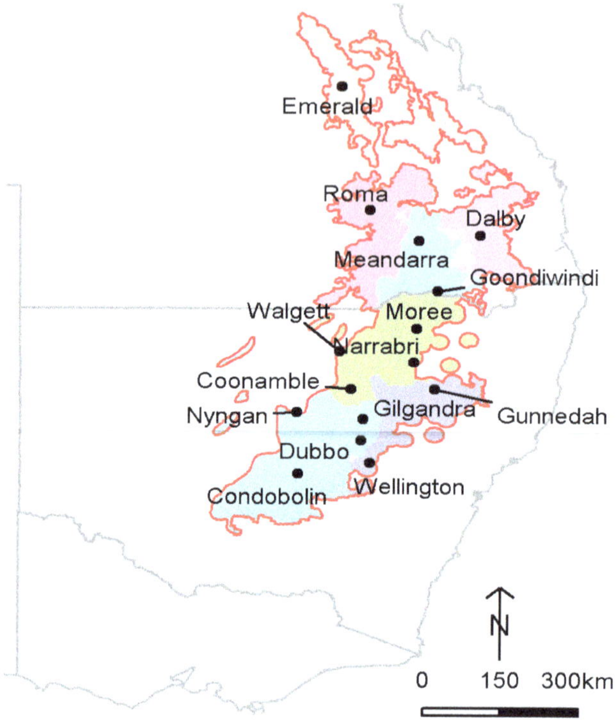

Figure 1. Map of the seven regions of the East Australian wheatbelt, with 15 sites chosen to represent those regions. Details on the locations can be found in Chenu et al., 2013 [22].

Table 1. Regions, locations, soil nitrogen at sowing and nitrogen fertilisation (in the baseline simulations), minimum pant available water (PAW) at sowing chosen to represent the East Australian wheatbelt. Initial and applied nitrogen (N) is indicated by 'x-y-z-a': x, initial N present in the soil at sowing; y, N applied at sowing as urea; z and a, N applied as nitrate at the stages 'beginning of stem elongation' and 'mid-stem elongation', respectively.

Region	Location	Lat.	Long.	Nitrogen (kg ha^{-1})	Minimum PAW at Sowing (mm)
Central Queensland	Emerald	−23.53	148.16	30-50-0-0	80
Eastern Darling Downs	Dalby	−27.18	151.26	30-130-0-0	80
Eastern NSW	Gunnedah	−30.98	150.25	50-70-60 *-0	80
	Wellington	−32.80	148.80	50-50-50 †-0	50
Northern NSW	Moree	−29.48	149.84	30-80-0-0	80
	Walgett	−30.04	148.12	30-80-0-0	80
	Narrabri	−30.32	149.78	30-130-0-0	80
	Coonamble	−30.98	148.38	50-70-60 †-0	50
Southern West Queensland	Roma	−26.57	148.79	30-50-0-0	80
Western Darling Downs	Meandarra	−27.32	149.88	30-80-0-0	80
	Goondiwindi	−28.55	150.31	30-80-0-0	80
Western NSW	Nyngan	−31.55	147.2	50-60-60 †-0	80
	Gilgandra	−31.71	148.66	50-50-50 †-0	50
	Dubbo	−32.24	148.61	50-50-50 †-0	50
	Condobolin	−33.07	147.23	50-60-60 †-0	80

* >80 mm of rainfall from sowing to the stage "end of tillering–beginning of stem elongation". † >100 mm of rainfall from sowing to the stage "end of tillering–beginning of stem elongation".

Monthly temperature and cumulated rainfall were calculated as the average for each month from 1900 to 2016. Daily minimum and maximum temperatures were used to determine occurrences of frost and heat events.

2.2. Crop Simulations and Gross Margins

Wheat yield (dry weight without moisture content) was simulated for the 15 sites (Figure 1, Table 1) from 1900 to 2016. The simulations were performed with the APSIM 7.5 model [27,30], which has been widely tested for wheat across Eastern Australia (e.g., Chenu et al., 2011; Holzworth et al., 2014; Christopher et al., 2016 [6,27,31]; http://www.apsim.info/APSIM.Validation/Main.aspx), and a wheat-phenology gene-based module [28], a heat impact module [10], and a frost impact module with a frost-stress threshold of −2 °C [12].

For each site and year, the simulations were begun with a summer fallow starting from 1 November with a soil containing 20% of its potential available soil water capacity (PAWC). Wheat crops were sown at two-day intervals within a fixed sowing window from the 1 April to 30 June for all 15 sites, when the soil held enough plant available water (PAW) at sowing (Table 1). Soil nitrogen and surface organic matter were reset at sowing. The base nitrogen fertilisation was chosen to reflect local farming practices, and therefore, varied with site and seasonal rainfall, as defined in Chenu et al., 2013 [22] (Table 1). Plants were grown at a density of 100 plants per m^2. Seasons with not enough soil water on 1 April (i.e., when management options were chosen for the tactical adaptation scenarios, see below) were excluded from the analysis.

Different management strategies were tested with a range of sowing dates (sowing every two days from 1 April to 30 June) and nitrogen applications. An extra 0 to 140 $kg{\cdot}ha^{-1}$ (at 20 $kg{\cdot}ha^{-1}$ interval) of nitrogen was applied to the base simulations. Nitrogen fertilisation was applied at the same stage(s) as in the base simulations (i.e., local farming practices) with the same proportions, i.e., at sowing and/or 'beginning of stem elongation' depending on the seasonal opportunities.

Simulations were performed for 208 genotypes including commercial varieties and virtual genotypes that could potentially be bred based on the flowering alleles present in the Australian germplasm pool (see Zheng et al., 2013 for details [28]). Virtual genotypes were created including all combinations of *VRN-A1*, *VRN-B1*, *VRN-D1*, and *PPD-D1* genes (two alleles for each gene), and the full range of values of additional thermal time requirement from floral to flowering (from 425 to 1025 °Cd [28]). Genotypes with the same phenology (from different allelic combinations) were disregarded, so that a total of 156 genotypes unique for their phenology were considered. Overall, the selected genotypes had APSIM parameters ranging from 0 to 1.2 for the photoperiod sensitivity (0.6 for the reference genotype Janz), 0.9 to 1.7 for the vernalisation sensitivity (0.9 for Janz), and 425 to 1025 °Cd for the additional thermal time requirement from floral to flowering (675 °Cd for Janz).

Odd and even years were first simulated separately as some crops matured after 1 November (date of the simulation initialisation). Odd- and even-year simulations were then merged together. Overall, 800 thousand simulations were run through the CSIRO HTCondor service using ClusterRun platform with the runs being completed in less than 4 h [32].

The gross margin was estimated for each simulation based on wheat and nitrogen prices. Other costs of wheat production were ignored, as only variations in gross margin were considered in this study (i.e., only the fertilisation costs varied among the tested management options). The wheat and nitrogen (as urea) prices were sourced from Australian Commodity Statistics [33], and calculated as median values from 2003 to 2012 (i.e., AU$ 269 and AU$ 547 per tonne for wheat and urea, respectively). Variation in grain quality was not considered in this study as the APSIM-wheat model is currently not able to accurately simulate changes in wheat protein content. Increase in gross margin at each site was multiplied with the planting area of the considered region (averaged data from 1975 to 2000, 2004 and 2006; source: Australian Bureau of Statistics), and all regional values were summed to obtain the total increase in gross margin for the Eastern region. Hence, gross margin estimations did not account for changes in fertilizer prices, changes in planting area, changes in wheat prices related to the harvested grain quality, nor changes in wheat prices related to fluctuations in the domestic and/or global market.

2.3. Fixed and Tactical Adaptation Options

Different scenarios exploring the GxExM interactions were evaluated to test their values for different pre-sowing levels of soil water and/or for the three different SOI classes at each studied site. The acceptable range of soil PAW (from minimum required PAW at sowing (Table 1) to the PAWC) calculated at 1 April was divided into three groups (0–33%, 33–67%, 67–100%) to represent Low, Median and High pre-sowing water levels. The 156 genotypes unique for their phenology were considered, as well as sowing dates from 1 April to 30 June and different nitrogen fertilisation options (see previous section).

To provide a reference against conventional practice, a baseline scenario was defined. In this scenario, the reference cultivar Janz was simulated from 1900 to 2016 for a sowing at 21 May using standard farmer practices for fertilisation (Table 1) [22]. A strategic 'fixed scenario' that considers the best management and best genotype for all years (in terms of highest average yield) at each location ('Fixed adaptation') was used as benchmark for best long-term practices (1900–2016). To investigate the potential tactical advantages of adapting, whereby a grower would modify planting decisions based on the SOI phases and/or soil PAW prior to sowing, scenarios with optimised genotypes and management practices specific to SOI classes and/or PAW groups were defined (at each location) through maximizing the average yield for crops grown within each SOI class and/or PAW group. For each location, the tactical adaptation scenarios consisted in either (1) the 'PAW' scenario, which considered the best overall genotype and management within situations from each PAW group, (2) the 'SOI' scenario, which considered the best overall genotype and management within situations from each SOI class, or (3) the 'PAW & SOI' scenario where both genotype and management were optimised for each combination of PAW group × SOI class. For the different scenarios, yield differences and changes in gross margin compared to the baseline and fixed adaptation scenarios were calculated for each year.

3. Results and Discussion

3.1. SOI Impacts on Seasonal Temperature and Rainfall

In the Eastern wheatbelt, SOI phases from the end of summer (calculated in March–April) were typically associated with temperature and cumulated rainfall recorded during the 'summer' fallow preceding the wheat crop (November to April). Higher temperatures were recorded for years with consistently negative SOI (phase I), and to a lesser extent, for years with rapidly falling SOI (III), while lower temperatures occurred in years with consistently positive SOI (II), and to a lesser extent, in years with rapidly increasing SOI (IV) (data not shown, Figure 2A and Figures S1–S3). For instance, the temperature in years from SOI phase I was up to 1.5 °C higher than the 'all years' data in February and March in Emerald, Roma and Gunnedah (data not shown). By contrast, summer rainfall tended to be lowest for SOI phases I & III, and highest for SOI phases II & IV (Figure 2B and Figure S7). As wheat crops in the Eastern wheatbelt heavily rely on soil-stored plant available water (PAW) [6,22], this implies that differences observed in summer rainfall for the different SOI phases are likely to impact crop water-stress pattern and yield, and also the type of genotype and management best suited for specific adaptation.

The impacts of SOI phases (calculated in March–April) for the upcoming 'winter' season (May to October) were weaker than climate variations observed during the previous 'summer' (Figure 2 and Figures S1–S8). In 'winter', differences in temperatures were forecasted for most sites with a tendency for greater differences in the northern sites (e.g., Emerald, Roma, and Meandarra), with highest temperatures for SOI phases I & III, and lowest for SOI phases II & IV (Figure 2B and Figures S4–S6). In any case, monthly temperatures of any SOI phases differed by less than 0.5 °C compared to 'all years' (data not shown, Figure 2A and Figures S4–S6). The impact of SOI phases on rainfall was only visible for a few sites, and mainly for higher rainfall in SOI phase IV years (data not shown, Figure 2B and Figure S8). As found in previous studies (e.g., [34]), ENSO had a substantial impact in northern sites, while a relatively weak impact in southern sites (data not presented for sites south of Condobolin).

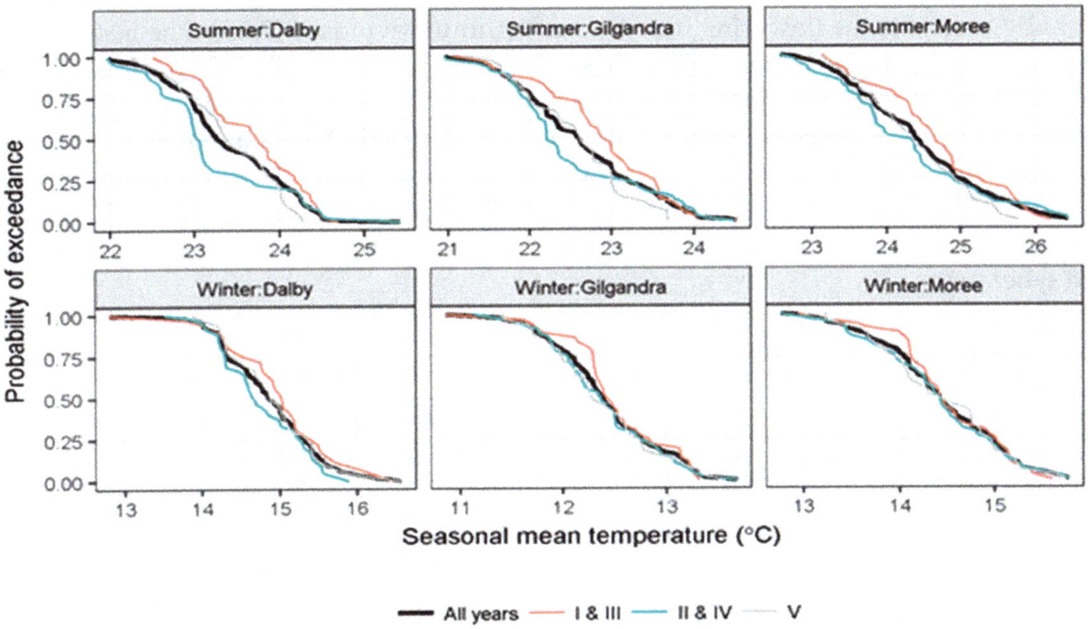

(A)

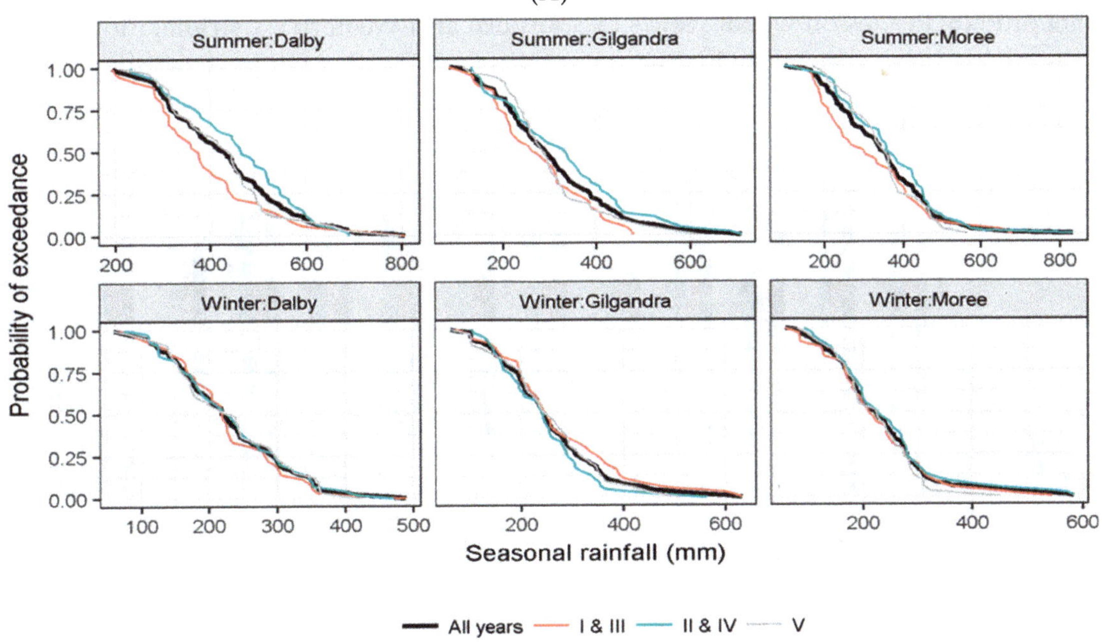

(B)

Figure 2. Cumulative probability distributions (probability of exceedance) of 'summer' and 'winter' average temperature (**A**) and cumulated rainfall (**B**) for the three SOI classes, singly (SOI phases I & III, SOI phases II & IV, and SOI phase V) and combined ('all years') for 1900–2016 at three sites in the Eastern wheatbelt. The three SOI classes correspond to SOI consistently negative and rapidly failing (phases I & III), SOI consistently positive and rapidly rising (phases II & IV), and SOI consistently near zero (phase V). SOI phases were determined in March–April, prior to sowing. The 'summer' data were recorded from the previous November to April, while 'winter' data are for the up-coming May to October period. See Figures S1–S8 for average, minimum and maximum temperature, and cumulated rainfall at other sites for both the 'summer' and 'winter' periods.

3.2. ENSO Impacts the Frequency of Occurrence of Frost and Heat Events around Flowering

Extreme temperatures can greatly decrease yield by affecting reproductive organs or impacting grain filling [11,12]. Australian wheat farmers manage their crops to minimise the risk of frost, heat, and drought by targeting the flowering time into an optimum window [7,25]. The last frost day with a 10% risk of frost tended to be earlier in SOI phases II & IV, and delayed in SOI phases I & III mostly in the eastern sites (Figure S9). By contrast, the first heat day with a 30% risk of heat tended to be earlier in SOI phases I & III, and delayed in SOI phases II & IV. Hence, in terms of temperature, the low-risk flowering window tended to last longer for SOI phases II & IV, while it tended to be reduced for SOI phases I & III. Using the three ENSO phases (i.e., El Niño, La Niña and Neutral), Alexander and Hayman [35] found similar trends for distribution and tails of last frost day in 15 sites across the Australian wheatbelt.

3.3. Variations in Yield across SOI Phases

For the reference cultivar and management (i.e., baseline simulations: Janz sown 21 May with farmer fertilisation practices), long-term average yield ranged from 0.93 to 2.71 t·ha^{-1} across sites and averaged 2.06 t·ha^{-1} among the 15 studied sites (Figure 3; 'all years'). Greatest yields were achieved in SOI phases II & IV, with long-term average yield ranging from 0.98 to 2.70 t·ha^{-1} across locations and averaging 2.10 t·ha^{-1} for Eastern Australia. By contrast, long-term average yield in the SOI phases I & III were commonly lower, ranging from 0.90 to 2.70 t·ha^{-1} across locations, and averaging 2.00 t·ha^{-1} for Eastern Australia. Strong links between wheat yield with ENSO were also found in the Eastern wheatbelt in other studies [18,36]. While the early study of Rimmington et al. [37] suggested little impact of ENSO types on wheat yields in Southern and Western Australia, more recent studies found wheat yields to be affected by ENSO in Southern and South-eastern Australia [38,39].

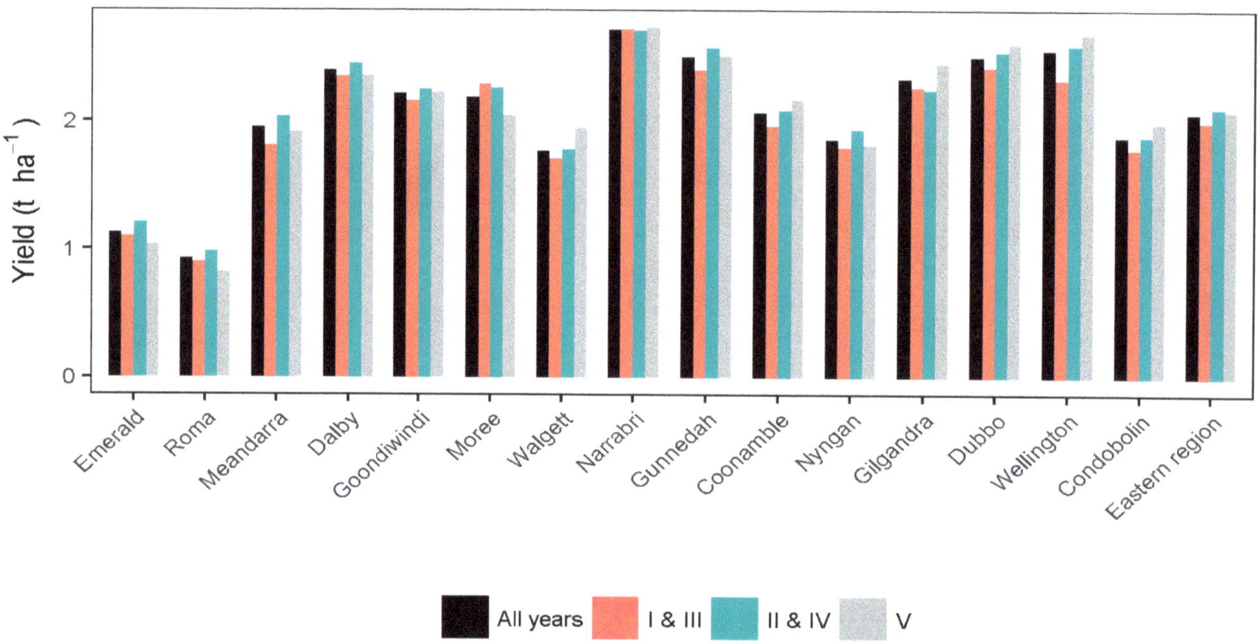

Figure 3. Simulated average yield in the baseline scenario for all years (1900–2016) and for years from each of the three SOI classes at 15 sites across the Eastern wheatbelt and for the whole Eastern wheatbelt region. Baseline simulations corresponded to a standard farmer practice (a medium-season cultivar Janz was sown at semi-optimum sowing date (21 May) with current fertilisation practice). The three SOI classes correspond to SOI consistently negative and rapidly failing (SOI phases I & III), SOI consistently positive and rapidly rising (SOI phases II & IV), and SOI consistently near zero (SOI phase V). SOI phases were determined in March–April, prior to sowing.

3.4. Optimising Genotype and Management across All Years Results in Consistent Yield Improvement and Higher Gross Margins

A large number of adaptation strategies were simulated, combining a wide range of genotypes (with all potential phenology range for Australian wheat) and diverse management practices (a broad range of sowing dates and nitrogen fertilisation options). These strategies were first applied to optimise average yield for a site across all years ('fixed adaptation') by selecting the top yielding genotype × management combination (Figure 4, Figure 5 and Figure S10).

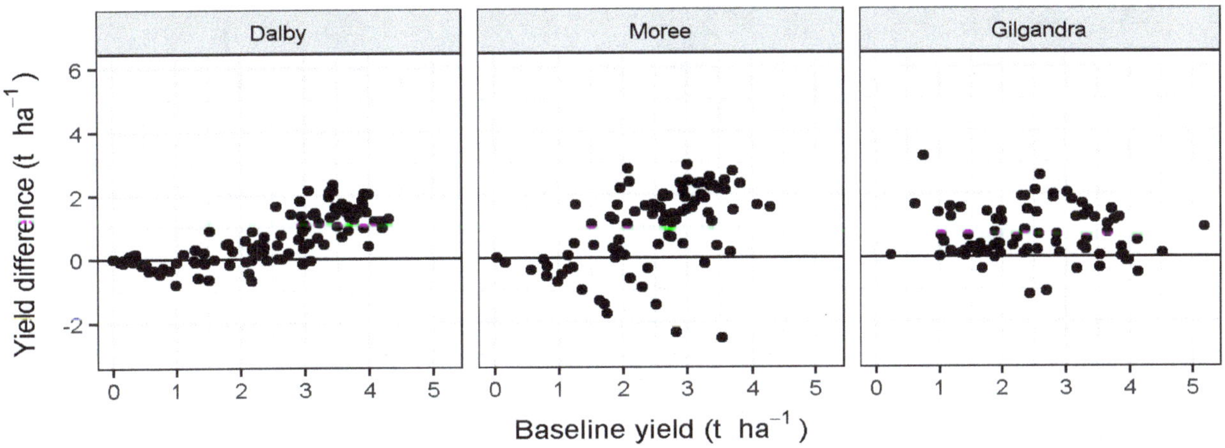

Figure 4. Yield advantage of fixed adaptation over the baseline scenario. The yield difference is calculated for each year. The baseline corresponds to simulated yield for a medium-season cultivar Janz sown at semi-optimum sowing date (21 May) with current fertilisation practice. See Figure S10 for other sites.

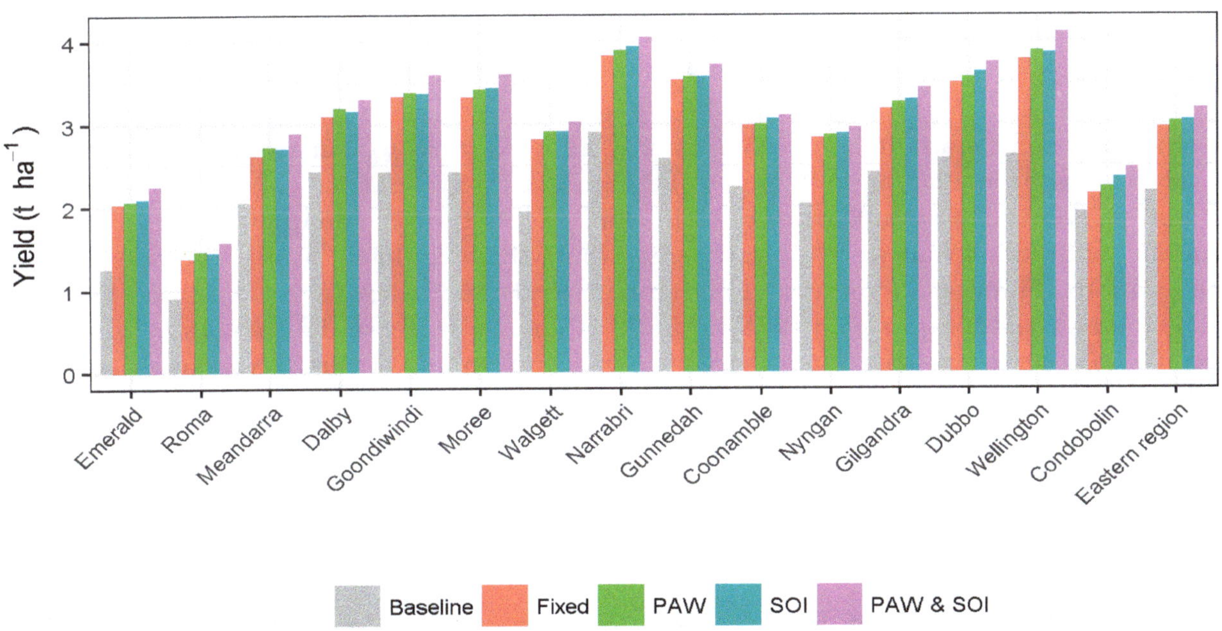

Figure 5. Simulated mean yields for the baseline, fixed adaption, and all the studied tactical adaptation scenarios related to pre-sown soil water and SOI forecast. The baseline corresponds to simulated yield for a medium-season cultivar Janz sown at semi-optimum sowing date (21 May) with no extra nitrogen input. The fixed adaptation scenario corresponds to optimised genotype and management across all years for each site. The three tactical adaptation scenarios include specific adaptation to soil PAW (plant available water) groups, SOI classes, and PAW & SOI groups. These adaptations correspond to optimised genotype and management for each PAW group and/or SOI class. SOI phases were determined in March–April, prior to sowing.

Compared to the baseline, the fixed adaptation scenario increased yields in the majority of years in all sites, although yield losses were also observed for a few years in all sites (Figure 4 and Figure S10). Regional yield (average across all sites) thus increased from 2.06 (baseline) to 2.96 t·ha^{-1} (fixed adaptation) (Figure 5). At the site level, long-term average yield ranged from 1.38 to 3.82 t·ha^{-1} for fixed adaptation compared to 0.91 to 2.90 t·ha^{-1} for the baseline, meaning a yield increase from 0.22 to 1.15 t·ha^{-1} (0.78 t·ha^{-1} on average for the whole region; Figure 3). Compared to the baseline, the fixed adaptation strategy corresponded to an earlier sowing with more nitrogen application of, in general, a shorter-maturing genotype in the northern part of the region, and a longer-maturing genotype in the southern part of the region (data not shown).

In terms of gross margin, site long-term increases from the baseline to fixed adaptation scenario ranged from AU$ 40.5 to 223.0 ha^{-1}, which corresponds to a regional increase of AU$ 153.00 ha^{-1} on average for the whole region. Across Eastern Australia, fixed adaptation resulted in an AU$ 440.00 M increase in gross margin compared to the standard current practice considered here.

3.5. Benefits of Tactical Compared to Fixed Adaptation Vary with the Location, the Soil Pre-Sowing Conditions and the SOI Forecast

To explore the potential of tactical adaptation over fixed adaptation, the genotype and management were optimised for pre-sowing PAW- and/or SOI-specific conditions, and then compared to the fixed adaptation scenario. For most sites (Figure 5), slight increases in long-term average yield were simulated when adapting the genotype and management to either pre-sown PAW or SOI solely. Substantial improvements occurred when optimising average yield for both the genotype and the management for each SOI class and PAW group together (PAW & SOI), rather than sole optimisation of either the PAW group or SOI class. It can nevertheless be noted that fewer of the 117 seasons were classified in each of the nine PAW & SOI groups than in each of three PAW groups or the three SOI classes, meaning that the optimised yield is prone to more uncertainty due to the likely reduction in environmental variations within groups of years considered.

However, tactical adaptation scenarios only allowed yield to increase for some of the years compared to the fixed adaptation (Figure 6 and Figures S11–S13). Actually, when adapting the PAW group only ('PAW adaptation'), losses in yield compared to the fixed adaptation occurred relatively frequently, with losses that were typically small, but which could be as substantial as 2 t·ha^{-1} in some locations (Figure 6 and Figure S11). Similar trends and extents were observed for adaption to SOI classes. Adapting to both PAW and SOI tended to increase the frequency of yield gains, especially in poor seasons (i.e., when yield was medium to low in the fixed adaptation scenario), mainly due to better tuning of the crop phenology (maturity type × sowing time) to the considered environmental conditions. However, yield losses compared to the fixed adaptation still occurred frequently in locations such as Wellington, while they were relatively rare in locations like Coonamble and Moree (Figure S13). Note that the unbalanced number of years among PAW groups, SOI classes, and PAW × SOI groups might cause bias in the adaptation values and risks.

Overall, when considering the optimised genotypes and management for each PAW group ('PAW'), long-term average yield and gross margin increased at each site from 0.012 to 0.13 t·ha^{-1}, and from AU$ 2.63 to 41.9 ha^{-1}, respectively, compared to the fixed adaptation scenario (Figure 7). When considering genotype and management practices optimised for each SOI class ('SOI'), long-term average yield and gross margin at each site increased from 0.037 to 0.17 t·ha^{-1}, and from AU$ 5.00 to 41.50 ha^{-1}, respectively. Finally, when considering optimised management and cultivar for each combination of PAW group and SOI class ('PAW & SOI'), average yield and gross margin at each site increased from 0.15 to 0.46 t·ha^{-1}, and from AU$ 34.2.00 to 108 ha^{-1}, respectively, compared to the fixed adaptation. Hammer et al. (1996) studied the adapted values of SOI phases through changing cultivars and nitrogen fertilisation at Goondiwindi, and also found a substantial, although more limited, increase in gross margin (\$26 ha^{-1}) compared with results in this study (\$48 ha^{-1}, Figure 7B) [1].

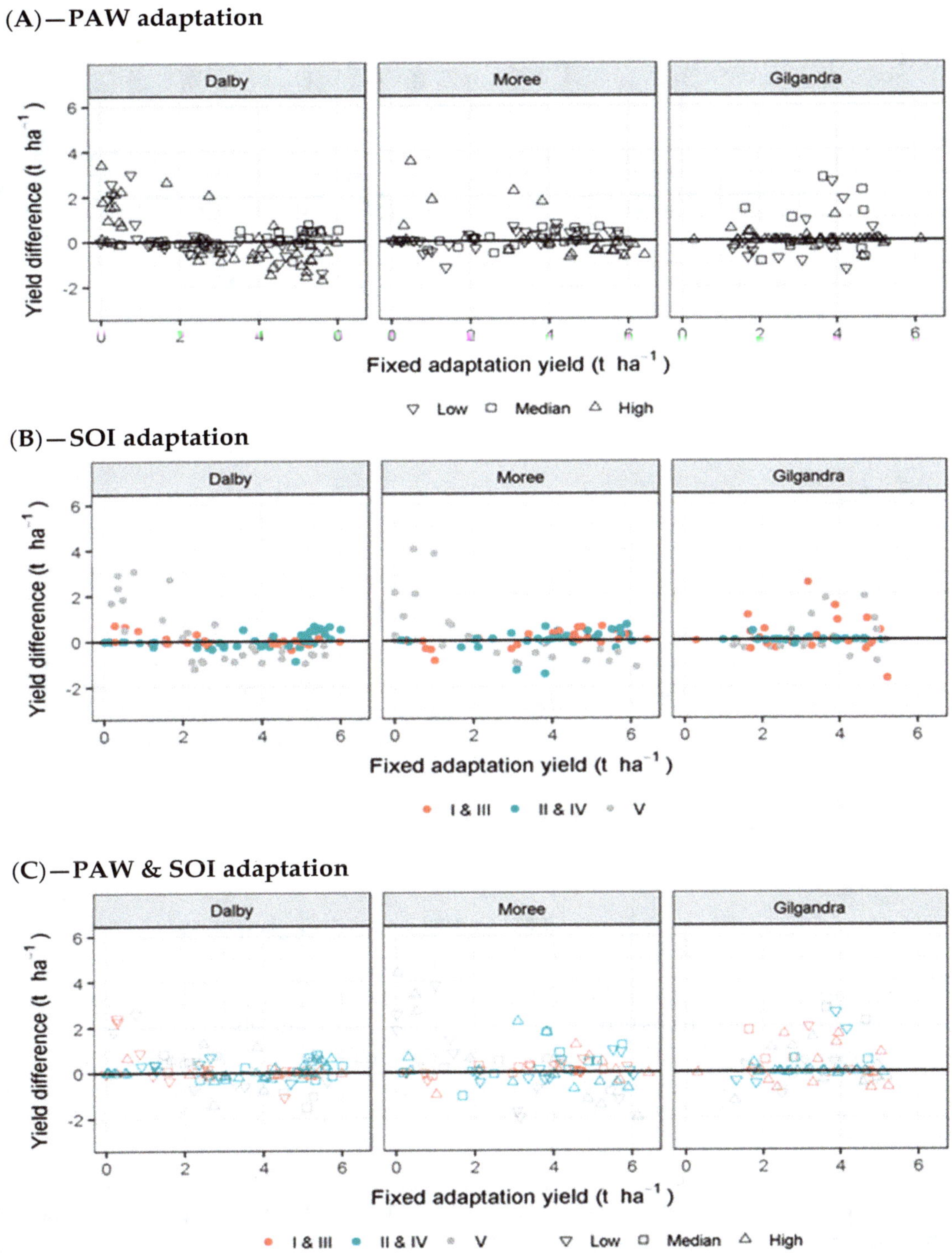

Figure 6. Yield advantage of tactical adaptation scenarios for (**A**) pre-sowing soil PAW (low, median and high), (**B**) SOI classes (I & III, II & IV and V), and (**C**) both PAW & SOI groups over fixed adaptation. The yield difference is calculated for each year. The fixed adaptation scenario corresponds to optimised genotype and management across all years. The three tactical adaptation scenarios correspond to optimised genotype and management for each PAW (plant available water) group and/or SOI class. SOI phases were determined in March–April, prior to sowing. See Figures S11–S13 for other sites.

(A)

(B)

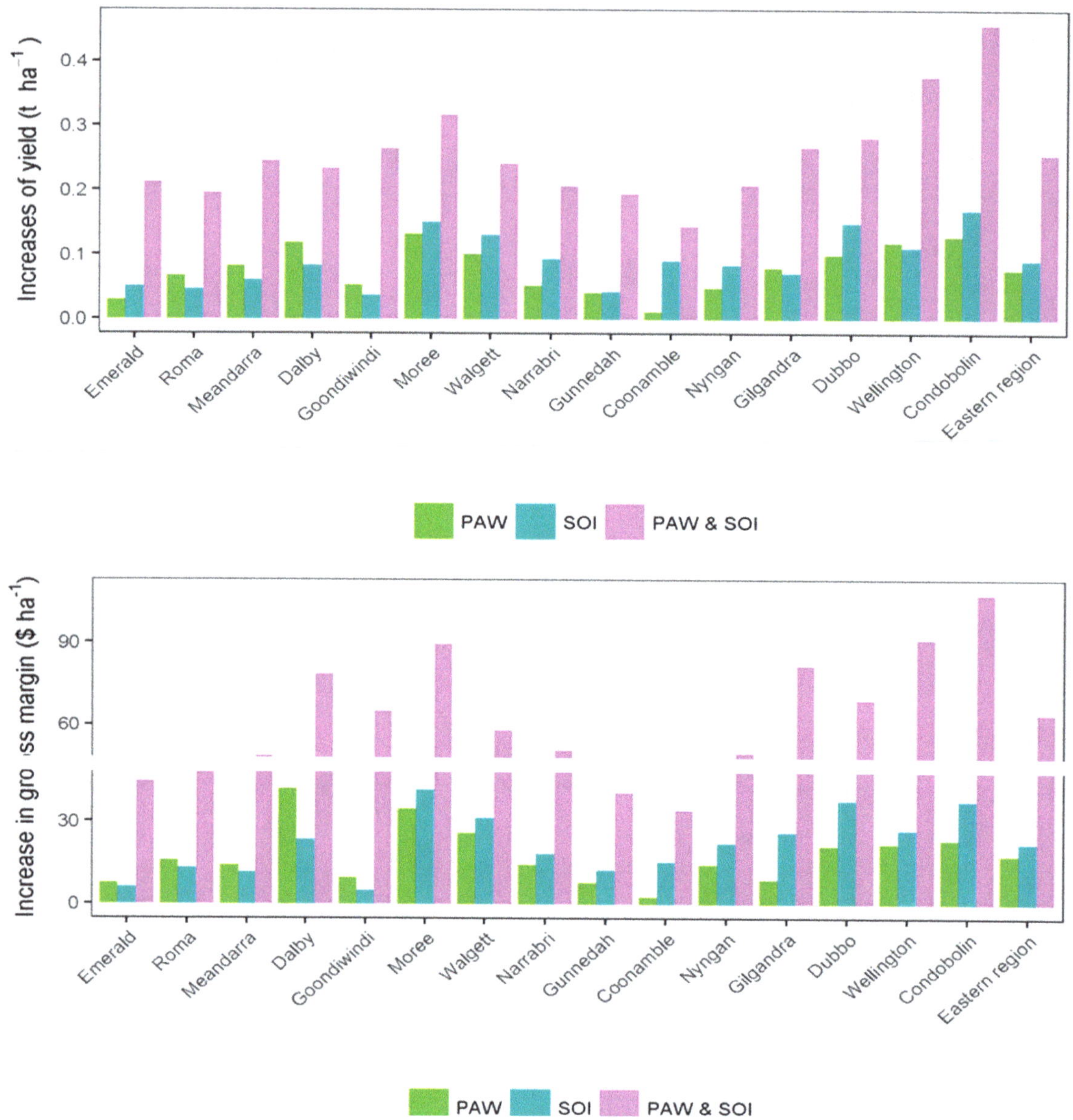

Figure 7. Increase in simulated yield (**A**) and gross margin (**B**) for tactical adaptation options compared to the fixed adaptation in each studied site and the whole Eastern wheatbelt. Increases in yield were averaged for all years from 1900 to 2016. The three tactical adaptation options correspond to optimise long-term yield for either (i) low/medium/high pre-sowing plant available water (PAW), (ii) each class of Southern Oscillation Index (SOI), or (iii) each combination of PAW group and SOI class.

At the regional scale, long-term mean increased in yield and gross margin of PAW & SOI tactical adaptation versus fixed adaptation were 0.26 t·ha^{-1} and AU$ 63.90 ha^{-1}, respectively. The cumulated regional increase in gross margin was AU$ 188 M for the Eastern wheatbelt. Note that the increase gross margin in this paper considered fixed nitrogen price, and accounted for neither changing wheat price related to the harvested grain quality, nor to the domestic and/or global market.

3.6. Should Eastern Australian Wheat Producers Adapt Their Decisions Based on SOI Phases?

In Australia, SOI phases impact the climate and crops mostly in the eastern part of the wheatbelt. In this region, wheat management occurs almost exclusively at sowing. While specific adaptation could be more accurate later in the season, i.e., when climate trends in regard to El Niño/La Niña are clearer, early forecast of the SOI phases is required to allow farmers to prepare seeds and plan suitable management. The potential gains of specific adaptation to such pre-sowing forecasts of SOI phases combined with knowledge on initial soil water appeared to be substantial (Figures 5 and 7). While the gains are variable depending on the location, the extra costs for seed companies and farmers to store large stocks of seeds required for specific adaptation may have to be considered, especially at the farm scale. Unpredictable rainfall in the autumn before sowing may also constrain sowing opportunities [1]. That said, the potential gains of specific adaptation appear to be well above what could be expected from most breeding innovations, at least in the short term [40].

Other methods exist to forecast short-term or seasonal climate. For instance, the Australian Bureau of Meteorology produced twice-weekly weather forecasts for a period of 270 days, with a dynamic model called POAMA (http://poama.bom.gov.au/info/poama-2.html) [41–43]. Such seasonal forecasts have been used to look at management strategies of crops [44–46]. To improve these forecasts, which use on a grid of about 250 km, the Bureau of Meteorology is now proposing seasonal forecasts (ACCESS-S) based on ACCESS (Australian Community Climate and Earth System Simulator; http://www.bom.gov.au/australia/charts/about/about_access.shtml) using a 60 km grid. Other indices than SOI or ENSO phases could also be used for specific adaptation, such as the Inter-decadal Pacific Oscillation (IPO) phases [47] or drought environmental types [20,22,48,49] if climate forecasts are sufficiently reliable. For instance, crop models have been used to assess the value of broad and specific adaptations to select sorghum varieties and managements for different types of drought environments [20].

4. Conclusions

In this study, we assessed the value of fixed adaptation (no distinction between the years) and tactical adaptations based on pre-sowing plant available water (PAW) and/or SOI forecasts to increase productivity at given sites. Overall, with our current knowledge, it appears that yield gains can be made from improving cultivar and management strategy both regardless of climate forecast (fixed adaptation), and as a tactical adaptation to pre-sowing soil water conditions and climate forecasts. The benefits of PAW and SOI tactical adaptation could be useful for farmers to adjust farm management practices according to the season, and may be improved with new forecasting climate methods such as the newly developed ACCESS-S model.

Supplementary Materials
Figure S1: Cumulative probability distributions of 'summer' average temperature for the three SOI classes, singly and combined ('all years'), Figure S2: Cumulative probability distributions of 'summer' maximum temperature for the three SOI classes, singly and combined ('all years'), Figure S3: Cumulative probability distributions of 'summer' minimum temperature for the three SOI classes, singly and combined ('all years'), Figure S4: Cumulative probability distributions of 'winter' average temperature for the three SOI classes, singly and combined ('all years'), Figure S5: Cumulative probability distributions of 'winter' maximum temperature for the three SOI classes, singly and combined ('all years'), Figure S6: Cumulative probability distributions of 'winter' minimum temperature for the three SOI classes, singly and combined ('all years'), Figure S7: Cumulative probability distributions of 'summer' total rainfall for the three SOI classes, singly and combined ('all years'), Figure S8: Cumulative probability distributions (probability of exceedance) of 'winter' total rainfall for the three SOI classes, singly and combined ('all years'), Figure S9: Probability of last frost day and first heat day for the three SOI classes, singly and combined ('all years'), Figure S10: Simulated yield advantage of fixed adaptation over the baseline scenario, Figure S11: Simulated yield advantage of PAW tactical adaptation scenario over fixed adaptation scenario, Figure S12: Simulated yield advantage of SOI tactical adaptation scenario over fixed adaptation scenario, Figure S13: Simulated yield advantage of PAW & SOI tactical adaptation scenario over fixed adaptation scenario.

Author Contributions: B.Z., S.C. and K.C. conceived and developed the ideas; B.Z. and K.C. performed the data analysis; B.Z., S.C. and K.C. wrote the paper.

Acknowledgments: This research was funded by the CSIRO and the University of Queensland.

References

1. Hammer, G.L.; Holzworth, D.P.; Stone, R. The value of skill in seasonal climate forecasting to wheat crop management in a region with high climatic variability. *Aust. J. Agric. Res.* **1996**, *47*, 717–737. [CrossRef]
2. Chapman, S.C.; Chakraborty, S.; Dreccer, M.F.; Howden, S.M. Plant adaptation to climate change? Opportunities and priorities in breeding. *Crop Pasture Sci.* **2012**, *63*, 251–268. [CrossRef]
3. Potgieter, A.B.; Hammer, G.L.; Butler, D. Spatial and temporal patterns in Australian wheat yield and their relationship with ENSO. *Aust. J. Agric. Res.* **2002**, *53*, 77–89. [CrossRef]
4. Meinke, H.; de Voil, P.; Hammer, G.L.; Power, S.; Allan, R.; Stone, R.C.; Folland, C.; Potgieter, A. Rainfall Variability at Decadal and Longer Time Scales: Signal or Noise? *J. Clim.* **2005**, *18*, 89–96. [CrossRef]
5. Yates, D.; Vervoort, R.W.; Minasny, B.; McBratney, A. The history of using rainfall data to improve production in the grain industry in Australia—From Goyder to ENSO. *Crop Pasture Sci.* **2016**, *67*, 467–479. [CrossRef]
6. Chenu, K.; Cooper, M.; Hammer, G.L.; Mathews, K.L.; Dreccer, M.F.; Chapman, S.C. Environment characterization as an aid to wheat improvement: Interpreting genotype–environment interactions by modelling water-deficit patterns in North-Eastern Australia. *J. Exp. Bot.* **2011**, *62*, 1743–1755. [CrossRef] [PubMed]
7. Zheng, B.Y.; Chenu, K.; Dreccer, M.F.; Chapman, S.C. Breeding for the future: What are the potential impacts of future frost and heat events on sowing and flowering time requirements for Australian bread wheat (*Triticum aestivum*) varieties? *Glob. Chang. Biol.* **2012**, *18*, 2899–2914. [CrossRef] [PubMed]
8. Nicholls, N.; Baek, H.J.; Gosai, A.; Chambers, L.E.; Choi, Y.; Collins, D.; Della-Marta, P.M.; Griffiths, G.M.; Haylock, M.R.; Iga, N.; et al. The El Niño–Southern Oscillation and daily temperature extremes in East Asia and the west Pacific. *Geophys. Res. Lett.* **2005**, *32*, L16714. [CrossRef]
9. Chambers, L.E.; Griffiths, G.M. The changing nature of temperature extremes in Australia and New Zealand. *Aust. Meteorol. Mag.* **2008**, *57*, 13–35.
10. Lobell, D.B.; Hammer, G.L.; Chenu, K.; Zheng, B.; McLean, G.; Chapman, S.C. The shifting influence of drought and heat stress for crops in Northeast Australia. *Glob. Chang. Biol.* **2015**, *21*, 4115–4127. [CrossRef] [PubMed]
11. Frederiks, T.M.; Christopher, J.T.; Harvey, G.L.; Sutherland, M.W.; Borrell, A.K. Current and emerging screening methods to identify post-head-emergence frost adaptation in wheat and barley. *J. Exp. Bot.* **2012**, *63*, 5405–5416. [CrossRef] [PubMed]
12. Zheng, B.; Chapman, S.C.; Christopher, J.T.; Frederiks, T.M.; Chenu, K. Frost trends and their estimated impact on yield in the Australian wheatbelt. *J. Exp. Bot.* **2015**, *66*, 3611–3623. [CrossRef] [PubMed]
13. Crimp, S.J.; Zheng, B.; Khimashia, N.; Gobbett, D.L.; Chapman, S.; Howden, M.; Nicholls, N. Recent changes in southern Australian frost occurrence: Implications for wheat production risk. *Crop Pasture Sci.* **2016**, *67*, 801–811. [CrossRef]
14. Timmermann, A.; Oberhuber, J.; Bacher, A.; Esch, M.; Latif, M.; Roeckner, E. Increased El Nino frequency in a climate model forced by future greenhouse warming. *Nature* **1999**, *398*, 694–697. [CrossRef]
15. Cai, W.; Wang, G.; Santoso, A.; McPhaden, M.J.; Wu, L.; Jin, F.-F.; Timmermann, A.; Collins, M.; Vecchi, G.; Lengaigne, M.; et al. Increased frequency of extreme La Nina events under greenhouse warming. *Nat. Clim. Chang.* **2015**, *5*, 132–137. [CrossRef]
16. McPhaden, M.J.; Zebiak, S.E.; Glantz, M.H. ENSO as an Integrating Concept in Earth Science. *Science* **2006**, *314*, 1740–1745. [CrossRef] [PubMed]
17. Ashok, K.; Yamagata, T. Climate change: The El Nino with a difference. *Nature* **2009**, *461*, 481–484. [CrossRef] [PubMed]
18. Stone, R.; Hammer, G.L.; Marcussen, T. Prediction of global rainfall probabilities using phases of the Southern Oscillation Index. *Nature* **1996**, *384*, 252–255. [CrossRef]
19. Hammer, G.L.; McLean, G.; Chapman, S.; Zheng, B.; Doherty, A.; Harrison, M.T.; van Oosterom, E.; Jordan, D. Crop design for specific adaptation in variable dryland production environments. *Crop Pasture Sci.* **2014**, *65*, 614–626.
20. Stone, R.; Nicholls, N.; Hammer, G. Frost in Northeast Australia: Trends and influences of phases of the Southern Oscillation. *J. Clim.* **1996**, *9*, 1896–1909. [CrossRef]
21. Woli, P.; Ortiz, B.V.; Johnson, J.; Hoogenboom, G. El Niño–Southern Oscillation Effects on winter wheat in the southeastern United States. *Agron. J.* **2015**, *107*, 2193. [CrossRef]

22. Chenu, K.; Deihimfard, R.; Chapman, S.C. Large-scale characterization of drought pattern: A continent-wide modelling approach applied to the Australian wheatbelt—Spatial and temporal trends. *New Phytol.* **2013**, *198*, 801–820. [CrossRef] [PubMed]
23. Dennett, M.D. Effects of sowing date and the determination of optimum sowing date. In *Wheat: Ecology and Physiology of Yield Determination*; Satorre, E.H., Slafer, G.H., Eds.; Food Products Press: Binghamton, NY, USA, 1999; pp. 123–140.
24. Pook, M.; Lisson, S.; Risbey, J.; Ummenhofer, C.C.; McIntosh, P.; Rebbeck, M. The autumn break for cropping in southeast Australia: Trends, synoptic influences and impacts on wheat yield. *Int. J. Climatol.* **2009**, *29*, 2012–2026. [CrossRef]
25. Flohr, B.M.; Hunt, J.R.; Kirkegaard, J.A.; Evans, J.R. Water and temperature stress define the optimal flowering period for wheat in south-eastern Australia. *Field Crops Res.* **2017**, *209*, 108–119. [CrossRef]
26. Chenu, K.; Porter, J.R.; Martre, P.; Basso, B.; Chapman, S.C.; Ewert, F.; Bindi, M.; Asseng, S. Contribution of Crop Models to Adaptation in Wheat. *Trends Plant Sci.* **2017**, *22*, 472–490. [CrossRef] [PubMed]
27. Holzworth, D.P.; Huth, N.; de Voil, P.G.; Zurcher, E.J.; Herrmann, N.I.; McLean, G.; Chenu, K.; Van Oosterom, E.; Murphy, C.; Moore, A.D.; et al. APSIM—Evolution towards a new generation of agricultural systems simulation. *Environ. Model. Softw.* **2014**, *62*, 327–350. [CrossRef]
28. Zheng, B.Y.; Biddulph, B.; Li, D.; Kuchel, H.; Chapman, S.C. Quantification of the effects of *VRN1* and *Ppd-D1* to predict spring wheat (*Triticum aestivum*) heading time across diverse environments. *J. Exp. Bot.* **2013**, *64*, 3747–3761. [CrossRef] [PubMed]
29. Jeffrey, S.J.; Carter, J.O.; Moodie, K.B.; Beswick, A.R. Using spatial interpolation to construct a comprehensive archive of Australian climate data. *Environ. Model. Softw.* **2001**, *16*, 309–330. [CrossRef]
30. Zheng, B.; Chenu, K.; Doherty, A.; Chapman, S. *The APSIM-Wheat Module (7.5 R3008)*; Agricultural Production Systems Simulator (APSIM) Initiative: Toowoomba, Australian, 2014.
31. Christopher, J.T.; Christopher, M.J.; Borrell, A.K.; Fletcher, S.; Chenu, K. Stay-green traits to improve wheat adaptation in well-watered and water-limited environments. *J. Exp. Bot.* **2016**, *67*, 5159–5172. [CrossRef] [PubMed]
32. Zheng, B.; Holland, E.; Chapman, S.C. A standardized workflow to utilise a grid-computing system through advanced message queuing protocols. *Environ. Model. Softw.* **2016**, *84*, 304–310. [CrossRef]
33. Australian Bureau of Agricultural and Resource Economics (ABARE). *Australian Commodity Statistics 2016*; ABARE: Canberra, Australia, 2016.
34. Stone, R.; Auliciems, A. SOI phase relationships with rainfall in eastern Australia. *Int. J. Climatol.* **1992**, *12*, 625–636. [CrossRef]
35. Alexander, B.; Hayman, P. Can we use forecasts of El Nino and La Nina for frost management in the Eastern and Southern grains belt? In Proceedings of the 14th Agronomy Conference, Adelaide, Australia, 21–25 September 2008.
36. Potgieter, A.B.; Hammer, G.L.; Meinke, H.; Stone, R.C.; Goddard, L. Three putative types of El Niño revealed by spatial variability in impact on Australian wheat yield. *J. Clim.* **2005**, *18*, 1566–1574. [CrossRef]
37. Rimmington, G.M.; Nicholls, N. Forecasting wheat yields in Australia with the Southern Oscillation Index. *Crop Pasture Sci.* **1993**, *44*, 625–632. [CrossRef]
38. Anwar, M.R.; Rodriguez, D.; Liu, D.L.; Power, S.; O'Leary, G.J. Quality and potential utility of ENSO-based forecasts of spring rainfall and wheat yield in south-eastern Australia. *Crop Pasture Sci.* **2008**, *59*, 112–126. [CrossRef]
39. Hayman, P.T.; Whitbread, A.M.; Gobbett, D.L. The impact of El Niño Southern Oscillation on seasonal drought in the southern Australian grainbelt. *Crop Pasture Sci.* **2010**, *61*, 528–539. [CrossRef]
40. Marshall, G.R.; Parton, K.A.; Hammer, G.L. Risk attitude, planting conditions and the value of seasonal forecasts to a dryland wheat grower. *Aust. J. Agric. Econ.* **1996**, *40*, 211–233. [CrossRef]
41. Zhao, M.; Hendon, H.H. Representation and prediction of the Indian Ocean dipole in the POAMA seasonal forecast model. *Q. J. R. Meteorol. Soc.* **2009**, *135*, 337–352. [CrossRef]
42. Marshall, A.G.; Hudson, D.; Hendon, H.H.; Pook, M.J.; Alves, O.; Wheeler, M.C. Simulation and prediction of blocking in the Australian region and its influence on intra-seasonal rainfall in POAMA-2. *Clim. Dyn.* **2014**, *42*, 3271–3288. [CrossRef]
43. Hudson, D.; Marshall, A.G.; Yin, Y.; Alves, O.; Hendon, H.H. Improving intraseasonal prediction with a new ensemble generation strategy. *Mon. Weather Rev.* **2013**, *141*, 4429–4449. [CrossRef]

44. Asseng, S.; McIntosh, P.C.; Wang, G.; Khimashia, N. Optimal N fertiliser management based on a seasonal forecast. *Eur. J. Agron.* **2012**, *38*, 66–73. [CrossRef]
45. Hayman, P.; Cooper, B.; Parton, K.; Alves, O.; Yong, G.; Henry, H.; Scheer, C. Can advances in climate forecasts improve the productive and environmental outcomes from nitrogen fertiliser on wheat? A case study using POAMA for topdressing wheat in South Australia. In Proceedings of the 17th Australian Agronomy Conference, Hobart, Tasmania, 20–24 September 2015; Australian Society of Agronomy: Hobart, Australia, 2015.
46. Rodriguez, D.; de Voil, P.; Hudson, D.; Brown, J.N.; Hayman, P.; Marrou, H.; Meinke, H. Predicting optimum crop designs using crop models and seasonal climate forecasts. *Sci. Rep.* **2018**, *8*, 2231. [CrossRef] [PubMed]
47. Power, S.; Casey, T.; Folland, C.; Colman, A.; Mehta, V. Inter-decadal modulation of the impact of ENSO on Australia. *Clim. Dyn.* **1999**, *15*, 319–324. [CrossRef]
48. Chapman, S.C.; Cooper, M.; Hammer, G.L. Using crop simulation to generate genotype by environment interaction effects for sorghum in water-limited environments. *Aust. J. Agric. Res.* **2002**, *53*, 379–389. [CrossRef]
49. Chenu, K. Characterizing the crop environment—Nature, significance and applications. In *Crop Physiology. Applications for Genetic Improvement and Agronomy*; Sadras, V.O., Calderini, D., Eds.; Academic Press: London, UK, 2015; pp. 321–348. ISBN 978-0-12-417104-6.

11

Resilience for Whom? The Problem Structuring Process of the Resilience Analysis

Hugo Herrera [1,2]

[1] Geography Department—System Dynamics Group, Bergen University, Fosswinckelsgate 6, 5007 Bergen, Norway; hugojhdl@gmail.com

[2] Department DEMS, University of Palermo, 90100 Palermo, Italy

Abstract: Resilience is a flexible concept open to many different interpretations. The openness of resilience implies that while talking about resilience, stakeholders risk talking past each other. The plurality of the interpretations has practical implications in the analysis and planning of resilience. This paper reflects on these implications that have so far not explicitly been addressed in the literature, by discussing the problem structuring process (PSP) of a modelling-based resilience analysis. The discussion is based on the analysis of food security resilience to climate change in Huehuetenango, Guatemala, jointly undertaken by the author, governmental authorities, small-scale farmers and academics of the national university. The aim of this discussion is to highlight the underestimated challenges and practical implications of the resilience concept ambiguity and potential avenues to address them. The contributions of the results presented in this paper are twofold. First, they show that, in practice, the resilience concept is constructed and subjective. Second, there remains a need for a participatory and contested framework for the PSP of resilience.

Keywords: food security; resilience; power; system dynamics; problem structuring process

---

1. Introduction

Climate change effects start to be recognised as threats to food system sustainability and food security [1]. Sustainability involves maintaining the functionality of the system without compromising its capacity to do so in the future [2]. However, undergoing effects of climate change compromises food system functionality by contributing to water scarcity and pest exacerbation [3]. Resilience is understood as the system adaptive ability of maintaining its functionality even when the system is being affected by a disturbance [4–6]. For this reason, resilience is a compelling framework for researchers and policymakers seeking to understand how socio-ecological systems (SESs) adapt and transform to withstand changes in the environment. In practice, resilience is often used as a measure of a SES's capability to respond and adapt to new conditions (e.g., climate change). Like Tendall et al. [2] (p. 18) describe, "sustainability is the measure of system performance, whereas resilience can be seen as a means to achieve it". Resilience has the potential to contribute to food security by enhancing farmers, and other stakeholders, capacity "for foreseeing and adapting to possible changes" [5] (p. 270). For instance, in the food systems literature, a number of studies have used resilience as framework for understanding how systems can adapt and transform in the presence of disturbances in the environment while still providing required amounts and quality of food [2,7].

Applications of resilience can be found in numerous disciplines, ranging from engineering to psychology to disaster risk management [8]. The increased popularity of resilience is due, at least partially, to the flexible meaning of the concept [8,9]. Resilience definitions have often been characterised as vague and unprecise in practical terms [2,9]. While the flexibility of resilience has moved it to the category of mainstream concepts and buzzwords, the same ambiguity represents a challenge to its application in prescriptive and normative settings. These challenges manifest when practitioners need to operationalise the concepts described in the literature to the context in which resilience will be applied. Unsurprisingly, different stakeholders of the analysed system have different and sometimes conflicting interpretations of what resilience means in practical terms.

Since each stakeholder interprets resilience differently, the scope of the analysis to be undertaken is not a given but is constructed through a problem structuring process (PSP). The term PSP is used in this paper to describe the "process by which a presented set of conditions is translated into a set of problems, issues sufficiently well-defined to allow specific research action" [10]. During the PSP, stakeholders interpret the available information in light of their values and knowledge and negotiate what is the purpose and the boundaries of the study to commence (referred to from now on as the "scope of the resilience analysis") [11,12]. The cognitive, social and political components, involved in the construction of the scope of analysis, condition its development and outcomes. The social and political nature of the PSP make it impossible to separate the conclusions and recommendations produced from the context in which they were produced. When talking about resilience, we cannot avoid the question: resilience for whom?

Literature has recently started to recognise some of the practical challenges of resilience ambiguity [2,9,13]; however, it still lags behind on recognising the political implications of resilience ambiguity in the analysis and its outcomes [8,14]. While some progress has been made by operationalising the definition of resilience (see for example [2,13]), resilience frequently continues to be presented as a "politically neutral approach" [9] (p. 134). The influence of stakeholders' agendas and power relationships are often overseen by practitioners [8,14]. Although these dimensions of the PSP have been discussed for a long time in the literature regarding problem structuring methods (PSMs), their implications for the resilience analysis are still unexplored.

This paper contributes to closing these gaps by discussing the political and social implications of resilience ambiguity in the PSP. To this purpose, this paper looks at the PSP of a modelling-based analysis of food security resilience to climate change. This case is used to discuss some of the cognitive and political challenges of resilience. This discussion is informed by the personal construct theory [15] and enriched by a post-normal science epistemology [16] for managing a wide range of perspectives. The aim of this discussion is to reflect on (a) the implications of having a diversity of resilience interpretations in the PSP and (b) the potential avenues to mediate stakeholder engagement and mitigate the challenges this diversity entails.

2. Case Study: Analysing the Resilience of Food Security to Climate Change in Guatemala

This research was conducted within the qualitative paradigm of case study research [17,18] and is part of an independent modelling-based discussion for the analysis of and planning for food security resilience to climate change in Guatemala. Specifically, this case study describes the PSP followed to define the scope of the resilience analysis undertaken in the district of Huehuetenango. As part of this PSP, the author conducted a series of semi-structured interviews among relevant stakeholders in the local maize production system.

2.1. Background

Guatemala, similar to other developing countries, faces food security challenges that will only increase as climate change affects small-scale farmers' capabilities to produce food. Guatemala's chronic malnutrition, an accepted measure of food insecurity, is one of the highest in the world [19], reaching 55% in rural areas [20]. Climate change effects, such as severe droughts and increased average temperatures, already compromise the food production in Guatemala, especially among small-scale farmers [21].

Recognising this as problematic, some studies that explore potential means to mitigate climate change effects have been commenced separately by academics, nongovernmental organisations (NGOs) and the local and central government in Guatemala. This research is part of these initiatives, independently conducted by the author with the cooperation of numerous stakeholders in the district of Huehuetenango.

Huehuetenango is located in the Northwest region of Guatemala, on the border with the South of Mexico. Huehuetenango is one of the poorest, most vulnerable districts in Guatemala. In 2014, its population was estimated at 1,150,000 people, with 67.6% of these people under the line of poverty [22]. Huehuetenango's main economic activities are the mining industry of silver and gold and the production of coffee [23]. Nevertheless, the production of maize is an important activity for self-consumption. The majority of the population is indigenous, from the ethnics of Mam and Quechi, with a cultural dependence on maize as the main source of calories. Among indigenous groups, maize represents a 71.2% of share in basic grains consumption).

2.2. Methodology

The intention of the study was to discuss potential policies to enhance food security resilience and to explore in an operational manner the impacts of these policies on different parts of the system. The author, with the support of two academics from the Universidad de San Carlos de Guatemala (national university in Guatemala), started by identifying (mapping) and engaging relevant stakeholders as early as possible and throughout the PSP. The following stakeholder groups accepted the invitation to participate in the PSP: (i) the central government; (ii) NGOs; (iii) farmers from Huehuetenango and (iv) academics and agronomists from the University. The number of delegates from each group and their backgrounds are presented in Table 1.

Table 1. Stakeholders' group representatives.

Stakeholder Group	Number of Delegates Participating	Background
Central Government (CG)	4	Agronomists Policymakers
Non-Governmental Organization (NGO)	3	Agronomist Project Managers
Farmers (F)	6	Maize Farmers
Academics (AC)	2	Agronomist Professor Researcher

During the PSP, the author conducted semi-structured interviews to gather stakeholders' perspectives about the food security resilience of the small-scale maize production system of the region. In the first part of the interviews, the author asked the delegates of the different stakeholder groups about the agendas they have for the local food system. Subsequently, causal loop diagrams (CLDs) were used to capture stakeholders' broad understanding of the underlying causes of system vulnerability (the extent to which the system will be affected by) climate change. Finally, the delegates were also asked to rank the stakeholders in the system in terms of influence on and interest in the local food system.

The elicitation of stakeholders' agendas for the local food system was done by discussing the following general questions with the delegates of each stakeholder group:

- What would you like to get from the small-scale maize production system?
- In this context, what does resilience of food security to climate change mean?
- What are the critical success factors of policies enhancing food security?

After the interviews, the author compiled and summarised the different answers. Similar answers were grouped in the same variable or short statement to simplify further analysis. The resulting statements were discussed in further interviews with each delegate to ensure they reflected their own perspectives. When needed, changes were made and again discussed with the specific delegate requesting the change.

Beside the narratives provided by the delegates, this paper uses CLDs as a means for capturing stakeholders' assumptions. CLDs are diagrams representing, in a simple manner, a possible set of causal relationships between different variables of the systems [24,25]. CLDs are particularly useful for identifying circular relationships known in the systems' literature as feedback loops. The rigor of diagramming forces the participants to "carefully and consistently" make their assumptions explicit and to "put their problem definition to test" [26] (p. 384). Thus, CLDs are a suitable way to represent and compare different interpretations of the problem and the causal explanations held by the stakeholder groups participating in the PSP.

CLDs might be employed in the PSP (also known as the conceptualisation stage of the modelling process) [27,28] to elicit participants' understanding of the problem. During the conceptualisation, the modeller focuses on "a verbal description of the feedback loops that are assumed to have caused the reference mode" [19] (p. 119). Namely, in this paper, the CLDs were used to diagrammatically represent the causal explanations for the lack of resilience of food security in the region. This elicitation might be done, as it was in the case of this paper, during one to one interviews with experts in the field, in our case an agronomist from the university, and stakeholders of the problem at hand.

During the semi-structured interviews, the author drafted CLDs representing what the delegates were describing. The author started by asking the delegates what were the main causes of the decrease and fluctuations of the affordability of maize (as a measure of food security [29]) experienced in the past 10 years in the region of Huehuetenango (see Figure 1). The causes stated by the delegates were summarized by the author in relevant variables while transcribing them to the diagram. Then, the author asked delegates to explain how those variables influenced each other. These causal links between different variables were represented in the diagram by arrows connecting the cause with its effects. When needed, new variables were added to the diagram.

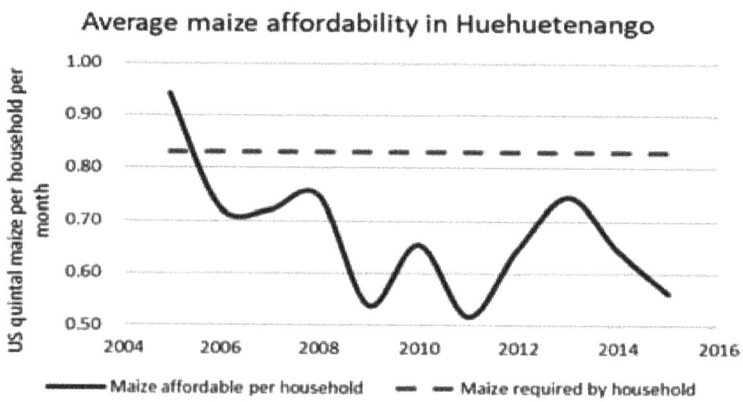

Figure 1. Maize affordability in Huehuetenango.

At the end of the interview, the delegates were asked to complete the CLDs drafted by the author by adding variables, causal relationships or any elements missing in the diagram. Later, the author worked on his own by summarising all the CLDs produced by each delegate into a single CLD per stakeholder group. The single CLDs were validated and discussed with the delegates of each stakeholder group in separate interviews to ensure all of their views were appropriately captured in the diagrams. If participants found important issues missing in the diagram, those issues were added to the final version.

Finally, delegates were asked to characterise the different stakeholders in the system. To be precise, participants were asked to rank from 1 (low) to 5 (high) the level of influence each stakeholder group has on the local food system. During this characterisation, participants were invited to consider in their assessment what resources each stakeholder can allocate for this purpose and the level of organisation and the reputation of each. Similarly, participants were asked to rank the stakeholders from 1 (low) to 5 (high) according to their interest in the problem (i.e., resilience of food security). The author tabulated the results into a single chart showing the average level of influence of each stakeholder group.

Analytical Framework

The results were analysed in light of personal construct theory (PCT) [15]. PCT is based on the assumption that a person needs to make sense of the problem to address it: "a person's processes are psychologically channelized by the ways in which he anticipates events" [30] (p. 7). Thus, to analyse resilience, stakeholders need first to make sense of what resilience means. To illustrate how this cognitive process unfolds, this paper adapts the simplified model proposed by Eden [31] to examine how stakeholders construct their own interpretations of resilience (see Figure 2). According to Eden's [31] model, stakeholders make sense of the concept of resilience by selecting particular elements that are applicable to the problem at hand and its context. This perception is then filtered through the individual system of values and beliefs to articulate its own interpretation of what resilience means in practical terms. This separation of selective perception and construal follows the personal construct theory of Kelly [15].

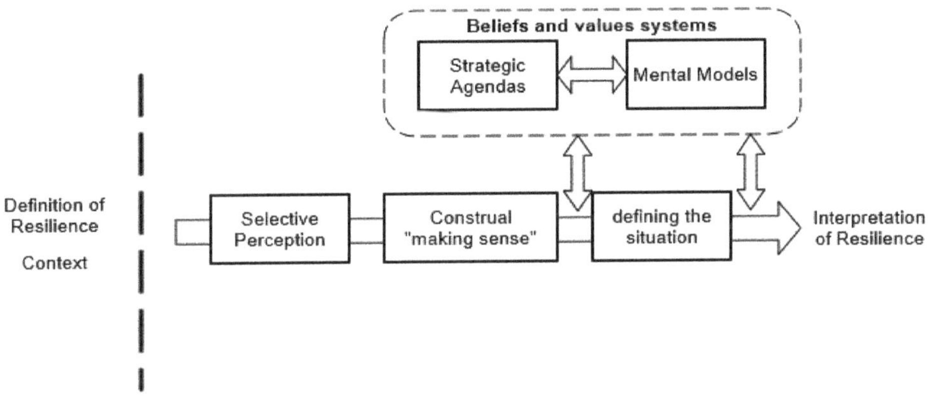

Figure 2. Construction of stakeholders' interpretation of resilience. Note: Adapted from Eden [31].

There is no clear distinction between values and beliefs, as they are closely interconnected [31]. However, for analysis purposes, this paper explores two separate interconnected aspects of the beliefs and values systems: strategic agendas and mental models. The term strategic agenda is used here to describe the set of goals each stakeholder has for the system. Similarly, the term mental model is used to describe the conceptual representations each stakeholder has about how the system works [32]. Strategic agendas and mental models are not separate entities. They support each other, and together, they are supported by wider individual value systems [31].

In policymaking settings, closely linked to the understanding of what resilience means in practical terms, is the concept of adaptability or the "the capacity of actors in the system to influence resilience" [33] (p. 5). Stakeholders' adaptive actions depend on how they perceive the disturbance is changing the conditions of their system. Since timing, magnitude and origin of the disturbance are, at least to some extent, unpredictable, the nature of the change that the disturbance produces deviates from the normal system-near-equilibrium analysis [34]. In these conditions of high uncertainty, identifying the mechanisms driving adaptation is not straight forward but depends on the stakeholder's mental models about how the system works.

To analyse how stakeholders' understand the system, this paper uses the reflections of Mayumi and Giampietro [35] about self-modifying systems and the theories of Funtowicz and Ravetz [36] on emergent systems. According to the aforementioned sources, the explanations each stakeholder group gave to the system behaviour were classified into:

(a) endogenously driven: the observed effects of disturbances affecting the system are the result of the functional links between its different elements. Adaptation emerges from the mechanisms the system has to regulate itself and can only be enhanced by strengthening them [35]. The solution to the problem is within the system boundaries.

(b) exogenously driven: the disturbance affecting the system comes from outside the system and, to adapt to the new conditions introduced, the system needs of external interventions that "push" it back to its equilibrium state. The solution is outside the system boundaries.

(c) chaos: the uncertainty about the disturbance affecting the system and complexity of the system itself are perceived so high that it is impossible to identify links between actions (outside or within the system) and their consequences. The solution is unknown.

This classification offers a helpful analytical framework to explain how delegates from different stakeholder groups understand the system and the differences in the policies they will propose in further stages.

Nonetheless, the agendas and mental models used by stakeholders to construct their own interpretation of resilience are only some of the ingredients for the scope of the resilience analysis. The manifestation of power in the PSP is indeed critical analytical lens to understand complications of resilience ambiguity. In fact, power effect on resilience is one of the most unexplored but most contested characteristics of resilience [14].

Case study research shows that in prescriptive settings, the PSP of resilience is predominantly a negotiation endeavour. For instance, Lebel et al. [37] describe that in many case studies undertaken by the Resilience Alliance, the scope of resilience analysis reflects, to a large extent, the interest of powerful stakeholders, undermining perspectives of ethnic minorities and small-villages (powerless stakeholders). Similarly, Larsen et al. [38] highlight the tensions regarding roles, control and ownerships between powerful stakeholders during the process of building resilience in Thailand tourism-dependent communities.

These cases studied in the literature show that during the PSP of resilience, stakeholders will try to persuade the others to join or accept their own interpretation of resilience and to articulate the scope of the resilience analysis accordingly. As illustrated in Figure 3, the scope of analysis is a negotiated outcome of the PSP that reflects not only the interpretations of each stakeholder in the system but also the power relationships between them.

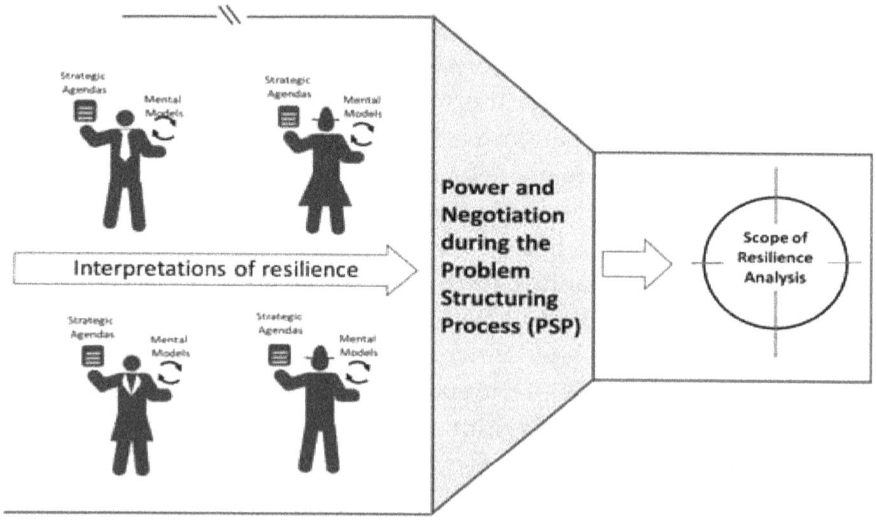

Figure 3. Simplified representation of the problem structuring process (PSP) of resilience analysis.

2.3. Results

2.3.1. Strategic Agendas

Tabulated results from the interview show that delegates from the same group coincide to a large extent in the answers they provided about their agendas. Table 2 summarizes these tabulated answers. In Table 2, it is noticeable that most of the delegates of the same group agreed on a similar answer.

Table 2. Summarized answers to the semistructured interviews.

	Delegate Code	CG1	CG2	CG3	CG4	NGO1	NGO2	NGO3	AC1	AC2	F1	F2	F3	F4	F5	F6
What would you like to get out from the small-scale maize production system?	Increase households' wealth	X	X		X										X	
	Produce revenues			X		X	X	X		X			X			
	Produce food								X		X				X	
	Produce food for locals						X			X		X	X	X		X
In this context, what resilience of food security to climate change means?	Being able to afford food even when droughts	X	X	X				X								
	Produce food constantly in despite of the droughts	X			X	X	X		X	X		X			X	
	Don't starve during the bad years								X							
	Have always enough food										X	X	X	X	X	X
What are the critical success factors of policies enhancing food security?	Money available for purchasing food	X	X	X												
	Crop productivity		X		X	X		X								
	Maize Yield				X		X		X	X			X			
	Maize reserve										X	X	X	X	X	X

Note: CG: Central Government; NGO: Non-Governmental Organization; AC: Academics, F: Farmers.

Based on the interviews results, the strategic agenda held by each stakeholder group can be summarized as follow:

Central Government (CG): The purpose of the analysis is to identify how to increase the household's wealth and particularly the money available to buy food so that households can afford enough food even when droughts reduce the yields of maize in the region.

Non-Governmental Organization (NGO): The purpose of the analysis is to identify how to enhance crop productivity so that households can produce food and revenues constantly despite the droughts. Note that in the words of the NGO delegates, crop productivity is understood as the amount of crop (not exclusively maize) produced from each Guatemalan Quetzal invested by the farmers.

Academics (AC): The purpose of the analysis is to identify how to increase maize yields and reserves as a mean to prevent starvation by increasing farmers revenues and food supply to the region.

Farmers (F): The purpose of the analysis is to identify how to increase food production (not limited to maize or crops in general) and maize reserves to have food year round.

2.3.2. Causal Loop Diagrams

Figure 4 presents the CLD's prepared jointly by the author and delegates of each group. In general, diagrams are relatively simple and focused (with the exception of diagram in Figure 4c) on one or two main causal explanations of the problem to address (decrease and fluctuations of maize affordability in the region).

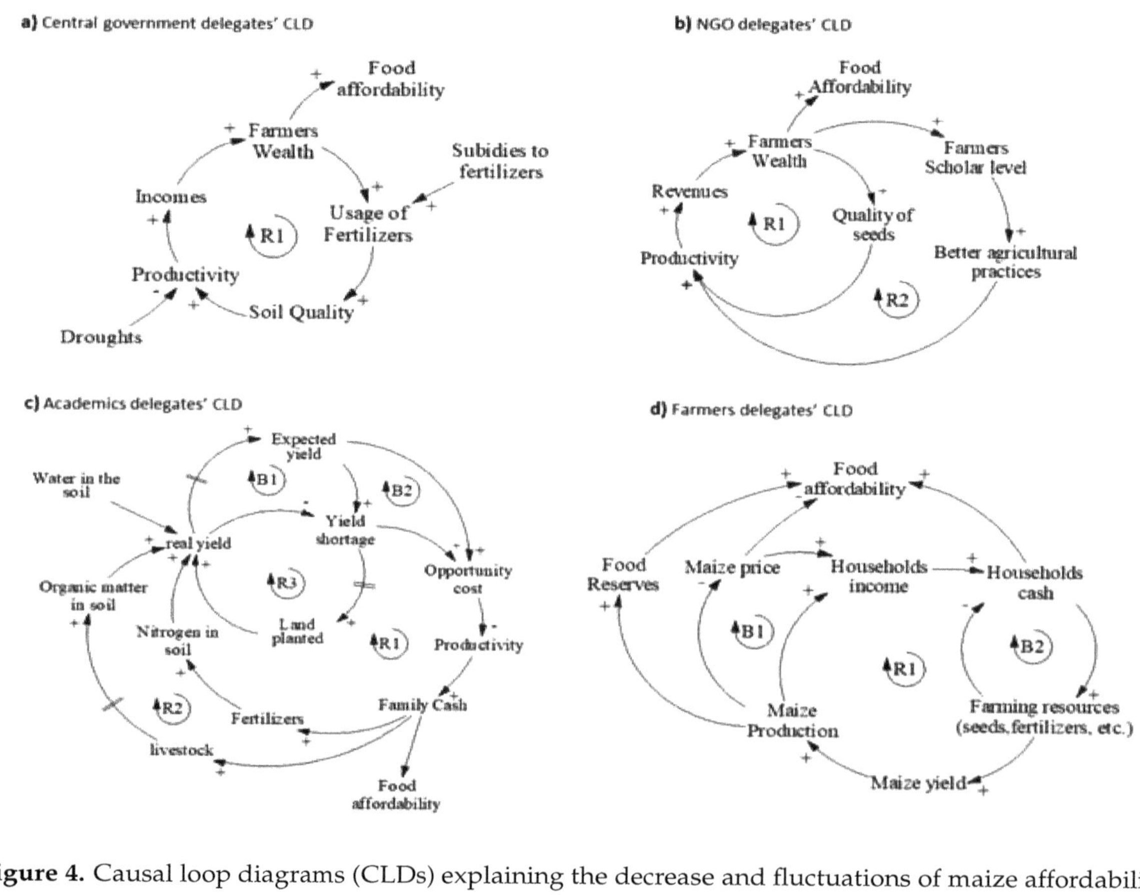

Figure 4. Causal loop diagrams (CLDs) explaining the decrease and fluctuations of maize affordability in Huehuetenango. CLDs were produced by (**a**) central government delegates; (**b**) NGO delegates; (**c**) academics delegates and (**d**) farmers delegates; during one-to-one semi structured interviews.

Next, there is a brief explanation of each diagram.

Central Government (CG): Farmers productivity increases the incomes and, therefore, the wealth of the farmers. Higher wealth increases farmers' capacity to use fertilizers (fertilizers are more affordable). Usage of fertilizers is directly related to the productivity and, therefore, the more fertilizers the farmers use the more productive they become in a virtuous cycle represented by the R1 feedback loop in Figure 4a. This loop, however, is perturbed by droughts (disturbances of the system) that reduce farmers productivity, reducing their overall wealth and hence their capacity to acquire food (food affordability).

Non-Governmental Organization (NGO): Farmers productivity increases the incomes and therefore the wealth of the farmers. Higher wealth increases farmers' capacity to access better seeds and formal education. Seeds of improved varieties, the ones that require less water, are assumed to increase crop productivities, especially during drought seasons, compared to seeds coming from informal sources (on farm save seeds for example). Better seeds increase wealth in the virtuous cycle represented by R1 in Figure 4b. Access to formal education is assumed to be linked to better agriculture practices (e.g., appropriate usage of fertilizers and land planning). Better agriculture practices increase revenues and wealth in the virtuous cycle represented by R2 in Figure 4b.

Academics (AC): The causal explanation represented in Figure 4c focuses on the variation of the real yield against the expected one (yield shortage in the diagram). Yield shortage results into lower productivities and opportunity costs that reduce families' cash and their capacity to invest in fertilizers and livestock (see feedback loops R1 and R2 in Figure 4c). Higher yield shortage also translates into a reduction of the land planted each season (see R3 in Figure 4c), because farmers need to spend more time on other activities (e.g., working on coffee plantations) and less time farming. The expected yield eventually gets adjusted, decreasing the yield shortage and opportunity costs (see loops B1 and B2 in Figure 4c). The increase in droughts occurrence increases yield shortage by affecting the maize system and its real yield, reducing at the same time the land planted and the cash available for the next season's harvest.

Farmers (F): Maize production increases incomes and households' cash, allowing farmers to acquire more resources needed in farming activities (seeds, fertilizers, etc.). This eventually increases the maize production. Higher production results in a) higher food reserves and b) higher incomes (see feedback loop R1 in Figure 4d). However, there are two drawbacks from the feedback loop R1. First, the acquisition of resources decreases households' cash (see feedback loop B1 in Figure 4d) thereby reducing the food affordability. Second, higher production will eventually translate into lower maize prices, reducing farmers' income and profit margins (see feedback loop B2 in Figure 4d).

2.3.3. Influence-Interest Grid

Figure 4 presents the stakeholders' grid produced by the delegates. The four stakeholder groups participating in this case study were consistently identified by all the delegates as those with the highest interest in the problem (see Figure 5). The central government and NGO working in the area were described as being the stakeholders in a better position to solve the problem or those with higher influence on the problem (see quadrant I in Figure 5). Other stakeholders, like the large-scale farmers producing food in the region and traders, were also recognised as highly influential. However, there was an agreement among delegates of all the stakeholder groups participating that, unfortunately, large-scale farmers and maize traders have no interest in enhancing food security in the region (quadrant IV in Figure 5). While recognized as those with higher interest in the problem, small-scale farmers were portraited as the group with the lowest influence on it (see quadrant II in Figure 5). Academics and stakeholders not participating in the PSP (local government) were also portrayed as interested parties with low influence.

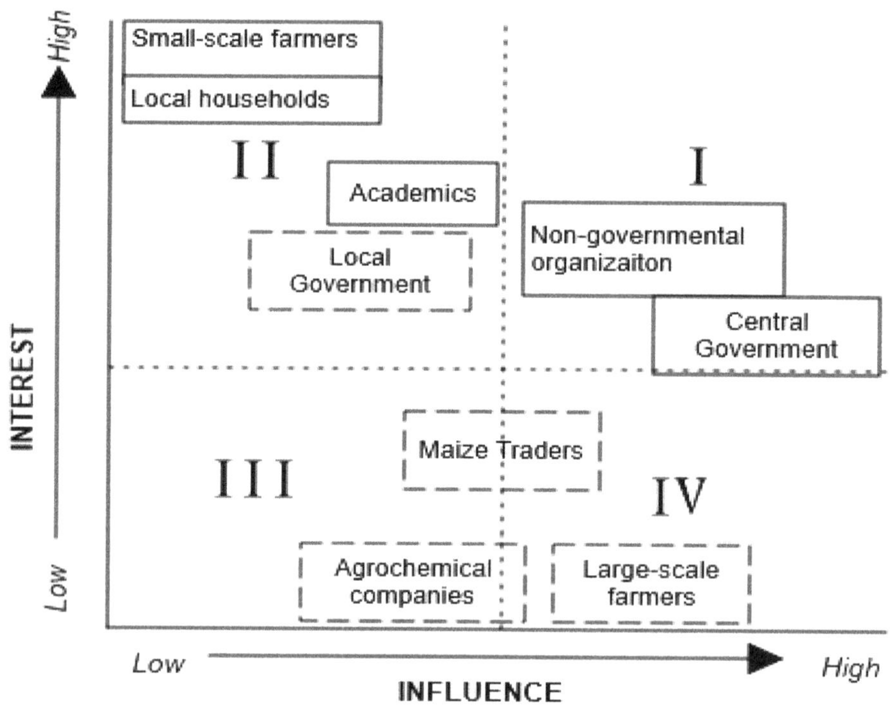

Figure 5. Influence/interest diagram summarizing stakeholders rank in the small-scale maize production system in Huehuetenango. Note: stakeholders in dotted lines did not take part in this research.

3. Complications of the PSP in the Analysis of Resilience

The results presented in Section 2 offer relevant evidence to discuss the ambiguity of resilience and its complications. The ambiguity of resilience, in this case, does not arise from the differences between many definitions of resilience [13], but from the way in which stakeholders interpret it for their specific context and problem. The differences that emerged during the PSP might already be noticeable for the reader, but the analytical lenses proposed in this paper offer a perspective of the deeper and more conflicting differences in the agendas and mental models held by each stakeholder group.

These cognitive differences set the scene for analysing the conflict that could unfold during the negotiation of a single scope of resilience. The more mutually exclusive agendas and mental models are, the harder it is to reach a scope of resilience that satisfies all the stakeholders. As Eriksen et al. [39] pointed out, adaptation will change social, political and economic relationships between stakeholders, "yet not all these changes are desirable for everybody".

This section concludes by discussing the practical implications of resilience ambiguity in the policymaking process. These implications are not only political but also methodological and require thoughtful planning of the PSP. While it might be possible to mitigate some drawbacks, more research is needed before outlining a comprehensive framework for addressing the political challenges that resilience entails.

3.1. Constructing an Interpretation of Resilience

The experience in the district of Huehuetenango in Guatemala shows that different stakeholders have different interpretations of resilience. These interpretations of resilience are context specific [40,41] and reflect the values and beliefs of the stakeholders involved. In other words, stakeholders make sense of what resilience means in their particular context and frame the analysis process accordingly. In this case study, different interpretations of resilience are reflected in (a) the different goals and desired outcomes (strategic agenda) stated during the interviews (see Table 2) and (b) the different descriptions of the causes of the problem (mental models) captured in the CLDs (see Figure 1).

When looking at the strategic agenda, stakeholders see the maize production system at different levels of aggregation (household level vs. regional level). As presented in the results section, delegates from the same stakeholder group share similar perspectives about the purpose of the system (see Table 2). With the exception of the farmers, the groups also share some alignment among themselves. The answers in Table 2 and summarized strategic agendas in the Results section show that most of the delegates have local/regional goals for the system, namely to promote local economic development. Alternatively, farmers focus on their current urgent problem of living in insecure food conditions.

In other words, there are two main strategic agendas for the system. One agenda (shared by many stakeholders) is seeking to use the system as a tool for local and/or regional economic development. The other agenda, held by the farmers, is to have food all year round. While there might be different arguments in favour of one agenda over the other one, it is unlikely that regional solutions will have any impact unless urgent issues challenging the farmers' own subsistence are addressed. Similarly, small-scale solutions, addressing farmers immediate needs, might prove to be unsustainable in the mid-term if the wider problem is not tackled.

Wider differences are found when looking at stakeholders' mental models reflected in the CLDs developed. Academics and NGO delegates describe the system in endogenous terms. This endogenous perspective is reflected in the feedback loops identified in the CLD they drafted (see Figure 4b,c). They look at the problem in a systemic way and try to find solutions within the system boundaries. They have, however, a different understanding of the vicious circles constraining food security. On the one hand, academics focus on the management of the water resources and reservoirs as a potential leverage point.

> *"The obvious cause of the problem is the deficiencies the communities face to access water .... This is why that, now that droughts are becoming more common, farmers face more problems."* (Academic delegate 1)

On the other hand, NGO delegates blame farmers' lack of technical skills and training as the cause of their poor productivity and, hence, food insecurity. The solution they propose is to increase training and to provide farmers with better seeds to increase their productivity in a sustainable way.

> *"You see, there are several complications in the situation of these poor people because their culture doesn't let them move forward. They use the same techniques they have been using since pre-colonial times. They have no formal education. You know that most of them cannot read. It is really difficult to teach them and change their minds. We need to make an effort to provide them with the right seeds and the proper instruction to use them well."* (NGO delegate 2)

The government delegates describe the system as exogenous driven. These delegates think the way to influence the system is through the artificial enhancement of farmers' productivity (see Figure 4a). Even though they identified a feedback loop in the system, their proposed solution focuses on ways to quickly boost the system performance, namely by using more fertilizers to increase productivity.

> *"The government is committed to provide a sustainable and plausible solution by providing the fertilizers they (farmers) need to increase their productivity and become more competitive .... Once they (farmers) level up with the market, the food affordability should be a natural condition."* (Central government delegate 2)

Farmers perceive the problem in a very different way. In their perspective, the increasing uncertainty about rainfall is transforming the system into a chaotic one. From their perspective, using more expensive seeds or more fertilisers will be useless if the weather conditions are not good. Farmers do not feel in control of the system. They feel they are victims of the uncertainty of the yields that they will get at the end of the season.

> *"The problem is you don't know if the yield is going to be good or not . . . . Now you never know . . . . If the yield goes bad, we lost the money we spent on seeds and fertilizers."* (Farmer delegate 4)

> *"The weather now cannot be predicted . . . . You gamble every time you plant."* (Farmer delegate 1)

Furthermore, the farmers do not see higher production as a means to increase their revenues but only as a means to increase their food reserves (see Figure 4d). In their view, the region is isolated, and they do not have access to other markets to trade. The benefit they perceive from higher production is in having more maize to build food reserves for the future.

Understanding and acknowledging different goals and mental models about the system will lead to a wider scope of analysis and might result in a more balanced decision-making process [34]. Short-term solutions and systemic interventions could provide a balanced view between achieving short-term outcomes and their long-term consequences. Farmers' chaotic view of the world challenges the mechanistic understanding other stakeholders might have and balances their deterministic view by the acknowledgement of uncertainty. The system cannot be assumed mechanistically following economic rules since human behaviour under stressful situations adapts in sometimes unexpected ways [7]. An oversimplified understanding about how different groups will react during a crisis might lead to policy failure [42]. For instance, while most of the stakeholders expect farmers to use a potential production surplus to increase their revenues, farmers will use it to increase their food reserves, affecting the policy's effectiveness.

3.2. Negotiating the Scope of Analysis

The power to influence the final outcome is not symmetrical among stakeholders, with those holding key resources being in an advantageous position to impose their own interpretations in the final scope. System adaptation will "influence social relations, governance and distribution of resources in any given population or place" [39] (p. 2). However, as shown in this case, there is not always agreement about the changes and the scale at which those changes should be made. Those with higher level of influence in the scope of analysis might not be those directly affected by its outcomes. For instance, the small-scale farmers in Huehuetenango are the stakeholders directly affected by potential decisions about how to enhance resilience, but they are also those with the least influence on the decision-making process (see Figure 5).

Power differences have contentious repercussions considering that those with a higher level of influence have different strategic agendas than those suffering the larger impacts of the policies implemented. This is particularly relevant since there is a clear difference between the farmers' interpretations and those held by the rest of the stakeholders. Considering the different interpretations of resilience, the power to set agendas about what issues are to be addressed needs to be an important consideration during the PSP.

Competitive agendas and mental models set the scenario for a game of power where different stakeholders seek to impose their own agendas on the scope of the analysis that will follow. The allocation and distribution of the access to natural resources have been, historically, an expression of power tension between different groups [14,39]. While building the resilience of the system outcomes, the resilience of the institutions and relationships defining those outcomes are also enhanced [43]. Many stakeholders perceive the resilience analysis as an opportunity to gain power or to influence the system towards their own interests [14]. This power might be exercised in many ways. For instance, stakeholders might scope the problem in isolation, ensuring their interpretations are the only ones represented. Alternatively, some groups could try to undermine those with competitive or opposite views by diminishing their credibility as shown in this case. For instance, note the comment above from NGO delegate 2 in which the delegate undermines farmers' practices because they have no formal education. Any analysis that does not account for these tensions would result in an incomplete understanding of the scope of potential responses [14,44].

In short, recognising that there might be different interpretations of resilience implies accepting the PSP as a negotiation and political process. Seeing the PSP as a negotiation forum means that practitioners need to acknowledge the social and political factors (e.g., inequality and legitimacy) shaping the scope of analysis and need to be transparent about the implications of these factors on their recommendations. Otherwise, the resilience analysis risks being used, possibly inadvertently, as a way to legitimise the power of particular groups and to impose particular means to manage natural resources [43].

3.3. *What Are the Potential Implications?*

There are at least two implications resulting from the flexibility of resilience to interpretation. First, it seems unlikely that a proper analysis would result in a PSP that does not account for the many different interpretations of resilience in each particular context. If the scope of analysis has been defined by only a few groups, it risks being too narrow, excluding important elements from the analysis and reducing the range of solutions explored. For instance, the analysis might focus on short-term solutions, ignoring important feedback loop mechanisms of the system. Alternatively, a pure systemic view of the problem might fail to recognise uncertainty and might oversimplify decision rules and human behaviours.

Second, stakeholders who have a different understanding of the problem will rarely support or get actively engaged in the implementation of a solution that is not addressing their initial understanding of the problem [45]. The contribution of any solution is null if those ultimately responsible for implementing them are not willing to do so [46]. For instance, stakeholders might sabotage the policies proposed at the end of the analysis by refusing to participate in the implementation (e.g., training and the introduction of new practices) or, even worse, by explicitly opposing them (e.g., demonstrations against the introduction of new seeds).

3.4. *Potential Avenues for Mitigation*

Recommendations are not conclusive, but it is possible to outline avenues for further development with the aim of reducing the potential drawback of power in the PSP. A possible avenue is to advocate for more participatory settings. So far, the SES literature has extensively discussed stakeholders' participation as a requirement for the enhancement of resilience in the SES. However, very little has been elaborated on the role of participation in the formulation of the problem as such. Facilitated modelling approaches, such as Group Model Building [47] or Cognitive Mapping [48], might contribute to mediating this process (e.g., by introducing the CLD as a transitional object that helps to leverage power differences) [49,50]. These methods contribute in leveraging the power between groups by forcing participants to make their assumptions explicit in a diagram that is challenged by the group [47,51,52]. In this case, the diagram is used to jointly represent the problem definition shared by and agreed upon by different stakeholders through a process of negotiation and dialogue [49].

Alternatively, another option is to aim for a broader perspective in the analysis of resilience and to consider possible trade-offs and asymmetries in resilience between different groups and communities within the system. A broader perspective might be particularly useful when there is a conflict between long-term and short-term goals or when the boundaries of the system are not clear [53]. By using computer simulations, for example, it is possible to uncover long-term unintended consequences that might result from short-term perspectives. Uncovering unintended effects is possible because computer simulations are especially useful when the delays between the policies and their results are too large to allow for assessment by simple intuition. Simulations might also uncover unexpected and unintended consequences of policies that are beneficial to one group but negative for others.

The latter is particularly important when analysing climate change problems because there are time lags or delays between policy measures (or non-action), and effects often extend beyond the normal period of analysis [54,55]. When important consequences of current policies materialise several years later (in some cases decades later), significant future stakeholders will not be present to voice their concerns and weigh in when preferences are aggregated into policy decisions. Present stakeholders might be willing to compromise the overall future detriment of the system for short-term benefits. Namely, in the resilience analysis, present stakeholders might favour policies that yield more efficiency in the short term but diminish the capability of the system to continue providing the desired outputs in the long term. The benefits for the few who are defining the problem now might be preferred over the benefits for the many tomorrow.

4. Conclusions

The ambiguity of resilience is a challenge for practitioners that want to implement it as an analytical and policymaking framework in real life problems. This paper addresses the ambiguity of resilience from a cognitive and political perspective by focusing on how resilience is interpreted in practice instead of its theoretical definition. This paper argues that the interpretation of what resilience means in a specific context (resilience of what?) and the ways to achieve it are results of the values and beliefs of those with a stake in the system. In this light, the case study presented methods to identify and highlight some of the challenges and practical implications of resilience ambiguity. Specifically, this paper focuses on strategic agendas and mental models as observable expressions of stakeholders' values, beliefs and knowledge about the system. The results discussed in this paper show that, in practice, different agendas and mental models compete during the PSP to be part of the scope of resilience analysis. The question of what outcome of the system needs to be resilient has many answers (revenues, yield, food supply).

The results presented in this paper show that stakeholders have different understandings of how the system works. For instance, while academics and delegates from the NGO participating in the study focused on enhancing virtuous cycles within the system, the central government delegates proposed solutions outsides the system's boundaries. All of these solutions, however, ignored the bounded rationality of the farmers and the premises of their decision-making process. Including only a few stakeholders in the process risks leaving many important aspects out of the scope of the analysis and therefore undermining its results.

It is also necessary to acknowledge the role of power shaping and filtering different interpretations of resilience into a formal scope of analysis. It is expected that those with more power will attempt to influence the PSP to reflect their views and agendas. In the case presented in this paper, farmers have little influence in the PSP and their agendas might, intentionally or accidentally, be bypassed by experts (e.g., academics and researchers) and policymakers. For instance, as discussed in this paper, farmers bounded rationality and socioeconomic position might be used as an argument for disregarding their knowledge and their claims.

In short, results show that the practical meaning of resilience is socially constructed by those participating in the PSP and the way this process is conducted will affect the result of the analysis. There are at least two practical implications of underestimating resilience ambiguity while structuring the scope of the resilience analysis. First, including only a few stakeholders in the process risks leaving many important aspects of the system out of the scope of the analysis to be undertaken. Second, poor stakeholder management also risks obstructing the implementation of proposed policies and, in the worst case, unintentionally harming those in more vulnerable positions. While literature starts to acknowledge the challenges and contentious implications of power in the resilience analysis (see for instance [7,14,39]), more research is needed toward defining a framework of how to facilitate negotiation during the PSP.

If resilience is to play a significant role in climate change adaptation, policymakers should be careful when structuring the scope of the resilience analysis and should seek for broader participation. Such broadening is not a simple case of bringing more perspectives. Instead, it is a "fundamental shift in how knowledge is understood to operate and consequences of this for the kinds of questions we formulate prior to our analyses" [14] (p. 484). Increasing participation is not a normatively uncontroversial route either, but at least it acknowledges that resilience-based policy solutions and institutions will have distributional and, thereby, moral consequences (as most other forms of public policy do).

Acknowledgments: To the stakeholders in Huehuetenango who participated in this research and kindly agreed to share their perspectives in this paper.

References

1. Food and Agriculture Organization of the United Nations (FAO). *2016 The State of Food and Agriculture. Climate Change, Agriculture and Food Security*; FAO: Rome, Italy, 2016.
2. Tendall, D.M.; Joerin, J.; Kopainsky, B.; Edwards, P.; Shreck, A.; Le, Q.B.; Kruetli, P.; Grant, M.; Six, J. Food system resilience: De fining the concept Resilience Sustainability. *Glob. Food Secur.* **2015**, *6*, 17–23. [CrossRef]
3. Campbell, B.M.; Vermeulen, S.J.; Aggarwal, P.K.; Corner-Dolloff, C.; Girvetz, E.; Loboguerrero, A.M.; Ramirez-Villegas, J.; Rosenstock, T.; Sebastian, L.; Thornton, P.K.; et al. Reducing risks to food security from climate change. *Glob. Food Secur.* **2016**, *11*, 34–43. [CrossRef]
4. Walker, B.; Carpenter, S.; Anderies, J.; Abel, N.; Cumming, G.; Janssen, M.; Norberg, J.; Peterson, G.D. Pritchard, R. Resilience Management in Social-ecological Systems: A Working Hypothesis for a Participatory Approach. *Conserv. Ecol.* **2002**, *6*, 14. [CrossRef]
5. Holling, C.S.; Gunderson, L.H. *Panarchy: Understanding Transformations in Human and Natural Systems*; Island Press: Washington, DC, USA, 2002.
6. Gallopín, G.C. Linkages between vulnerability, resilience, and adaptive capacity. *Glob. Environ. Chang.* **2006**, *16*, 293–303. [CrossRef]
7. Maleksaeidi, H.; Karami, E. Social-ecological resilience and sustainable agriculture under water scarcity. *Agroecol. Sustain. Food Syst.* **2013**, *37*, 262–290. [CrossRef]
8. Duit, A. Resilience Thinking: Lessons for Public Administration. *Public Adm.* **2015**, *92*, 364–380. [CrossRef]
9. Pizzo, B. Problematizing resilience: Implications for planning theory and practice. *Cities* **2015**, *43*, 133–140. [CrossRef]
10. Woolley, A.R.N.; Pidd, M. Problem Structuring? A Literature Review. *J. Oper. Res. Soc.* **1981**, *32*, 197–206.
11. Shaw, D.; Westcombe, M.; Hodgkin, J.; Montibeller, G. Problem structuring methods for large group interventions. *J. Oper. Res. Soc.* **2004**, *55*, 453–463. [CrossRef]
12. Weingart, L.R.; Bennett, R.J.; Brett, J.M. The impact of consideration of issues and motivational orientation on group negotiation process and outcome. *J. Appl. Psychol.* **1993**, *78*, 504–517. [CrossRef]
13. Quinlan, A.E.; Berbés-Blázquez, M.; Haider, L.J.; Peterson, G.D. Measuring and assessing resilience: Broadening understanding through multiple disciplinary perspectives. *J. Appl. Ecol.* **2016**, *53*, 677–687. [CrossRef]
14. Cote, M.; Nightingale, A.J. Resilience thinking meets social theory, Situating social change in socio-ecological systems (SES) research. *Prog. Hum. Geogr.* **2012**, *36*, 475–489. [CrossRef]
15. Kelly, G. The nature of personal constructs. *Psychol. Pers. Constr.* **1955**, *1*, 105–183.
16. Funtowicz, S.; Ravetz, J.R. Uncertainty, complexity and post-normal science. *Environ. Toxicol. Chem.* **1994**, *13*, 1881–1885. [CrossRef]
17. Merriam, S.B. Introduction to qualitative research. In *Qualitative Research in Practice: Examples for Discussion and Analysis*; John Wiley and Sons: New York, NY, USA, 2002.
18. Stake, R.E. *The Art of Case Study Research*; Sage: London, UK, 1995.
19. World Food Programme. *Countries-Guatemala*; World Food Programme: Guatemala, Guateamala, 2016.
20. Guardiola, J.; Cano, V.G.; Pol, J.L.V. *La Seguridad Alimentaria: Estimación de Índices de Vulnerabilidad en Guatemala*; VIII Reunión de Economía Mundial; Universidad de Alicante: Alicante, Spain, 2006.
21. Bouroncle, C.; Imbach, P.; Läderach, P.; Rodirguez, B.; Medellin, C.; Fung, E.; Martinez-Rodriguez, R.; Donatti, C.I. *La Agricultura de Guatemala y el Cambio Climático: ¿Dónde Están Lasprioridades Para la Adaptación?* CGIAR Research Program on Climate Change, Agriculture and Food Security (CCAFS): Copenhague, Dinamarca, 2015.

22. Instituto Nacional de Estadistica (INE). *Caracterización Estadística República de Guatemala 2012*; INE: Guatemala, Guatemala, 2012.
23. Camposeco, M.; Thomas, M.; Kreynmar, W. *Huehuetenango en Cifras*; Centro de Estudios y Documentación de la Frontera Occidental de Guatemala (CEDFOG): Guatemala, Guatemala, 2008.
24. Lane, D.C. The Emergence and Use of Diagramming in System Dynamics: A Critical Account. *Syst. Res. Behav. Sci.* **2008**, *25*, 3–23. [CrossRef]
25. Richardson, G.P. Problems with causal-loop diagrams. *Syst. Dyn. Rev.* **1986**, *2*, 158–170. [CrossRef]
26. Vennix, J.A.M. Group model-building: Tackling messy problems. *Syst. Dyn. Rev.* **1999**, *15*, 379–401. [CrossRef]
27. Luna-Reyes, L.F.; Andersen, D.L. Collecting and analyzing qualitative data for system dynamics: Methods and models. *Syst. Dyn. Rev.* **2003**, *19*, 271–296. [CrossRef]
28. Randers, J. Guidelines for model conceptualization. In *Elements of the System Dynamics Method*; MIT Press: Cambridge, MA, USA, 1980; pp. 117–139.
29. Ingram, J.; Ericksen, P.; Liverman, D. *Food Security and Global Environmental Change*; Earthscan: London, UK, 2009.
30. Kelly, G. A Brief Introduction to Personal Construct Theory. In *International Handbook of Personal Construct Psychology*; John Wiley & Sons: Chichester, UK, 2003; pp. 3–20.
31. Eden, C. Cognitive mapping and problem structuring for system dynamics model building. *Syst. Dyn. Rev.* **1994**, *10*, 257–276. [CrossRef]
32. Doyle, J.K.; Ford, D.N. Mental models concepts revisited: Some clarifications and a reply to Lane. *Syst. Dyn. Rev.* **1999**, *15*, 411–415. [CrossRef]
33. Walker, B.; Holling, C.S.; Carpenter, S.R.; Kinzig, A. Resilience, adaptability and transformability in social-ecological systems. *Ecol. Soc.* **2004**, *9*, 5. [CrossRef]
34. Darnhofer, I.; Bellon, S.; Dedieu, B.; Milestad, R. Adaptiveness to enhance the sustainability of farming systems. A review. *Sustain. Agric.* **2011**, *2*, 45–58.
35. Mayumi, K.; Giampietro, M. The epistemological challenge of self-modifying systems: Governance and sustainability in the post-normal science era. *Ecol. Econ.* **2006**, *57*, 382–399. [CrossRef]
36. Funtowicz, S.; Ravetz, J.R. Emergent complex systems. *Futures* **1994**, *26*, 568–582. [CrossRef]
37. Lebel, L.; Anderies, J.M.; Campbell, B.; Folke, C. Governance and the Capacity to Manage Resilience in Regional Social-Ecological systems. *Ecol. Soc.* **2006**, *11*, 1.
38. Larsen, R.K.; Calgaro, E.; Thomalla, F. Governing resilience building in Thailand's tourism-dependent coastal communities: Conceptualising stakeholder agency in social-ecological systems. *Glob. Environ. Chang.* **2011**, *21*, 481–491. [CrossRef]
39. Eriksen, S.H.; Nightingale, A.J.; Eakin, H. Reframing adaptation: The political nature of climate change adaptation. *Glob. Environ. Chang.* **2015**, *35*, 523–533. [CrossRef]
40. Carpenter, S.; Walker, B.; Anderies, M.J.; Abel, N. From metaphor to measurement: Resilience of what to what? *Ecosystems* **2001**, *4*, 765–781. [CrossRef]
41. Marshall, N.A.; Marshall, P.A. Conceptualizing and Operationalizing Social Resiliance within Commercial Fisheries in Northern Australia. *Ecol. Soc.* **2007**, *12*, 1. [CrossRef]
42. Nightingale, A.J. "The experts taught us all we know": Professionalisation and knowledge in Nepalese community forestry. *Antipode* **2005**, *37*, 581–603. [CrossRef]
43. Peterson, G. Political ecology and ecological resilience: An integration of human and ecological dynamics. *Ecol. Econ.* **2000**, *35*, 323–336. [CrossRef]
44. Adger, W.N.; Arnell, N.W.; Tompkins, E.L. Successful adaptation to climate change across scales. *Glob. Environ. Chang.* **2005**, *15*, 77–86. [CrossRef]
45. Größler, A. System Dynamics projects that failed to make an impact. *Syst. Dyn. Rev.* **2007**, *23*, 437–452. [CrossRef]
46. Ackermann, F. Problem structuring methods 'in the Dock': Arguing the case for Soft OR. *Eur. J. Oper. Res.* **2012**, *219*, 652–658. [CrossRef]
47. Vennix, J.A.M. *Group Model Building*; John Willey and Sons Ltd.: Chichester, UK, 1996.
48. Eden, C.; Ackermann, F. Group Decision and Negotiation in Strategy Making. *Gr. Decis. Negot.* **2001**, *10*, 119–140. [CrossRef]

49. Franco, L.A.; Montibeller, G. Facilitated modelling in operational research. *Eur. J. Oper. Res.* **2010**, *205*, 489–500. [CrossRef]
50. Davies, K.K.; Fisher, K.T.; Dickson, M.E.; Thrush, S.F.; Le Heron, R. Improving ecosystem service frameworks to address wicked problems. *Ecol. Soc.* **2015**, *20*, 2. [CrossRef]
51. Rouwette, E.A.J.A.; Vennix, J.A.M.; Felling, A.J.A. On evaluating the performance of problem structuring methods: An attempt at formulating a conceptual model. *Gr. Decis. Negot.* **2009**, *18*, 567–587. [CrossRef]
52. Akkermans, H.A.; Vennix, J.A.M. Clients' opinions on group model-building: An exploratory study. *Syst. Dyn. Rev.* **1997**, *13*, 3–31. [CrossRef]
53. Duit, A.; Galaz, V.; Eckerberg, K.; Ebbesson, J. Governance, complexity, and resilience. *Glob. Environ. Chang.* **2010**, *20*, 363–368. [CrossRef]
54. Young, O.R. Institutional dynamics: Resilience, vulnerability and adaptation in environmental and resource regimes. *Glob. Environ. Chang.* **2010**, *20*, 378–385. [CrossRef]
55. Warner, K. Global environmental change and migration: Governance challenges. *Glob. Environ. Chang.* **2010**, *20*, 402–413. [CrossRef]

12

Impact of Climate Change on Twenty-First Century Crop Yields in the U.S.

Lillian Kay Petersen

Los Alamos High School, Los Alamos, NM 87544, USA; lilliankay.petersen@gmail.com

Abstract: Crop yields are strongly dependent on the average climate, extreme temperatures, and carbon dioxide concentrations, all of which are projected to increase in the coming century. In this study, a statistical model was created to predict US yields to 2100 for three crops using low and high-emissions future scenarios (RCP 4.5 and 8.5). The model is based on linear regressions between historical crop yields and daily weather observations since 1970 for every county in the US. Yields were found to be most strongly dependent on heat waves, summer average temperatures, and killing degree days; these relationships were hence used to predict future yields. The model shows that warming temperatures will significantly decrease corn and soybean yields, but will not have as strong of an influence on rice. Before accounting for CO_2 fertilization, crops in the high-emissions scenario are predicted to produce 77%, 85%, and 96% of their expected yield without climate change for corn, soybeans, and rice, respectively. When a simple CO_2 fertilization factor is included, corn, a C4 plant, increases slightly, while the yields of the C3 plants (soybeans and rice) are actually predicted to increase compared to today's yields. This study exhibits the wide range of possible impacts of climate change on crop yields in the coming century, and emphasizes the need for field research on the combined effects of CO_2 fertilization and heat extremes.

Keywords: future crop yields; climate change impacts; CO_2 fertilization; corn; rice; soybeans

---

1. Introduction

1.1. Impacts of Climate Change and the Social Cost of Carbon

On 8 October 2018, the Intergovernmental Panel on Climate Change (IPCC) released a new report titled *Global Warming of 1.5 °C*, which concludes that drastic action must be taken to limit global temperature rise and avoid serious negative impacts. It finds that natural, managed and human systems have a high risk of permanent damage as the climate warms. Extreme weather events will occur more often, including droughts, floods, coastal storms, and heat waves, increasing mortality and property damage. As warmer temperatures combine with more extreme weather events, cereal yields will decrease [1,2]. If we fail to limit fossil fuel combustion, all of the effects will increase [3].

Also on 8 October, the Nobel Prize in Economic Science was awarded to Paul Romer and William Nordhaus for their research on using economics as a driving factor to reduce greenhouse gas emissions [4]. Nordhaus, recognized as the founder of climate change economics, developed economic models to weigh the cost of reducing carbon footprints today against future costs of current emissions [5]. Romer focused on how market factors influence technological growth. Both advocate carbon taxes to employ market forces to reduce emissions and spur innovation in energy efficiency.

The correct amount to tax carbon may be found through the social cost of carbon, or the external cost of carbon emissions. The social cost of carbon is calculated by integrating all future economic losses due to climate change discussed in the IPCC. In total, warming temperatures cause a loss of annual national average gross domestic product (GDP) of 1.0% to 3.0% at the end of the century [6]. Agriculture is a substantial portion of the economy, and crop yields are highly dependent on temperature. Other impacts are more difficult to put a price on, such as biodiversity and ecosystem loss. The currently accepted cost of carbon dioxide when considering these externalities varies between $37 and $220 per ton emitted [7].

1.2. Warming Temperatures Impact Agriculture

The United States produces 41% of the world's corn and 38% of the world's soybeans, two of the four largest crop sources of caloric energy [8]. These crops are thus crucial to food security, and understanding how their yields will change in the next century could help drive more informed policy decisions.

The growing world population requires a larger food supply. Historical improvements in crop yields from agricultural technology (e.g., pesticides, fertilizers, farm machinery, gene modification, and shifting of production to large corporations [9]) have kept up with increasing demand for several decades, but it is doubtful that yields will continue to grow at the same rate as they have since 1970. Population is unlikely to stop growing this century, and by 2100 there will be between 9.6 and 12.3 billion people on earth [10]. Research has shown that yields are projected to drop in coming decades due to warming temperatures and the potential emergence of virulent crop diseases [11,12].

Crop yields are strongly dependent on the weather and may be predicted from observed weather events during the growing season [13–16]. Over 60% of yield variability in global breadbaskets can be explained by climate variation [17], particularly temperature extremes during crucial phases of the growing season, such as the grain fill stages [18]. Some research suggests that yields decrease exponentially as temperatures warm [8]. Therefore, a warming climate could harm crop yields and global food security. In fact, corn and wheat yields have already decreased by 1–2% per decade since 1980 relative to the expected harvest without warming [19].

There are many different approaches to identifying the the impacts of climate change on crop yields. Statistical models, including this study, use historical correlations from observations to develop empirical relationships between yields and weather. These relationships are then applied to climate model output to predict future crop yields. Process models are based on the mechanisms of an individual plant's physiology and then are scaled up to large domains. Each type of model has its own advantages and disadvantages. Statistical models are accurate for the specific locations and conditions of their training data sets, and are a direct way to model yields within those constraints [20,21]. Process models offer a deeper understanding of the cause and effect of the environmental impacts on yields, and they can potentially model future yields outside of historical observations. Process models have become more sophisticated in recent years [22], but still have difficulty reproducing historical yields in certain circumstances [23].

Many previous studies have analyzed the relationships between crop yields and emperature [13,24,25], precipitation [14,26,27], or radiation measurements [28,29], and have predicted future yields based on these relationships [13,26,30]. These models may be statistical [17,31], process-based [30,32], or both [33], and have focused on US [14,24,33], China [23,25,29,31,32], Europe [28,30], or global bread baskets [15,17,27]. All of these studies conclude that climate change will have a negative impact on future crop yields, and bread basket failures could pose a threat to food security.

The purpose of this study is to evaluate the future economic losses or gains of three crops through 2100 for different climate scenarios. It examines the historical relationship between crop yields and extreme weather to better understand which factors affect yields, and then projects crop yields into the future for every county in the United States. This study also incorporates CO_2 fertilization to show the range of possible future impacts from these processes. Finally, a monetary value of changing crop yields is calculated, an integral part to the social cost of carbon.

2. Methods

A statistical model was created to evaluate historical weather and crop data, compute correlations and linear regressions for each county, and project crop yields to 2100 based on two future climate model scenarios. Annual crop yield data of corn, soybeans, and rice was obtained for every US county for 1970 through 2015 from the United States Department of Agriculture (USDA, [34]). In addition, 1970 was chosen as a start date because yields were more variable and the farming practices were not as standardized before then (e.g., irrigation, pesticides, fertilizers). Daily weather station observations, provided by the Daily Global Historical Climatology Network [35], were downloaded for all weather stations in the US with data since 1970. Daily minimum and maximum air temperature and precipitation were computed for each county from the average of the two weather stations closest to the center of that county. This provided redundancy—if one station was missing data, the other station's data was used.

Next, various means and extremes were computed for each county and year. Most of these are standard measures reported by the Intergovernmental Panel on Climate Change (IPCC, [36], Box 2.4 p. 221). Table 1 lists all extremes computed from the daily temperature. Values of the 10th and 90th percentile for each variable and county were computed from the daily data from the years 1970 to 1990. The extreme measures were computed over the growing season, which varies for each crop and state, and were obtained from the USDA [37].

Table 1. Temperature measures computed to find correlations to crop yields. Summer average temperature, heat waves, and killing degree days had the highest correlation, and thus were used as predictors of crop yields. All statistics after the first four are summed over the growing season for each crop and location. Here, highs and lows refer to the recorded daily high and low temperature at each site.

Measurement	Definition	Units
Average Yearly high	Average of all daily highs in a year	°C
Average Yearly low	Average of all daily lows in a year	°C
Summer Average	Average of all daily max temps over months June, July, and August	°C
Warmest Day	The warmest high in the growing season	°C
Coldest Day	The coldest high in the growing season	°C
Warmest Night	The warmest low in the growing season	°C
Coldest Night	The coldest low in the growing season	°C
Heat Waves of highs	Frequency of 3 daily highs in a row >90th percentile	#/year
Heat Waves of lows	Frequency of 3 daily lows in a row >90th percentile	#/year
Cold Spells of lows	Frequency of 3 daily lows in a row <10th percentile	#/year
Cold Spells of highs	Frequency of 3 daily highs in a row <10th percentile	#/year
Warm Days	Days when daily high >90th percentile	days/year
Cold Days	Days when daily high <10th percentile	days/year
Warm Nights	Days when daily low >90th percentile	days/year
Cold Nights	Days when daily low <10th percentile	days/year
Tropical Nights	Frequency of daily lows >20 °C (68 °F)	days/year
Frost Nights	Frequency of daily lows <0 °C (32 °F)	days/year
Growing Degree Days	Summation of daily highs above 10 °C (50 °F)	°C*days
Killing Degree Days	Summation of daily highs above 29 °C (68 °F) [13]	°C
Precipitation	Total precipitation	mm/year

Correlations between the detrended crop yield and each of the weather statistics were then computed for each county and crop. In order to account for increasing crop yields due to improvements in agricultural technology, the yields were first detrended for each county. Figure 1a shows the increase in corn yields since 1970 in an example county of Champaign, Illinois, and Figure 1b shows the correlation between the detrended corn yield and summer average temperature for that same county. Correlations with a p-value less than 0.05 are considered significant and ones under 0.01 are highly significant [38]. The three with the highest correlations were summer average temperature, heat waves, and killing degree days. Thus, these three statistics were used to predict future yields. Although regressions were also computed with precipitation, it was found to have little to no correlations with yields.

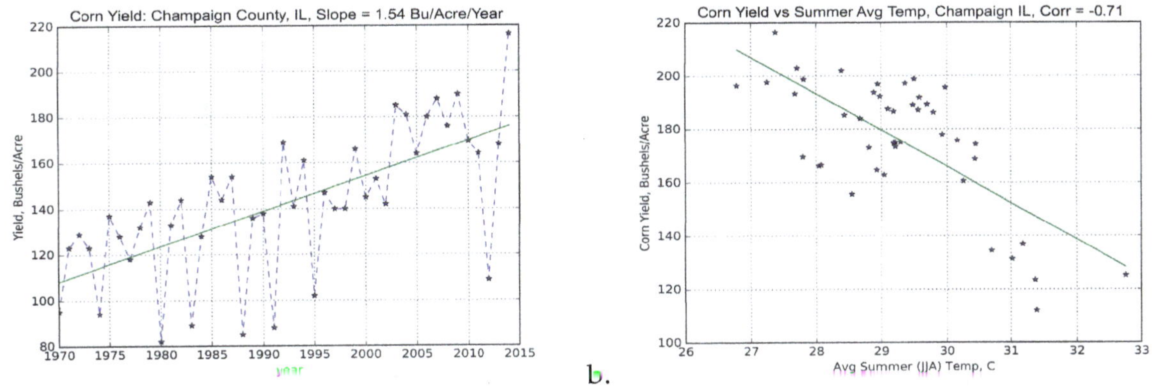

Figure 1. The corn yield over time for an example county (**a**) and detrended corn yield plotted against summer average temperature (**b**). The correlation of −0.71 is highly significant. Data is from the USDA [34].

Future climate model data was obtained from the Coupled Model Intercomparison Project Version 5 (CMIP5), using the Community Climate System Model (CCSM4) [39], and was obtained from the MACA data portal. I downloaded datasets for two IPCC scenarios: a high emission future with a Representative Concentration Pathway (RCP) that induces an extra 8.5 W/m^2 of radiative forcing (RCP 8.5) and a low emissions scenario with an RCP of 4.5 W/m^2 (RCP 4.5). Climate forcings in the MACAv2-METDATA were drawn from a statistical downscaling of global climate model data from CMIP5 [40] utilizing a modification of the Multivariate Adaptive Constructed Analogs (MACA) [41] method with the METDATA [42] observational dataset as training data. The climate data had high resolution in space (one-tenth of a degree) and time (daily). Summer average temperature, heat waves, and killing degree days were computed for each county for every year until 2100, using data from the closest model grid-cell to the center of each county. The histograms displaying these future heat measurements (Figure 2) were computed using a latitude/longitude rectangle around the dominant corn-growing region, with corners at (40N, 100W), (44N, 85W).

Crop yields were then predicted by applying the historical linear regressions to the future projections of summer average temperature, heat waves, and killing degree days. For each county, the crop predictions from the three statistics were averaged, as each measure predicted the yields slightly differently. National averages of crop yields were computed by averaging all counties that either consistently grew their crop over the past 10 years or grew at least 10% as much as the highest-producing county for that crop.

After future crop yields were predicted from temperature projections, the yield was multiplied by the expected yield factor from CO_2 fertilization. Future carbon dioxide concentrations to 2100 for RCP8.5 and RCP4.5 scenarios were obtained from [43]. The yield factor is the change in yield due to carbon dioxide fertilization, where 0.8 is a 20% reduction and 1.2 is a 20% increase in yield. The yield factor for C3 and C4 crops under different CO_2 concentrations (Figure 3) was acquired from [44], in which results from the DSSAT4 models [45] were interpolated based on CO_2 enrichment experiments, and were then normalized at 2015 CO_2 concentrations. Figure 4 displays the yield factor due to carbon dioxide fertilization to year 2100 for C3 and C4 crops under low and high-emissions scenarios.

Crop yields in the US have improved in recent decades due to better technologies. In fact, corn yields have doubled since 1970. It is not known whether these trends will continue to hold in the future or if biological constraints will impose maximum achievable crop yields. In this study, best and worst case scenarios were computed as a proxy for all possible futures. For the best case scenario of continuous technology improvement, the historical trend was added to the future yield predictions. For the worst case scenario of no future technology improvement, no trend was added to the yield predictions.

Ultimately, future yield forecasts were computed based on three sets of indicators: (1) only temperature changes, (2) only CO_2 fertilization, and (3) both combined. Each of these predictions were then split into two technology scenarios: (1) no technology improvement and (2) continuous technology improvement, resulting in six forecasts for each crop.

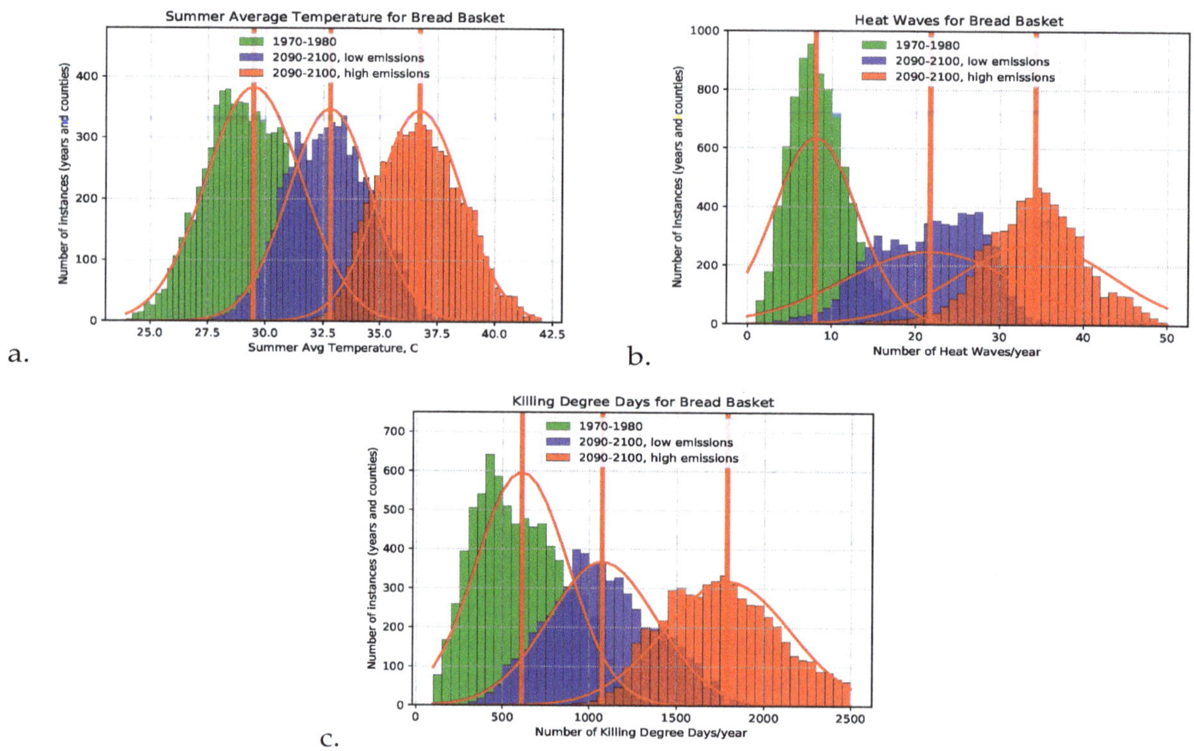

Figure 2. Distributions of summer average temperature (**a**), heat waves (**b**), and killing degree days (**c**) for historical (green), future low-emissions RCP 4.5 scenario (blue) and future high-emissions RCP 8.5 scenario (red). Results are for the US corn growing region. Historical data from [35] and future projections from [41].

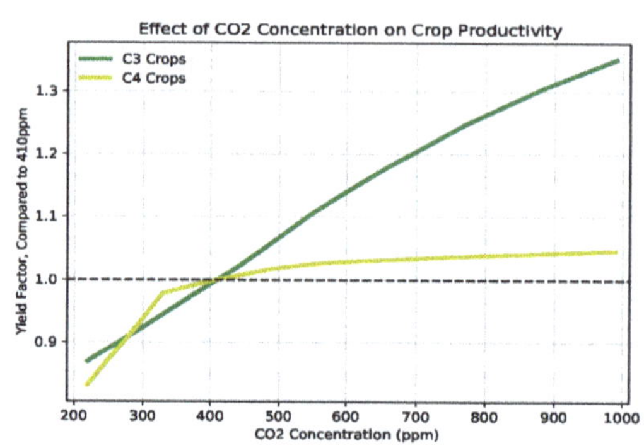

Figure 3. Yield factor for C3 (green) and C4 (yellow) crops versus CO_2 concentration. Crop productivity was acquired from [44], in which results from the DSSAT4 models [45] were interpolated based on CO_2 enrichment experiments.

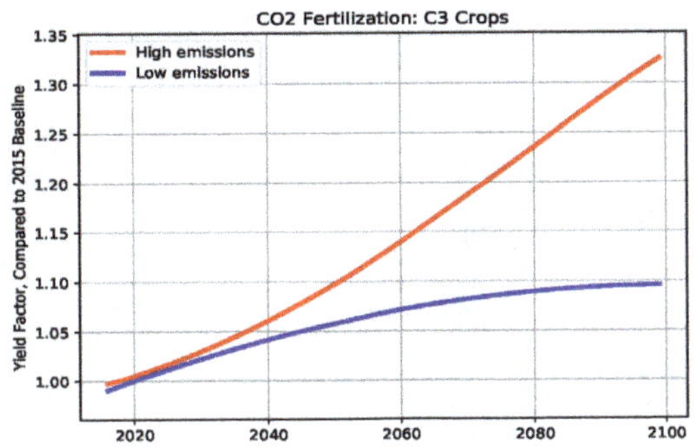

a.

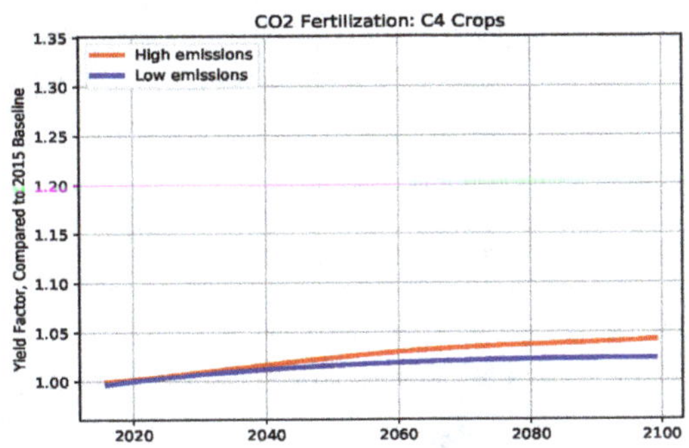

b.

Figure 4. Yield factor for C3 (**a**) and C4 (**b**) crops through 2100 for a low (blue) and high (red) emissions scenario. Crop productivity under CO_2 concentrations was obtained from [44], and CO_2 concentrations for RCP4.5 and RCP8.5 are from [43].

3. Results

3.1. Historical Correlations and Regressions

The slopes of the linear regressions computed between crop yields and weather statistics for every county in the US can be seen in Figure 5. Almost all of the slopes are negative, meaning that higher temperatures result in lower crop yields. Corn has average correlations of -0.44, -0.46, and -0.41 to summer average temperature, heat waves, and killing degree days, respectively (Figure 6), and three in every five counties have a significant correlation. All temperature indices have similar correlations, indicating that all three have similar predictive power. Corn and soybeans observe large geographical distributions in the eastern and central US, while rice is mostly grown along the Mississippi River. Spatial variations of slopes and correlations can be examined for corn and soybeans. In southern growing regions (Missouri, southern Illinois, and Indiana), slopes are very negative with highly significant correlations (Figures 5 and 6). Crops here are thus extremely sensitive to heat extremes. In contrast, slopes and correlations farther north (Minnesota and South Dakota) are about zero, indicating that the yields are not affected by temperatures classified as extreme events in these states. The correlations with heat waves, summer average temperature, and killing degree days all follow similar geographic distributions. These results indicate that the places where crops are grown will likely shift north over time, where average temperatures are cooler.

Heat waves have the highest impact on all three crops. Corn has the strongest correlation to heat measurements, soybeans have a slighly weaker correlation, and rice has little to no correlation to temperatures. When averaged across crop-growing counties, soybeans have a correlation of -0.37 with heat waves, and about half of the counties have significant correlations. Rice has an average correlation -0.22 with heat waves and no counties have significant correlations (Figure 6).

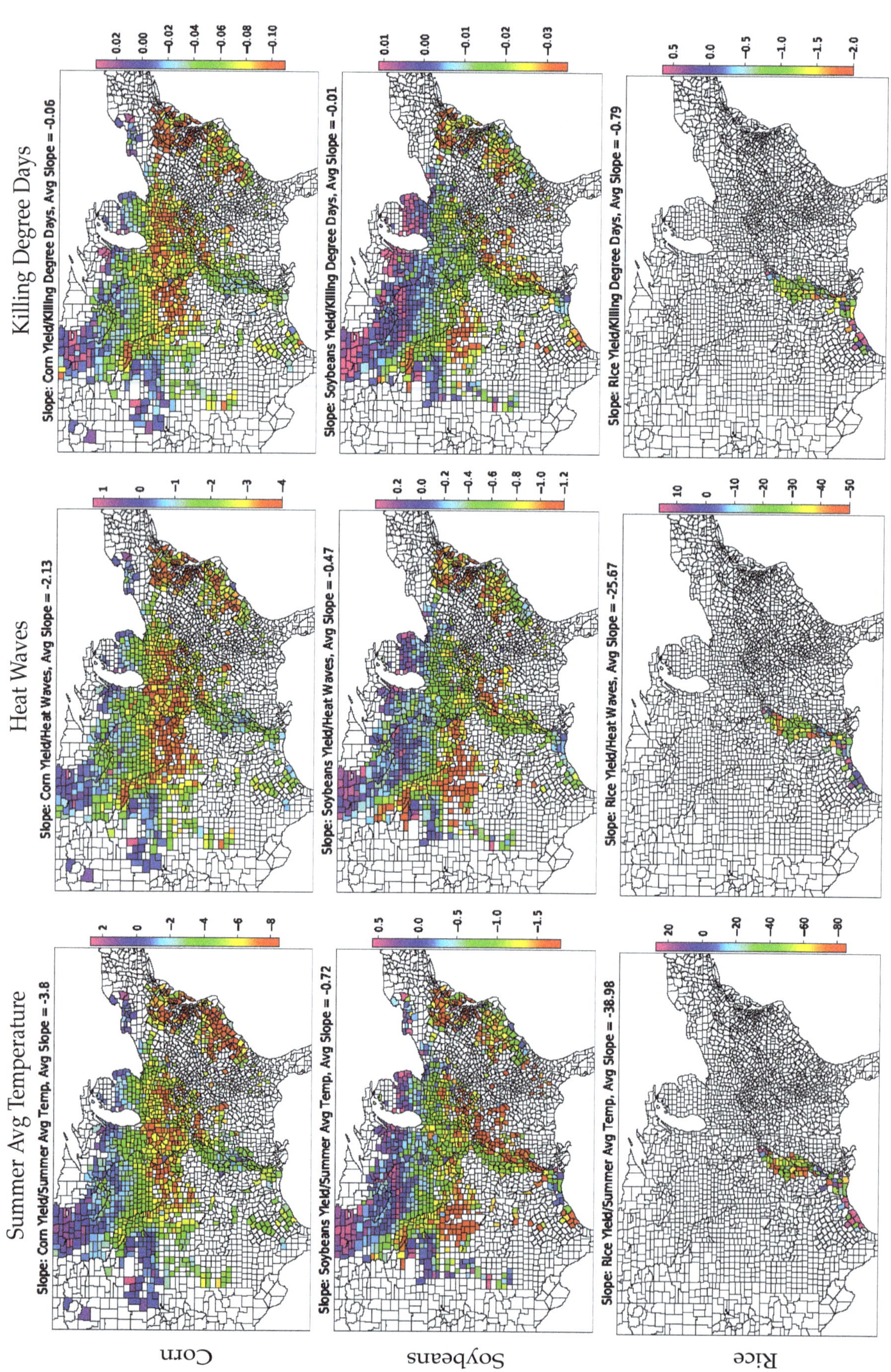

Figure 5. The slopes of the linear regressions between corn (top), soybean (middle), and rice (bottom) yields and summer average temperature (left), heat waves (center), and killing degree days (right), in each US county with reported crop yields.

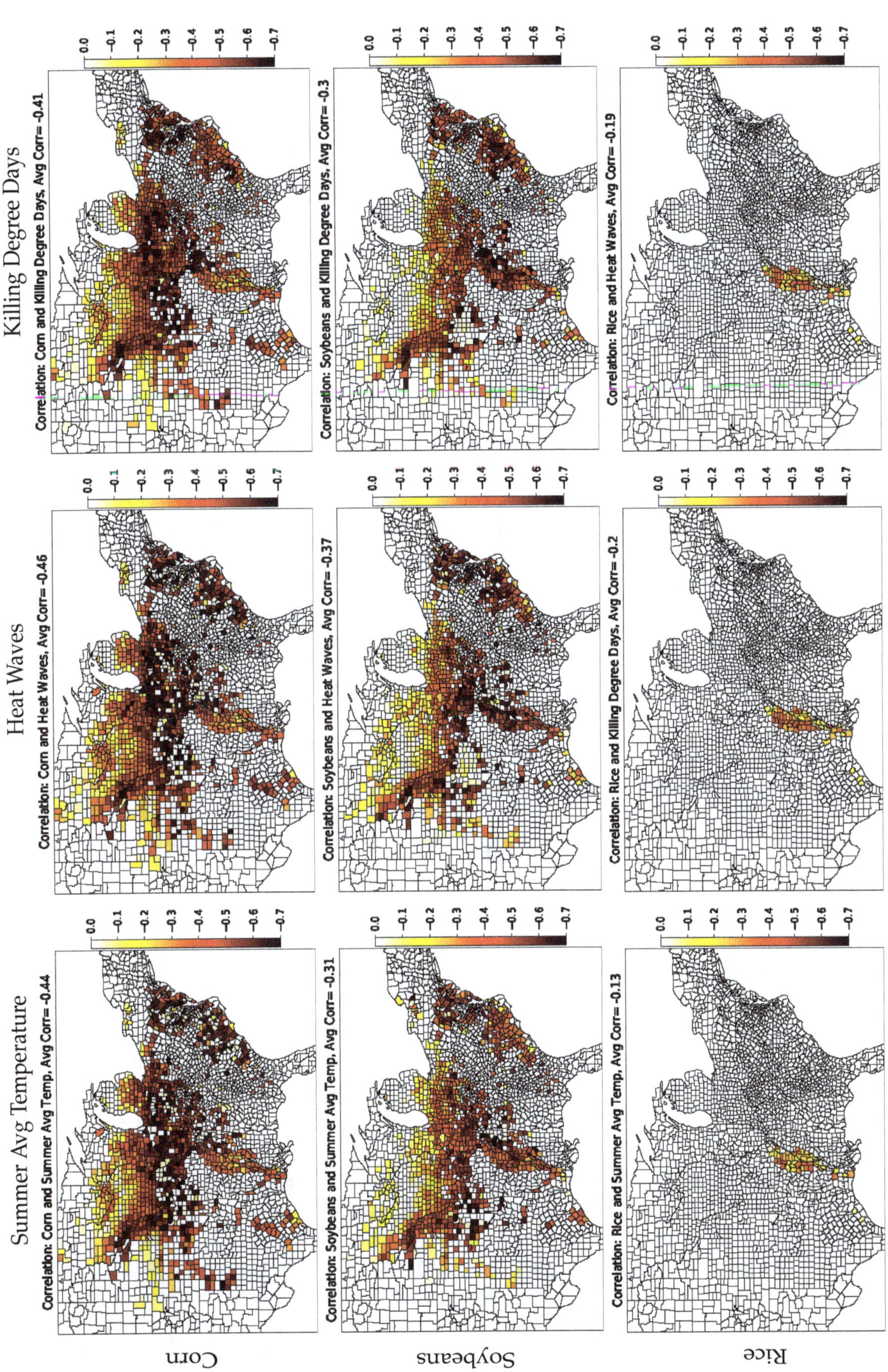

Figure 6. The same as Figure 5, but for correlations between the crop yield and the temperature metric. Correlations below −0.49 are significant ($p < 0.05$) and below −0.59 are highly significant ($p < 0.01$).

3.2. Prediction of Future Crop Yield

Heat extremes are expected to increase in the coming century. Histograms of the three heat measurements are shown for three different times and scenarios: 1970–1980 observed, 2090–2100 low emissions, and 2090–2100 high emissions, over the corn growing counties (Figure 2). Both mean and extreme temperatures dramatically increase in the future, with high emission scenarios increasing more than low emission. For example, the average summer daily high temperature was 29 °C (85 °F) in 1970 to 1980. In 2090 to 2100, the summer average temperature is projected to be 33 °C (91 °F) for RCP 4.5 and 36 °C (97 °F) for RCP 8.5. Heat waves and killing degree days are also expected to increase dramatically (Figure 2b,c). Interestingly, the tails of histograms in the future are much wider, indicating higher probabilities of extreme weather events.

Historical regressions and future climate extremes were used to predict future yields to 2100 for each county, year, and crop. First, results without CO_2 fertilization will be presented, followed by those including CO_2 fertilization.

Before accounting for CO_2 fertilization, crop yields are projected to dramatically drop in the coming century. Forecasted crop yields, with and without future technology improvements, can be seen in Figure 7. In addition to having the highest correlations, corn is also affected the most by the warming climate. Average US corn yields doubled from 80 bushels/acre in 1970 to 170 bushels/acre in 2015. Predicted yields in 2096 through 2100 drop to 76% (86%) of expected yields without warming for a high (low) emissions scenario. This translates to a 3.8% decrease in corn yields per decade for a high emissions scenario and a 1.8% decrease per decade for a low emissions scenario, compared to a historical 24% increase in corn yields per decade due to agricultural technology improvements. For more details, refer to Table 2.

Table 2. Statistics on future yield predictions, for forecasts without and with CO_2 fertilization. Future yield change per decade are in comparison to expected yields without climate change. Monetary losses are calculated from the acres harvested and the crop prices in 2016 [46,47].

	Corn		**Soybeans**		**Rice**	
Historical yield change/decade (%)	23.7		17.7		17.4	
Future: low emissions	no CO_2	with CO_2	no CO_2	with CO_2	no CO_2	with CO_2
Yield change/decade (%)	−1.77	−1.50	−1.20	+0.002	−0.367	0.860
Projected yield diff to steady climate (%)	87.9	90.0	92.3	101	98.6	108
Monetary loss or gain (billion 2019 US$)	−12.7	−10.5	−5.00	+0.717	−0.343	+1.99
Future: high emissions	no CO_2	with CO_2	no CO_2	with CO_2	no CO_2	with CO_2
Yield change/decade (%)	−3.78	−3.41	−2.40	0.96	−0.830	3.09
Projected yield diff to steady climate (%)	76.8	80.0	84.6	111	95.7	126
Monetary loss or gain (billion 2019 US$)	−24.3	−21.0	−10.0	+7.29	−1.06	+6.42

Even with the optimistic conditions of continuous technology improvement, there is a huge loss in yields below expected yields with a steady climate. In 2100, there is a loss of $24 billion per year for high emissions and $13 billion per year for a low emissions scenario. This estimate assumes the acres harvested and the cost of corn in 2016, and does not account for inflation [46,47]. Soybeans are affected by temperature extremes slightly less than corn, with losses of $5 and $10 billion per year in 2100 for high and low emission scenarios. Rice, being least sensitive to climate change, only has losses of $0.34 and $1.06 billion per year (Table 2). Rice is the least affected by heat, likely because it is grown in flooded conditions. These computations assume the current market prices of crops and do not include CO_2 fertilization.

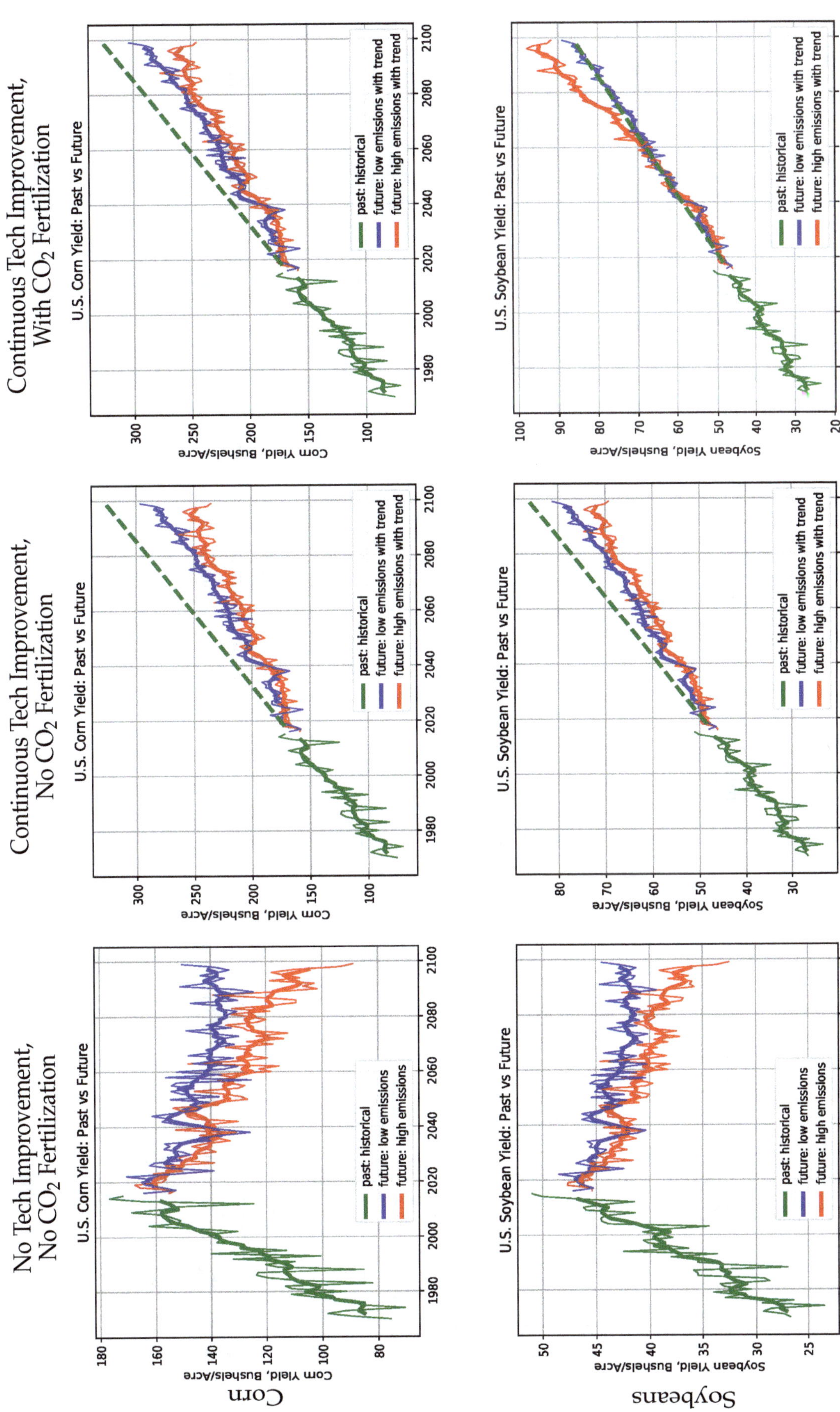

Figure 7. *Cont.*

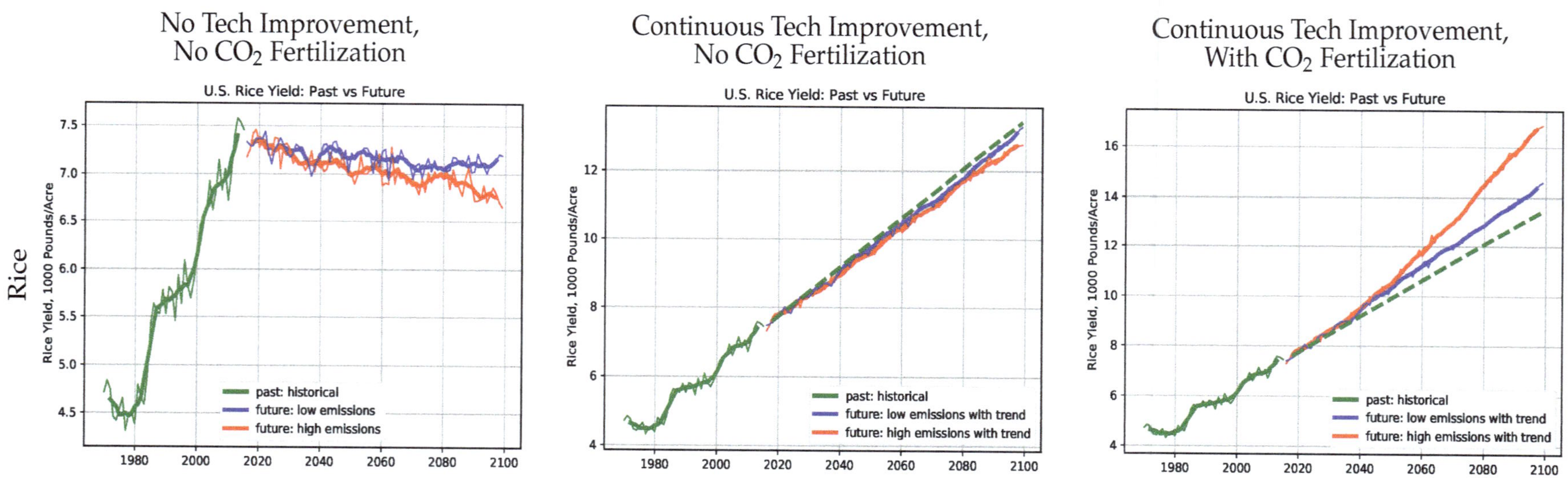

Figure 7. Projected US corn (top), soybean (middle), and rice (bottom) yields to 2100. The prediction scenarios include: (left) no technology improvement and no CO_2 fertilization; (center) continuous technology improvement and no CO_2 fertilization; and (right) continuous technology improvement with CO_2 fertilization. The green dashed line is a linear extension of the 1970–2015 trend. Thin lines are yearly data, solid is the five-year running average.

Although these results display a dismal future, the story changes when we account for CO_2 fertilization. In RCP 8.5, carbon dioxide concentrations break 900 ppm by 2100. That translates to almost 135% productivity for C3 crops (here: soybeans and rice) compared to productivity in 2015. Even for a low emissions scenario, C3 crops gain 10% productivity by the end of the century. CO_2 concentrations have a much smaller effect on C4 crops (here: corn), which reach 104% and 102% productivity in 2100 for a high and low-emissions scenario.

The projections of corn, a C4 crop, are very similar before and after accounting for CO_2 fertilization (Figure 7). Projections of soybeans and rice with CO_2 fertilization, however, are considerably higher. In fact, rice reaches 126% expected productivity by 2100 with a high emissions scenario. A summary of future crop yield estimates is shown in Figure 8.

The spatial distributions of projected crop yields may be examined (Figure 9). In 2005 through 2015, corn and soybean yields are relatively uniform across the US. In contrast, their yields are spatially disparate in 2100, with very low yields in the south and much higher yields in the northern Midwest. This distribution likely is a product of cooler climates farther north, and holds true with and without CO_2 fertilization. Few spatial differences are predicted in rice yields, as rice is grown in a relatively small geographic range and has weaker correlations to temperature.

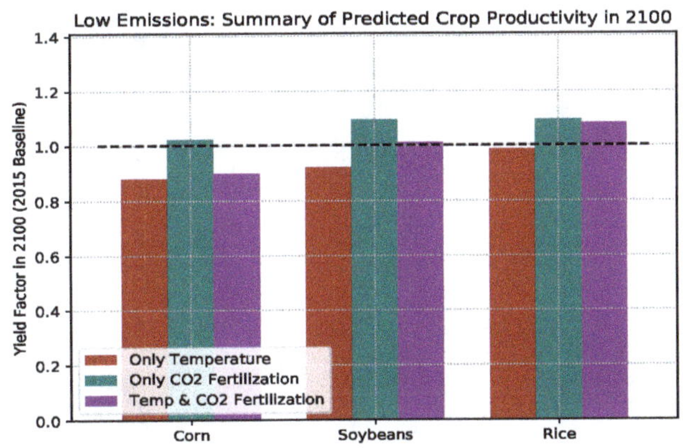

a.

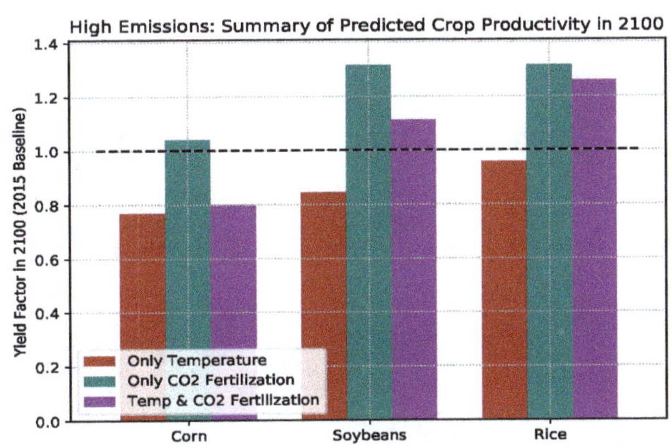

b.

Figure 8. Crop productivity averaged over 2096–2100, compared to a 2011 through 2015 baseline for a low (**a**) and high (**b**) emissions scenario. Estimates are based on only temperatures (red), only CO_2 fertilization (teal), and both assuming compounding effects (purple).

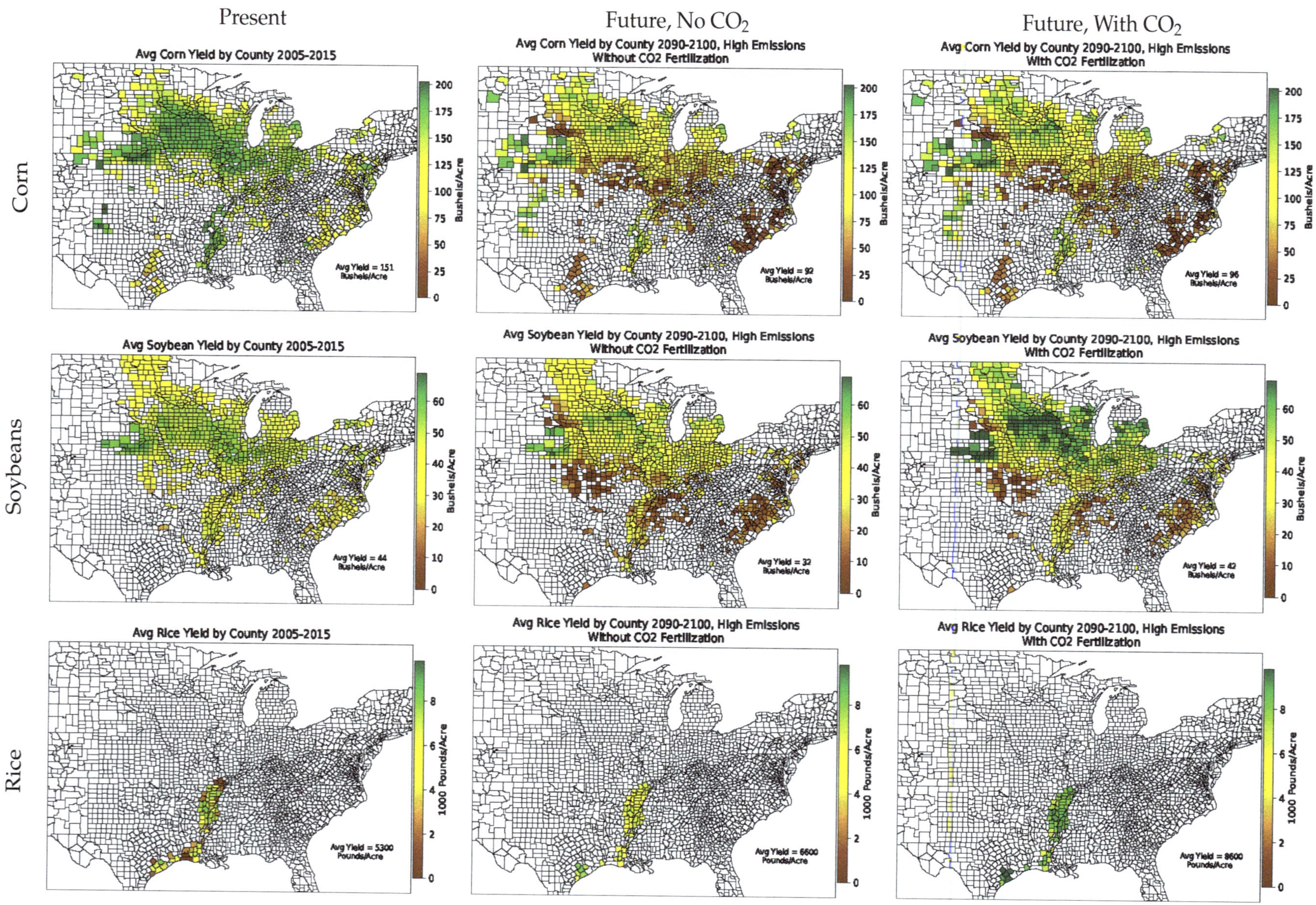

Figure 9. Crop yield of corn (top), soybeans (middle), and rice (bottom) for 2005–2015 (left), 2090–2100 without CO_2 fertilization (center), and 2090–2100 with CO_2 fertilization (right). Crop predictions are for a high-emissions scenario without future technology improvements.

4. Discussion

The purpose of this study was to analyze the potential economic losses or gains of US crop yields due to climate change by applying historical relationships between yields and heat extremes to multiple future climate scenarios, while accounting for CO_2 fertilization. Historical records give insights into factors that affect yields. The most dominant influence on crop yields since 1970 is the secular trend due to improving farming practices and technologies, where yields nearly double over that period. On top of this trend, there is year-to-year variability that can be explained by local weather. Corn, soybean, and rice yields were correlated to several measures of mean and extreme weather, and all were most strongly dependent on heat waves, summer average temperature, and killing degree days. This indicates that hot temperatures have the strongest effect on crop growth, while moderate or cold temperatures have little effect. Interestingly, precipitation had insignificant correlations with crop yields, possibly because of the prevalence of drought-resistant breeds in the US or the confounding influence of irrigation.

This study evaluates future crop yields with and without CO_2 fertilization. Without CO_2 fertilization, increasing temperatures significantly decrease crop production, with yields reaching as low as 77% of their expected productivity without climate change. In all cases, corn is most sensitive to heat, soybeans slightly less so, and rice the least, likely because rice is grown in flooded conditions. When carbon dioxide fertilization is added into the future projections, crop yields increase dramatically. Carbon dioxide concentrations alone are expected to increase C3 crop yields by 35% by the end of the century for a high emissions scenario. When the effects of CO_2 and temperatures are combined in a simple way, rice and soybean yields are actually shown to increase over the next century.

In this study, the effects of temperature extremes on crop yields were measured during historical CO_2 concentrations, while the effects of increased CO_2 concentrations were measured under laboratory conditions, independently of heat extremes. To project the effects of both together, the two factors were multiplied, but this may in fact not be how plants respond. It is unknown whether CO_2 fertilization and temperatures will have compounding physiological effects: the positive influence of CO_2 fertilization may be severely curbed at sufficient heat. Thus, more data on the effects of these combined influences is required for more accurate yield forecasts.

Projected yield losses due to climate change may be compared to past studies, shown in Table 7-2 of the IPCC Working Group II [36]. Numbers in this study compare well with the past corn and rice studies (10 and 13 studies, respectively). The soybean losses here are much greater than the 10 past publications when only accounting for temperatures, but are comparative when including CO_2 fertilization. These comparisons are complicated by the mixture of scenarios and model types in the IPCC summary, and most previous studies do not include CO_2 fertilization.

This study is very similar to [44], which predicts global crop yields for different low-emission climate scenarios, with and without CO_2 fertilization. They find that accounting for CO_2 fertilization mitigates the effect of warming temperatures. The results of this study are comparable to those of [44], since both studies find that C3 crops benefit strongly from higher carbon dioxide concentrations. The results of this study are also similar to those of [15], who use temperature, precipitation, and CO_2 fertilization to predict global corn and wheat yields. They also find clear benefits of reducing greenhouse gas emissions for corn yields, but fewer for wheat yields, as wheat is a C3 crop and benefits from CO_2 fertilization [48].

The statistical model developed here includes several assumptions. The three seasonal climate statistics only involve temperature, but crop yields may also be correlated with other conditions, such as precipitation, soil moisture, and radiation. Some other statistical models include these, but this study found much higher correlations with temperature than precipitation. Soil moisture and radiation were not available from weather station data. In order to project into the future, the model assumes that temperature continues to influence crop yields as they have historically, despite potential changes in other factors such as precipitation, soil conditions, or more advanced technologies. Another assumption is that linear regressions may be used as a predictive model for temperatures much hotter than those recorded historically, even though previous papers have found crop growth to respond nonlinearly to climate [8,49]. Despite these shortcomings, the high correlations in Figure 6 show that the climate statistics used for prediction are reasonable predictors of crop yields.

This study indicates that land in the northern United States may be more suitable for crop production as temperatures rise. Crops in the northern counties are less susceptible to events classified as extreme temperatures in those counties (Figure 6), and are forecasted to have much higher yields in 2100 (Figure 9). A limitation of a statistical model is that it can only predict crop yields in areas with sufficient crop data. Thus, this model cannot judge whether crops will have high yields in counties further north where crops have not been historically grown. A process model would be required to predict crop performance in these counties. A large unknown in this study is whether agricultural technology will continue to improve or whether crop yields will hit a fundamental biological limit. Given the difficulties of predicting future technologies, this study has instead projected a best case and a worst case scenario. Most likely, technology will have decreasing effects on crop yields due to biological constraints on production, and crop yields will increase at slower rates than have been in the last 50 years. In order to prepare for climate change, we should develop farming practices and crop breeds that are resistant to stronger and more frequent heat extremes.

This study highlights the trade-offs of climate change, where CO_2 fertilization is a potential benefit even while average temperatures and heat waves are increasing. Research to date has shown that the positive influences of CO_2 fertilization will increase yields for much of the 21st century, but will be countered by increasingly hot and dry conditions [50–52]. The results in this study exhibit the wide range of possible future impacts of climate change in the next century, and emphasize the need for continued research on the compounding effects of carbon dioxide fertilization and heat extremes.

Acknowledgments: L.K.P. is grateful to Mark Petersen and Phillip Wolfram for discussions on statistical methods and scientific presentation.

References

1. Shrestha, S.; Deb, P.; Bui, T.T.T. Adaptation strategies for rice cultivation under climate change in Central Vietnam. *Mitig. Adapt. Strateg. Glob. Chang.* **2016**, *21*, 15–37. [CrossRef]
2. Deb, P.; Shrestha, S.; Babel, M.S. Forecasting climate change impacts and evaluation of adaptation options for maize cropping in the hilly terrain of Himalayas: Sikkim, India. *Theor. Appl. Climatol.* **2015**, *121*, 649–667. [CrossRef]
3. IPCC. Summary for Policymakers. In *Global Warming of 1.5 °C*; An IPCC Special Report on the Impacts of Global Warming of 1.5 °C above Pre-Industrial Levels and Related Global Greenhouse Gas Emission Pathways, In the Context of Strengthening the Global Response to the Threat of Climate Change, Sustainable Development, and Efforts to Eradicate Poverty; Intergovernmental Panel on Climate Change (IPCC): Geneva, Switzerland, 2018.
4. Schiermeier, Q. Economists Who Changed Thinking on Climate Change Win Nobel Prize. *Nature*, 8 October 2018. [CrossRef]
5. Nordhaus, W.D. Economic aspects of global warming in a post-Copenhagen environment. *Proc. Natl. Acad. Sci. USA* **2010**, *107*, 11721–11726. [CrossRef] [PubMed]
6. Hsiang, S.; Kopp, R.; Jina, A.; Rising, J.; Delgado, M.; Mohan, S.; Rasmussen, D.J.; Muir-Wood, R.; Wilson, P.; Oppenheimer, M.; et al. Estimating economic damage from climate change in the United States. *Science* **2017**, *356*, 1362–1369. [CrossRef] [PubMed]
7. Moore, F.C.; Diaz, D.B. Temperature impacts on economic growth warrant stringent mitigation policy. *Nat. Clim. Chang.* **2015**, *5*, 127–131. [CrossRef]
8. Schlenker, W.; Roberts, M.J. Nonlinear temperature effects indicate severe damages to U.S. crop yields under climate change. *Proc. Natl. Acad. Sci. USA* **2009**, *106*, 15594–15598. [CrossRef] [PubMed]

9. Edgerton, M.D. Increasing Crop Productivity to Meet Global Needs for Feed, Food, and Fuel. *Plant Physiol.* **2009**, *149*, 7–13. [CrossRef] [PubMed]
10. Gerland, P.; Raftery, A.E.; Ševčíková, H.; Li, N.; Gu, D.; Spoorenberg, T.; Alkema, L.; Fosdick, B.K.; Chunn, J.; Lalic, N.; et al. World population stabilization unlikely this century. *Science* **2014**, *346*, 234–237. [CrossRef] [PubMed]
11. Alston, J.M.; Beddow, J.M.; Pardey, P.G. Agricultural Research, Productivity, and Food Prices in the Long Run. *Science* **2009**, *325*, 1209–1210. [CrossRef]
12. Singh, R.P.; Hodson, D.P.; Huerta-Espino, J.; Jin, Y.; Njau, P.; Wanyera, R.; Herrera-Foessel, S.A.; Ward, R.W. Will Stem Rust Destroy the World's Wheat Crop? *Adv. Agron.* **2018**, *98*, 271–309. [CrossRef]
13. Butler, E.E.; Huybers, P. Adaptation of US maize to temperature variations. *Nat. Clim. Chang.* **2013**, *3*, 68–72. [CrossRef]
14. Liang, X.Z.; Wu, Y.; Chambers, R.G.; Schmoldt, D.L.; Gao, W.; Liu, C.; Liu, Y.A.; Sun, C.; Kennedy, J.A. Determining climate effects on US total agricultural productivity. *Proc. Natl. Acad. Sci. USA* **2017**, *114*, E2285–E2292. [CrossRef] [PubMed]
15. Tebaldi, C.; Lobell, D. Estimated impacts of emission reductions on wheat and maize crops. *Clim. Chang.* **2015**, 1–13. [CrossRef]
16. Petersen, L.K. Real-Time Prediction of Crop Yields From MODIS Relative Vegetation Health: A Continent-Wide Analysis of Africa. *Remote Sens.* **2018**, *10*, 1726. [CrossRef]
17. Ray, D.K.; Gerber, J.S.; MacDonald, G.K.; West, P.C. Climate variation explains a third of global crop yield variability. *Nat. Commun.* **2015**, *6*, 5989. [CrossRef]
18. Ortiz-Bobea, A.; Just, R.E. Modeling the Structure of Adaptation in Climate Change Impact Assessment. *Am. J. Agric. Econ.* **2013**, *95*, 244–251. [CrossRef]
19. Lobell, D.B.; Schlenker, W.; Costa-Roberts, J. Climate Trends and Global Crop Production Since 1980. *Science* **2011**, *333*, 616–620. [CrossRef] [PubMed]
20. Kukal, M.S.; Irmak, S. Climate-Driven Crop Yield and Yield Variability and Climate Change Impacts on the U.S. Great Plains Agricultural Production. *Sci. Rep.* **2018**, *8*, 3450. [CrossRef]
21. Najafi, E.; Devineni, N.; Khanbilvardi, R.M.; Kogan, F. Understanding the Changes in Global Crop Yields Through Changes in Climate and Technology. *Earths Future* **2018**, *6*, 410–427. [CrossRef]
22. Peng, B.; Guan, K.; Chen, M.; Lawrence, D.M.; Pokhrel, Y.; Suyker, A.; Arkebauer, T.; Lu, Y. Improving maize growth processes in the community land model: Implementation and evaluation. *Agric. For. Meteorol.* **2018**, *250–251*, 64–89. [CrossRef]
23. Sheng, M.; Liu, J.; Zhu, A.X.; Rossiter, D.G.; Zhu, L.; Peng, G. Evaluation of CLM-Crop for maize growth simulation over Northeast China. *Ecol. Model.* **2018**, *377*, 26–34. [CrossRef]
24. Butler, E.E.; Huybers, P. Variations in the sensitivity of US maize yield to extreme temperatures by region and growth phase. *Environ. Res. Lett.* **2015**, *10*, 034009. [CrossRef]
25. Zhang, Q.; Zhang, J.; Guo, E.; Yan, D.; Sun, Z. The impacts of long-term and year-to-year temperature change on corn yield in China. *Theor. Appl. Climatol.* **2015**, *119*, 77–82. [CrossRef]
26. Lobell, D.B.; Tebaldi, C. Getting caught with our plants down: the risks of a global crop yield slowdown from climate trends in the next two decades. *Environ. Res. Lett.* **2014**, *9*, 074003. [CrossRef]
27. Tebaldi, C.; Lobell, D.B. Towards probabilistic projections of climate change impacts on global crop yields. *Geophys. Res. Lett.* **2008**, *35*, L08705. [CrossRef]
28. Gornott, C.; Wechsung, F. Statistical regression models for assessing climate impacts on crop yields: A validation study for winter wheat and silage maize in Germany. *Agric. For. Meteorol.* **2016**, *217*, 89–100. [CrossRef]
29. Tao, F.; Zhang, Z.; Zhang, S.; Ratter, R.P.; Shi, W.; Xiao, D.; Liu, Y.; Wang, M.; Liu, F.; Zhang, H. Historical data provide new insights into response and adaptation of maize production systems to climate change/variability in China. *Field Crops Res.* **2016**, *185*, 1–11. [CrossRef]
30. Ummenhofer, C.C.; Xu, H.; Twine, T.E.; Girvetz, E.H.; McCarthy, H.R.; Chhetri, N.; Nicholas, K.A. How Climate Change Affects Extremes in Maize and Wheat Yield in Two Cropping Regions. *J. Clim.* **2015**, *28*, 4653–4687. [CrossRef]
31. Wang, P.; Zhang, Z.; Song, X.; Chen, Y.; Wei, X.; Shi, P.; Tao, F. Temperature variations and rice yields in China: Historical contributions and future trends. *Clim. Chang.* **2014**, *124*, 777–789. [CrossRef]

32. Wang, P.; Zhang, Z.; Chen, Y.; Wei, X.; Feng, B.; Tao, F. How much yield loss has been caused by extreme temperature stress to the irrigated rice production in China? *Clim. Chang.* **2016**, *134*, 635–650. [CrossRef]
33. Anderson, C.J.; Babcock, B.A.; Peng, Y.; Gassman, P.W.; Campbell, T.D. Placing bounds on extreme temperature response of maize. *Environ. Res. Lett.* **2015**, *10*, 124001. [CrossRef]
34. Hamer, H.; Picanso, R.; Prusacki, J.J.; Rater, B.; Johnson, J.; Barnes, K.; Parsons, J.; Young, D.L. *National Agricultural Statistics Service*; United States Department of Agriculture: Washington, DC, USA, 2017.
35. Menne, M.J.; Durre, I.; Vose, R.S.; Gleason, B.E.; Houston, T.G. An Overview of the Global Historical Climatology Network-Daily Database. *J. Atmos. Ocean. Technol.* **2012**, *29*, 897–910. [CrossRef]
36. Hartmann, D.; Tank, A.K.; Rusticucci, M. *Climate Change 2013: The Physical Science Basis. Contribution of Working Group I to the Fifth Assessment Report of the Intergovernmental Panel on Climate Change*; Technical Report; IPCC: Geneva, Switzerland, 2013.
37. USDA. *Field Crops: Usual Planting and Harvesting Dates*; Technical Report; National Agricultural Statistics Service: Washington, DC, USA, 2010.
38. Crow, E.L.; Davis, F.A.; Maxfield, M.W. *Statistics Manual*; Dover Publications, Inc.: Mineola, NY, USA, 1960.
39. Gent, P.R.; Danabasoglu, G.; Donner, L.J.; Holland, M.M.; Hunke, E.C.; Jayne, S.R.; Lawrence, D.M.; Neale, R.B.; Rasch, P.J.; Vertenstein, M.; et al. The Community Climate System Model Version 4. *J. Clim.* **2011**, *24*, 4973–4991. [CrossRef]
40. Taylor, K.E.; Stouffer, R.J.; Meehl, G.A. An Overview of CMIP5 and the Experiment Design. *Bull. Am. Meteorol. Soc.* **2011**, *93*, 485–498. [CrossRef]
41. Abatzoglou, J.T.; Brown, T.J. A comparison of statistical downscaling methods suited for wildfire applications. *Int. J. Climatol.* **2012**, *32*, 772–780. [CrossRef]
42. Abatzoglou, J.T. Development of gridded surface meteorological data for ecological applications and modelling. *Int. J. Climatol.* **2011**, *33*, 121–131. [CrossRef]
43. Hayhoe, K.; Edmonds, J.; Kopp, R.; LeGrande, A.; Sanderson, B.; Wehner, M.; Wuebbles, D. Climate models, scenarios, and projections. **2017**, *1*, 133–160. [CrossRef]
44. Tebaldi, C.; Lobell, D. Differences, or lack thereof, in wheat and maize yields under three low-warming scenarios. *Environ. Res. Lett.* **2018**, *13*, 065001. [CrossRef]
45. Jones, J.W.; Hoogenboom, G.; Porter, C.H.; Boote, K.J.; Batchelor, W.D.; Hunt, L.A.; Wilkens, P.W.; Singh, U.; Gijsman, A.J.; Ritchie, J.T. The DSSAT cropping system model. *Eur. J. Agron.* **2003**, *18*, 235–265. [CrossRef]
46. USDA, NASS. *Agricultural Prices*; USDA, NASS: Washington, DC, USA, 2018.
47. USDA, NASS. *Crop Production 2016 Summary*; USDA, NASS: Washington, DC, USA, 2017.
48. Högy, P.; Wieser, H.; Köhler, P.; Schwadorf, K.; Breuer, J.; Franzaring, J.; Muntifering, R.; Fangmeier, A. Effects of elevated CO_2 on grain yield and quality of wheat: Results from a 3-year free-air CO_2 enrichment experiment. *Plant Biol.* **2009**, *11*, 60–69. [CrossRef] [PubMed]
49. Lobell, D.B.; Bänziger, M.; Magorokosho, C.; Vivek, B. Nonlinear heat effects on African maize as evidenced by historical yield trials. *Nat. Clim. Chang.* **2011**, *1*, 42–45. [CrossRef]
50. Jin, Z.; Ainsworth, E.A.; Leakey, A.D.B.; Lobell, D.B. Increasing drought and diminishing benefits of elevated carbon dioxide for soybean yields across the US Midwest. *Glob. Chang. Biol.* **2018**, *24*, e522–e533. [CrossRef]
51. Jin, Z.; Zhuang, Q.; Wang, J.; Archontoulis, S.V.; Zobel, Z.; Kotamarthi, V.R. The combined and separate impacts of climate extremes on the current and future US rainfed maize and soybean production under elevated CO_2. *Glob. Chang. Biol.* **2017**, *23*, 2687–2704. [CrossRef] [PubMed]
52. Obermeier, W.A.; Lehnert, L.W.; Kammann, C.I.; Müller, C.; Grünhage, L.; Luterbacher, J.; Erbs, M.; Moser, G.; Seibert, R.; Yuan, N.; et al. Reduced CO_2 fertilization effect in temperate C3 grasslands under more extreme weather conditions. *Nat. Clim. Chang.* **2016**, *7*, 137. [CrossRef]

13

Estimation of the Virtual Water Content of Main Crops on the Korean Peninsula using Multiple Regional Climate Models and Evapotranspiration Methods

Chul-Hee Lim [1,2], Seung Hee Kim [2], Yuyoung Choi [1], Menas C. Kafatos [2] and Woo-Kyun Lee [1,*]

1 Department of Environmental Science and Ecological Engineering, Korea University, Seoul 02841, Korea; limpossible@korea.ac.kr (C.-H.L.); cuteyu0@korea.ac.kr (Y.C.)
2 Center of Excellence in Earth Systems Modeling and Observations, Chapman University, Orange, CA 92866, USA; sekim@chapman.edu (S.H.K.); kafatos@chapman.edu (M.C.K.)
* Correspondence: leewk@korea.ac.kr

Abstract: Sustainable agriculture in the era of climate change needs to find solutions for the retention and proper utilization of water. This study proposes an ensemble approach for identifying the virtual water content (VWC) of main crops on the Korean Peninsula in past and future climates. Ensemble results with low uncertainty were obtained using three regional climate models, five potential evapotranspiration methods, and the Environmental Policy Integrated Climate (EPIC) crop model. The productivity results of major crops (rice and maize) under climate change are likely to increase more than in the past based on the ensemble results. The ensemble VWC is calculated using three types of crop yields and fifteen consumptive amounts of water use in the past and the future. While the ensemble VWC of rice and maize was 1.18 $m^3\ kg^{-1}$ and 0.58 $m^3\ kg^{-1}$, respectively, in the past, the future amounts were estimated at 0.76 $m^3\ kg^{-1}$ and 0.48 $m^3\ kg^{-1}$, respectively. The yields of both crops showed a decline in future projections, indicating that this change could have a positive impact on future water demand. The positive changes in crop productivity and water consumption due to climate change suggest that adaptation to climate change can be an opportunity for enhancing sustainability as well as for minimizing agricultural damage.

Keywords: virtual water content; ensemble result; crop yield; regional climate models; PET methods

1. Introduction

Agriculture is the most climate-dependent production sector; thus, it is necessary to accurately assess the impacts of climate change on the agriculture sector to achieve sustainability [1,2]. The agriculture sector is also linked to the largest number of indicators of sustainable development [3]. The demand for agriculture-related predictions is expected to rise in the future with population expansion and a greater need for food security [4,5].

Agriculture and crop production are the sectors that have the highest water demand [6]. While it is necessary to manage multiple stress factors to achieve agricultural productivity, an abundant supply of water is essential. Several global studies have predicted significant changes in crop productivity and hydrological circulation in the mid-latitude regions during the 21st century [7,8]. The Korean Peninsula has a temperate, monsoon climate with a large annual variation in precipitation, and is classified as a region of water shortage by the UN (United Nations); therefore, it is necessary to study the agricultural water use in the region [9,10].

The concept of virtual water or water footprint has been presented by several international organizations and studies as an effective way to estimate water demand [11,12]. Virtual water refers to the amount of water used per unit product and is synonymous with water footprint. Many studies have estimated the water demand of the agricultural sector by calculating the virtual water of crops [13–15]. By estimating the virtual water per crop, we can also analyze the influence of regional differences and climate change on the water demand.

While estimates of the virtual water content of crops in East Asian cases can be easily found [11–16], the crop-hydrological model-based estimation approach for production and water consumption is not common. In particular, there were no cases which estimated the virtual water of a crop with a model-based approach on the Korean Peninsula, and only Zhao et al. [15] suggested the virtual water of several crops in China using a modeling approach. The information from a model-based virtual water content analysis in the Korean Peninsula can help solve problems of food and water security [4,10].

Previous model-based virtual water studies have calculated the amount of water used, crop yields, and evapotranspiration rates to estimate the virtual water content of crops [13,15,16]. The most basic data used in the estimation of evapotranspiration and crop yield is climatic data, which has significant uncertainties [10,17], depending on the climate model used [9,18].

To estimate and reduce uncertainties in a model's predictions, multiple methods have recently been employed by the crop modeling community [17,19–21]. The major source of a model's error is from the uncertainty of the input data, such as climate data. Another source is due to a model's characteristics having various approaches and parameterizations to determine the growth and the phenological development of the crop [20].

In this study, we propose a multi-input and multi-model super ensemble approach for the virtual water content of main crops based on the past and future climate data of the Korean Peninsula. Specifically, we use five evapotranspiration methods to reduce model uncertainty and three regional climate models to reduce the uncertainty in the climate data. In particular, we present virtual water content that represents the past and the future by averaging the number of virtual water contents based on multiple data and methodologies. The average of the multiple results is defined as the ensemble average, and the uncertainty reduction is analyzed by considering the difference in each value without considering the variability. Given the importance and increasing emphasis of agricultural water use under climate change, it is useful to provide the estimation of the ensemble virtual water of main crops.

2. Data and Methods

2.1. Research Area and Crops

The research area covers the whole cropland of the Korean Peninsula, which includes both South and North Korea (Figure 1). The Korean Peninsula is located on the eastern end of Asia, covers an area of about 221,000 km^2, and is located between the 33.23° N and 43.01° N latitudes and the 124.14° E and 130.93° E longitudes. The Korean Peninsula is located in the temperate zone. It is largely influenced by the temperate monsoon climate, and has large annual variations in precipitation, with high temperatures and humidity levels in the summer. Under the Köppen climate classification, a warm and dry winter climate (Cwa) and a cold and dry winter climate (Dwa) are the most common, with an average annual temperature of 10–16 °C, and an average cumulative precipitation of 1200 mm [22]. Geographically, there are high mountains in the north and east and plains in the south and west.

The cropped area of South Korea is about 20,000 km^2 and that of North Korea about 30,000 km^2, which is over 20% of the total area of the peninsula. Rice paddies and maize fields appear to be mixed, but rice is widely distributed on a broad plain and the number of maize fields increases in the north [23]. In this study, considering the resolution of the available climate data, the cropland of the peninsula was mapped with a 5 km^2 grid resolution, and all the data were processed accordingly.

In this study, rice and maize, staple crops in the Korean Peninsula, are selected for the analysis. Rice production is dominant throughout the region, where precipitation is concentrated in the summer. Maize is mainly produced in the northeastern region, where low precipitation and temperature conditions are dominant. Note that maize accounts for 35% of the total food production in North Korea [23,24].

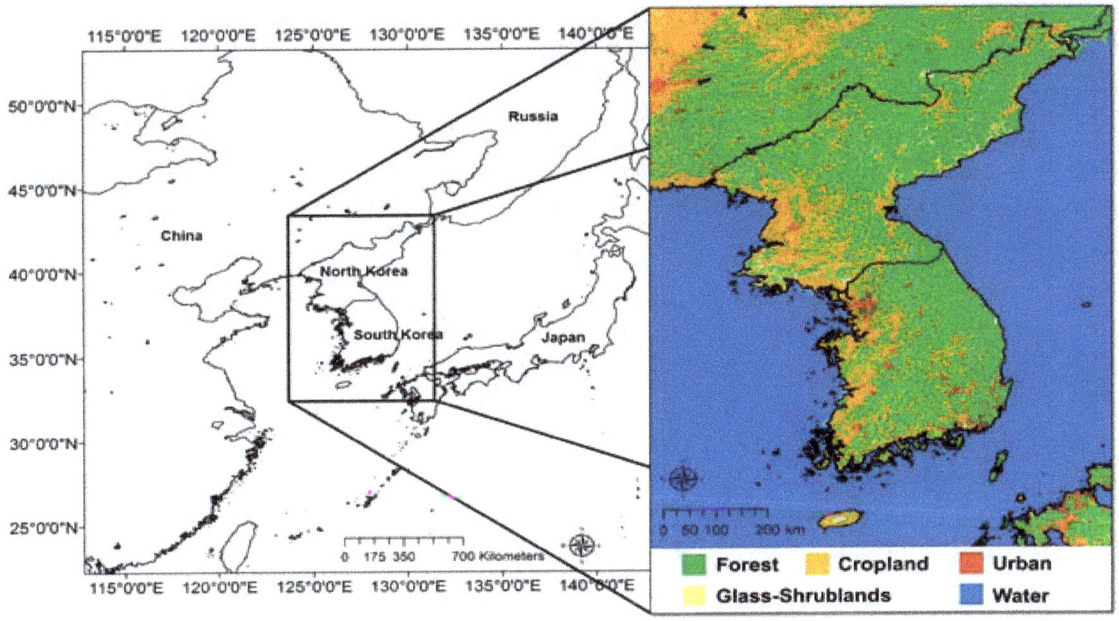

Figure 1. Land cover map of the research area.

2.2. *Assessment Model Hierarchy*

Figure 2 shows a data flow of this study. We have selected three regional climate models (RCM) for climate data, and each climate data are used in a simulation in an EPIC model with five different potential evapotranspiration (PET) methods. In the research framework, fifteen ensemble members are produced and are used to estimate the virtual water contents (VWC) of crops over the study region. A detailed explanation of each component is described in the following sections.

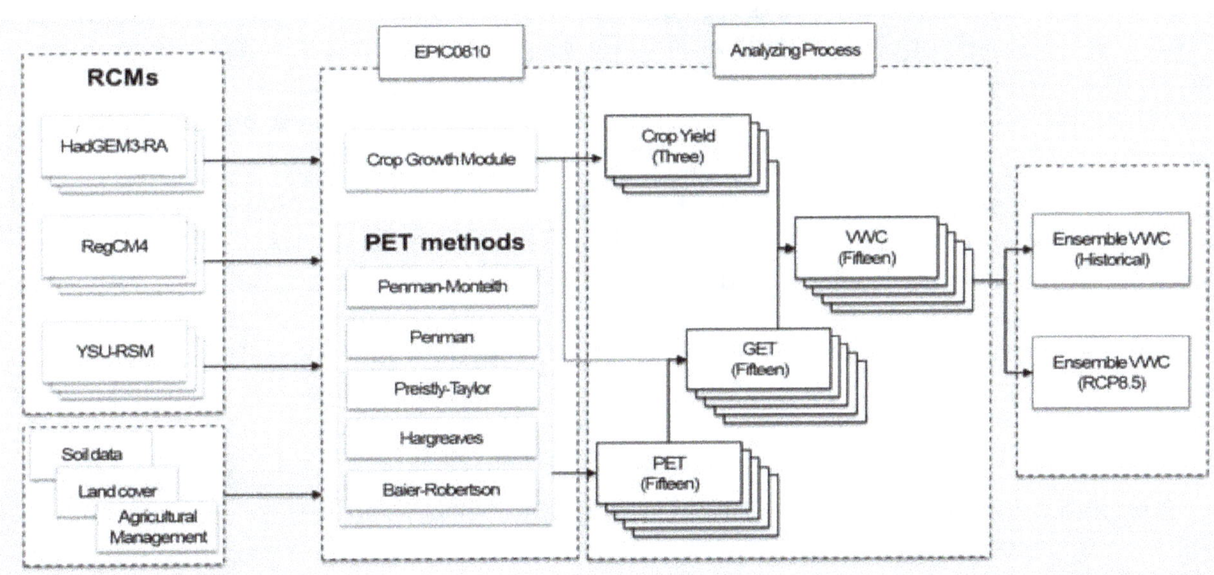

Figure 2. Schematic diagram of the applied methodology. RCM, regional climate model; PET, potential evapotranspiration; VWC, virtual water content; growing season evapotranspiration (GET).

2.3. *Multiple Regional Climate Models*

To reduce the uncertainty in the climate data, the data of three regional climate models (RCM) were used. The climate model data for the region are obtained through the COordinated Regional climate Downscaling EXperiment (CORDEX)-East Asia. The CORDEX initiative was launched by the Task Force for Regional Climate Downscaling (TFRCD), which was established in 2009 by the World Climate Research Program (WCRP).

The climate data of the RCMs acquired through CORDEX-East Asia are HadGEM3-RA, RegCM4, and YSU-RCM, and the three models were verified using previous studies [10,17]. These models provide dynamically downscaled data with a spatial resolution of 12.5 km^2 based on the HadGEM2-AO model, a large-scale climate model.

HadGEM3-RA is a regional version of the Hadley Center Global Environment Model, a non-hydrostatic regional climate model following the Arakawa-C horizontal grid and a terrain-following vertical coordinate [9]. RegCM4, the fourth version of the RegCM regional climate model system, is a hydrostatic, compressible model with sigma-p vertical coordinates and an Arakawa-B horizontal grid [25]. YSU-RSM, a disturbance model, is defined by a two-dimensional sine series for the perturbation of vorticity and by a two-dimensional cosine series for perturbations of pressure, divergence, temperature, and mixing ratio [26]. More detailed information about these models can be found on the CORDEX-East Asia web page (https://cordex-ea.climate.go.kr/main/aboutCordexPage.do).

We used RCMs for the representative concentration pathways (RCP) 8.5 scenario, which is the lowest greenhouse gas reduction scenario, and for the analysis of the data for 1981–2000 and 2031–2050.

2.4. EPIC Model and PET Methods

The EPIC model was developed in the 1980s to assess soil erosion and soil productivity, followed by a model on plant growth and hydrological parameters [27,28]. Since the EPIC model was first published, many components have been added, such as GLEAMS [29], Century [30], and RUSLE [31]. This model was renamed the Environmental Policy Integrated Climate model (EPIC) with the addition of environmental assessment functions for pesticides and water quality [32].

The EPIC model has a structure that converts daily energy and biomass to simulate crop growth [28]. The daily potential biomass increase is calculated using climate variables such as the solar radiation and biomass–energy conversion rates of the crop. The daily response to plant stress variables (water, nutrient, temperature, aeration, and salinity) decreases the potential biomass. Crop yields are ultimately estimated based on the crop harvest index and the actual biomass accumulation [33].

The estimation of crop productivity through the EPIC model has been successfully applied to the whole of the Korean Peninsula, as well as Eastern Asia, by previous studies [34–36]. In this study, calibration was performed to estimate the crop productivity on the Korean Peninsula. Some of the key crop parameters were modified through calibration. For rice, the biomass–energy ratio was set at 30 kg MJ^{-1}, the harvest index at 0.55 mg mg^{-1}, the optimum temperature at 25 °C, the base temperature at 10 °C, and the potential heat unit (PHU) ranged from 1300 to 1500 °C, depending on the climate of the specific grid cells. For maize, the biomass–energy ratio was set at 43 kg MJ^{-1}, the harvest index at 0.45 mg mg^{-1}, the optimum temperature at 25 °C, the base temperature at 8 °C, and the PHU range at 1000–1200 °C, depending on the climate of the specific grid cells [34,37,38].

After calibration, a statistical validation was conducted to evaluate the model's performance for estimating crop productivity using the results from 1981 to 2000 and the statistical data. The HadGEM3-RA climate model was used as a standard in the calibration and validation process. HadGEM3-RA has been used most commonly for the Korean Peninsula and shown to average between the RegCM4 and YSU-RSM [17,18,39]. Since only the amount of production by country is known and the statistics for North Korea are very limited, the model was only verified for the South Korean data. The statistical data on rice and maize production were retrieved from the Korea Statistical Information System (KOSIS). Three statistical indicators were used to validate the model's performance: (i) the Root Mean Square Error (RMSE), (ii) the Nash–Sutcliffe Efficiency Coefficient (NSEC), and (iii) the Relative Error (RE) [40,41]. These statistical measures have been described in detail in the literature [35,37,42].

Five representative methods (Penman–Monteith (PM) [43], Penman (P) [44], Preistly–Taylor (PT) [45], Hargreaves (H) [46], and Baier–Robertson (BR) [47]) were used to estimate the PET. All of the PET methods available in the EPIC model were used, and these five methods are commonly used in equations in many previous studies [12–15,21].

$$PET_{PM} = (RN \times \delta + 86.66 \times AD \times EA\,(1 - RH) \times U / 350) / ((2.51 - 0.0022 \times T) \times (\delta + \gamma)) \quad (1)$$

$$PET_{P} = (RN \times \delta / (2.051 - 0.0022 \times T) + \gamma \times (2.7 + 1.63 \times U) \times EA\,(1 - RH)) / (\delta + \gamma) \quad (2)$$

$$PET_{PT} = 1.28 \times (RN \times (1.0 - AB) / (2.501 - 0.0022 \times T)) \times (\delta / (\delta + \gamma)) \quad (3)$$

$$PET_{H} = 0.0032 \times (RAMX / (2.501 - 0.0022 \times T)) \times (T + 17.8) \times (T_{max} - T_{min})^{0.6} \quad (4)$$

$$PET_{BR} = 0.288 \times T_{max} - 0.144 \times T_{min} + 0.139 \times RAMX - 4.931 \quad (5)$$

PET_{PM}, PET_{P}, PET_{PT}, PET_{H}, and PET_{BR} simulate the daily PET by each method. T_{min} and T_{max} are the daily minimum and maximum temperatures (°C), and RAMX is the solar irradiance on a clear day (MJ m^{-2} d^{-1}). T is the slope of the daily average temperature (°C), RN is the total solar irradiance (MJ m^{-2} d^{-1}), δ is the slope of the saturation vapor pressure curve (kPa °C^{-1}), γ is the psychrometric constant (kPa °C^{-1}), U is the average daily wind speed (ms^{-1}), and EA is the saturation vapor pressure at mean air temperature (kPa). RH is the average relative humidity per day, AD is air density (Kg m^{-3}), and AB is the soil albedo. The required meteorological data are different for each PET method. PET_{PM} and PET_{P} require the temperature, solar radiation, relative humidity, and wind speed variables; PET_{PT} requires the relative humidity and wind speed values; and PET_{H} and PET_{BR} require only the temperature-related variables. The climate requirement of each PET method is described in Table 1.

Table 1. Climate variables required by each PET method (○: Required climate variables).

Method	Temperature (T, T_{min}, T_{max})	Solar Radiation	Relative Humidity	Wind Speed	Reference
Penman–Monteith	○	○	○	○	Monteith (1965)
Penman	○	○	○	○	Penman (1948)
Priestley–Taylor	○	○			Priestley and Taylor (1972)
Hargreaves	○				Hargreaves and Samani (1985)
Baier–Robertson	○				Baier and Robertson (1965)

2.5. Method to Calculate Virtual Water Content of Crops

The method proposed by Zhao et al. [15] to estimate the virtual water content (VWC) of crops was used in this study. The VWC of crops represents the amount of water used per unit of production, defined as the ratio of the Consumptive Water Use (CWU) to crop production [48]. The CWU of the crop is estimated from the actual evapotranspiration during the growing season of each crop. The VWC of the crop is an indicator of the agricultural demand, which can be used to estimate the total demand for water in the region and the supply requirements. The VWC is calculated as follows.

$$VWC = CWU/P = 10 \times GET/Y \quad (6)$$

where CWU is the volume of water (m^3) used by the crop during the growing season, P is the amount of crop (kg) produced during the same period, GET is the actual evapotranspiration in the growing season (mm), Y is the crop yield in kg ha^{-1} units, and the number 10 was used for the conversion of mm to m^3 ha^{-1}.

2.6. Other Data and Management Description

The EPIC model requires various soil-related parameters (OC (%), pH, cation exchange capacity (cmol kg^{-1}), sand (%), silt (%), bulk density (tm^{-3}), and electrical conductivity (mS cm^{-1})). The input form was completed using the Digital Soil Map of the World [49], constructed spatially using the ISRIC-WISE database [50].

We used the 2010 Global Land Cover 30 (GLC30) data to reflect the current land cover [51] (accessed from http://www.globallandcover.com). GLC30, a 30 m land cover map based on Landsat 7 satellite imagery, was extracted from the cropland of the Korean peninsula and constructed with 5 km × 5 km grid cells.

The amount of fertilizer and irrigation water required for each crop and area in the EPIC model is calculated to determine the spatially required amount. Since the planting and harvest dates for crops are different for each region, these are automatically assigned based on the temperature by setting the first farming start date to reflect agricultural activities on the Korean Peninsula. The starting date of the first farming for rice was set to 1 March and for corn, to 1 April.

3. Results

3.1. Estimation of Crop Yield

3.1.1. Evaluation of the Model's Performance

The HadGEM-EPIC results were used for evaluating the model's performance (Table 2). The statistical data for the rice yield showed an average range of 4.0–5.0 t ha^{-1} with a low standard deviation. The maize yield data showed relatively high standard deviations and an average of 3.6–5.1 t ha^{-1}. The non-main production area for maize had a very low yield, which led to high standard deviations. The estimates showed low standard deviations and were similar to the reported values. The RMSE values ranged from 0.2 to 0.9 for rice and 0.3 to 1.7 for maize, rice data showing higher accuracy (Table 2). The NSEC and RE values showed that the rice yield data were more accurate, which contributed to the high standard deviations of the maize production data (Table 2). The accuracy assessment of the statistical data for the two crops showed an overall high degree of accuracy; therefore, the data can be used in past climate models and in future climate change research.

Table 2. Evaluation of the model's performance in estimating rice and maize yield. SD, standard deviation; RMSE, root mean square error, NSEC, Nash–Sutcliffe efficiency coefficient; RE, relative error.

Year	Reported (t ha^{-1})		Estimated (t ha^{-1})		RMSE	E (NSEC)	RE (%)
	Mean	SD	Mean	SD			
			Rice Yield				
1981	4.03	0.39	3.96	0.36	0.28	−0.41	−1.12
1982	4.21	0.38	5.16	0.33	0.59	−0.71	3.35
1983	4.20	0.39	5.12	0.37	0.51	−1.01	4.28
1984	4.38	0.40	4.77	0.54	0.40	−0.01	1.79
1985	4.31	0.42	4.81	0.45	0.57	−0.65	5.21
1986	4.25	0.68	4.85	0.43	0.63	−0.72	2.56
1987	4.19	0.42	4.08	0.27	0.31	−0.08	−1.93
1988	4.64	0.45	4.98	0.31	0.44	−0.48	1.52
1989	4.56	0.34	4.67	0.41	0.32	−0.09	0.69
1990	4.34	0.42	4.87	0.56	0.56	−1.08	2.87
1991	4.28	0.34	4.69	0.27	0.48	−0.57	1.36
1992	4.44	0.34	5.44	0.63	0.87	−0.82	3.82
1993	4.08	0.58	4.84	0.33	0.74	−1.16	2.61
1994	4.45	0.32	4.56	0.45	0.27	−0.25	0.42
1995	4.35	0.28	4.90	0.29	0.59	−0.89	1.93
1996	4.94	0.26	4.56	0.36	0.41	−0.12	−1.25
1997	5.00	0.32	4.89	0.42	0.37	−0.08	−0.73
1998	4.62	0.36	5.01	0.43	0.45	−0.33	1.13
1999	4.78	0.37	3.91	0.27	0.77	−1.02	−3.27
2000	4.80	0.31	4.99	0.44	0.29	−0.04	1.11
			Maize Yield				
1981	4.38	1.58	4.32	0.18	0.42	−0.51	−0.81
1982	4.12	1.37	5.20	0.38	0.87	−1.12	3.67
1983	3.66	1.24	5.36	0.28	1.21	−2.33	6.15
1984	4.44	1.61	5.05	0.33	0.75	−0.62	2.49

Table 2. *Cont.*

Year	Reported (t ha^{-1})		Estimated (t ha^{-1})		RMSE	E (NSEC)	RE (%)
	Mean	SD	Mean	SD			
1985	5.04	1.76	5.26	0.33	0.38	−0.09	1.22
1986	4.79	1.71	5.47	0.43	0.73	−0.43	5.76
1987	4.85	1.63	4.60	0.29	0.42	−0.11	−1.93
1988	4.80	1.93	5.25	0.25	0.55	−0.91	7.32
1989	4.88	1.84	4.38	0.22	0.65	−1.15	−1.86
1990	4.61	1.78	5.07	0.29	0.71	−1.32	2.83
1991	3.41	1.34	5.57	0.37	1.63	−2.78	12.32
1992	4.40	1.52	5.63	0.38	0.88	−1.18	8.17
1993	4.18	1.54	5.15	0.29	0.91	−1.36	5.86
1994	4.09	1.53	4.39	0.27	0.37	−0.12	1.04
1995	4.25	1.54	5.19	0.35	0.82	−0.58	3.63
1996	4.03	1.60	5.83	0.29	1.44	−2.30	8.29
1997	4.11	1.48	5.19	0.27	1.01	−0.91	5.59
1998	3.98	1.31	4.68	0.30	0.71	−1.21	5.15
1999	3.94	1.37	5.30	0.30	1.09	−0.81	10.28
2000	4.06	1.16	5.06	0.24	0.91	−1.11	9.11

3.1.2. Estimation of Crop Yields Using Multiple RCMs

The estimated results of rice and maize production using three RCMs showed significant differences between the models. Based on the RCP 8.5 scenario, the productivity is predicted to increase in the future.

Rice yields showed a gradual increase from 1981 to 2000, and the RegCM4-EPIC result was the lowest (mean 3.1 t ha^{-1}) while the YSU-RSM-EPIC result was the highest (mean 4.25 t ha^{-1}) (Figure 2). The estimation results of HadGEM3-RA-EPIC were moderate (mean 4.05 t ha^{-1}), and the values of the three ensembles (mean 4.25 t ha^{-1}) were similar to those of HadGEM3-RA-EPIC (Figure 2). The results of the ensemble were similar to the agricultural statistics, and more accurate than the individual model results (Figure 2 and Table 2). The spatial distribution of productivity calculated by RegCM4-EPIC showed that the whole of North Korea and the central region of South Korea had low productivity. YSU-RSM-EPIC showed high productivity in the entirety of South Korea, while HadGEM3-RA-EPIC showed high productivity in the western plains of North Korea (Figure 3). The rice cultivation on the Korean Peninsula was more accurately expressed in the ensemble results than the individual model results, compared to the existing studies or the actual production status [52,53].

In the RCP 8.5 scenario for 2031–2050, the rice productivity estimates are predicted to increase by 15–25% compared to the historical data. In the historical data, the YSU-RSM-EPIC results were the highest (mean 5.6 t ha^{-1}), while the RegCM4-EPIC results were slightly higher than the HadGEM3-RA-EPIC results. The ensemble results showed an average rice yield of 4.7 t ha^{-1}, with annual variability, but not with a time series increase or decrease (Figure 3). Differences in spatial distribution between the models were observed, but overall, rice productivity improved in most regions except the northeast mountainous region of the Korean Peninsula (Figure 4).

The maize yield results showed no significant increase or decrease from 1981 to 2000, but the YSU-RSM-EPIC results (mean 6.4 t ha^{-1}) were the highest, similar to those for the rice data. RegCM4-EPIC showed moderate levels (mean 5.25 t ha^{-1}), while the lowest results were calculated by HadGEM3-RA-EPIC (mean 4.85 t ha^{-1}). The ensemble results showed an average value of 5.5 t ha^{-1} (Figure 3), which was similar to the existing Korean Peninsula maize productivity data [34]. Most regions, except the high plateau region of the Korean Peninsula, showed high productivity, which was more evident in the ensemble results (Figure 5).

The estimates of the future maize productivity, from 2031 to 2050, under climate change showed a slight increase in productivity compared to past productivity. The YSU-RSM-EPIC results were the highest, and the RegCM4-EPIC results were slightly higher than the HadGEM3-RA-EPIC results (Figure 3). The ensemble results showed an average maize yield of 5.75 t ha^{-1}, with annual variability, but no significant increase in the time series. There was no significant difference in spatial distribution between the models, and maize productivity was found to increase across the Korean Peninsula (Figure 5).

Three RCM and EPIC models were used to estimate the change in crop productivity, which was then used as an ensemble result to reduce the uncertainty in the climate models. This ensemble result considers the fertilization and irrigation required for crop growth, and is estimated for the whole cultivation area without distinguishing between rice paddies and maize fields.

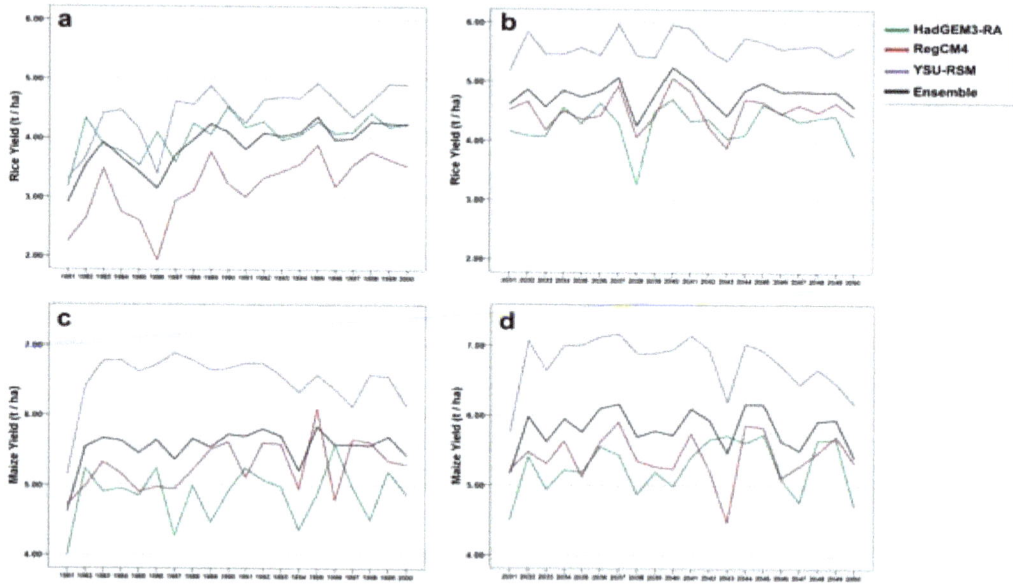

Figure 3. Estimated crop yield using multiple regional climate models (RCMs) in the past and future (**a**) Historical-rice; (**b**) representative concentration pathways (RCP) 8.5-rice; (**c**) Historical-maize; (**d**) RCP8.5-maize.

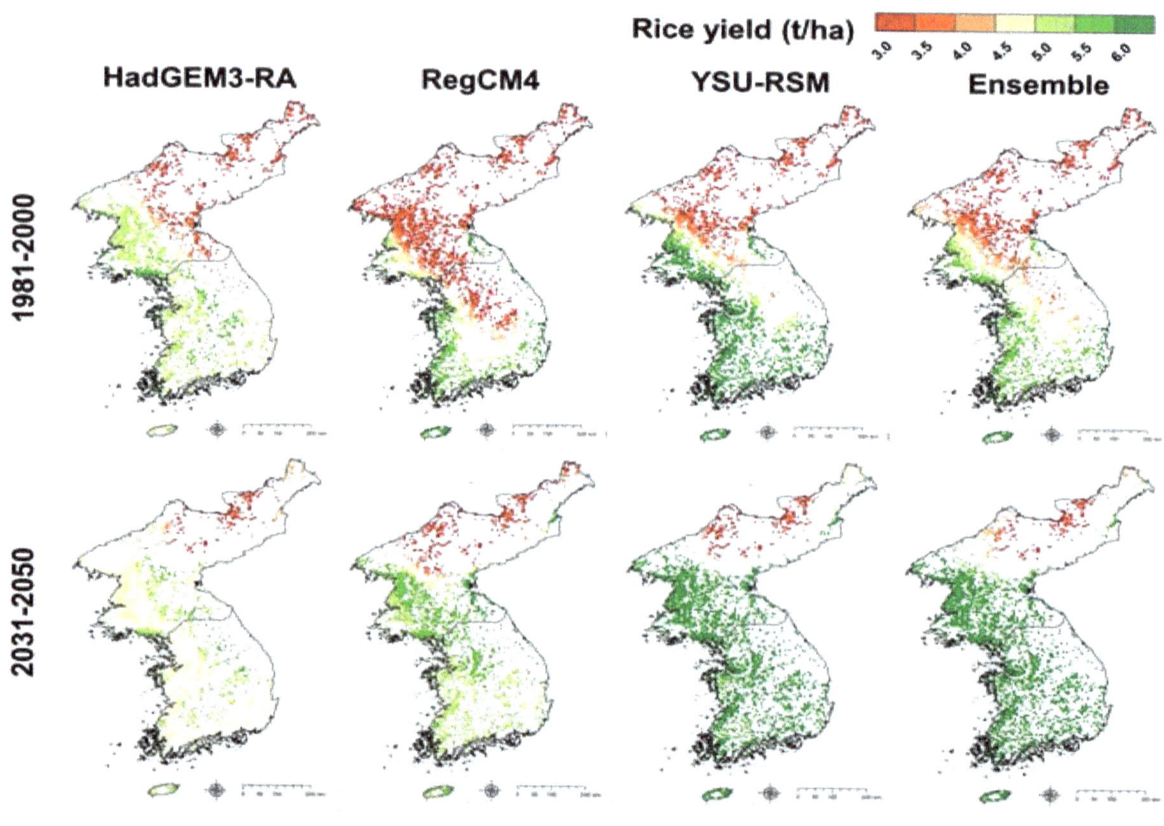

Figure 4. Spatial distribution of rice yield in the past and future for each climate model.

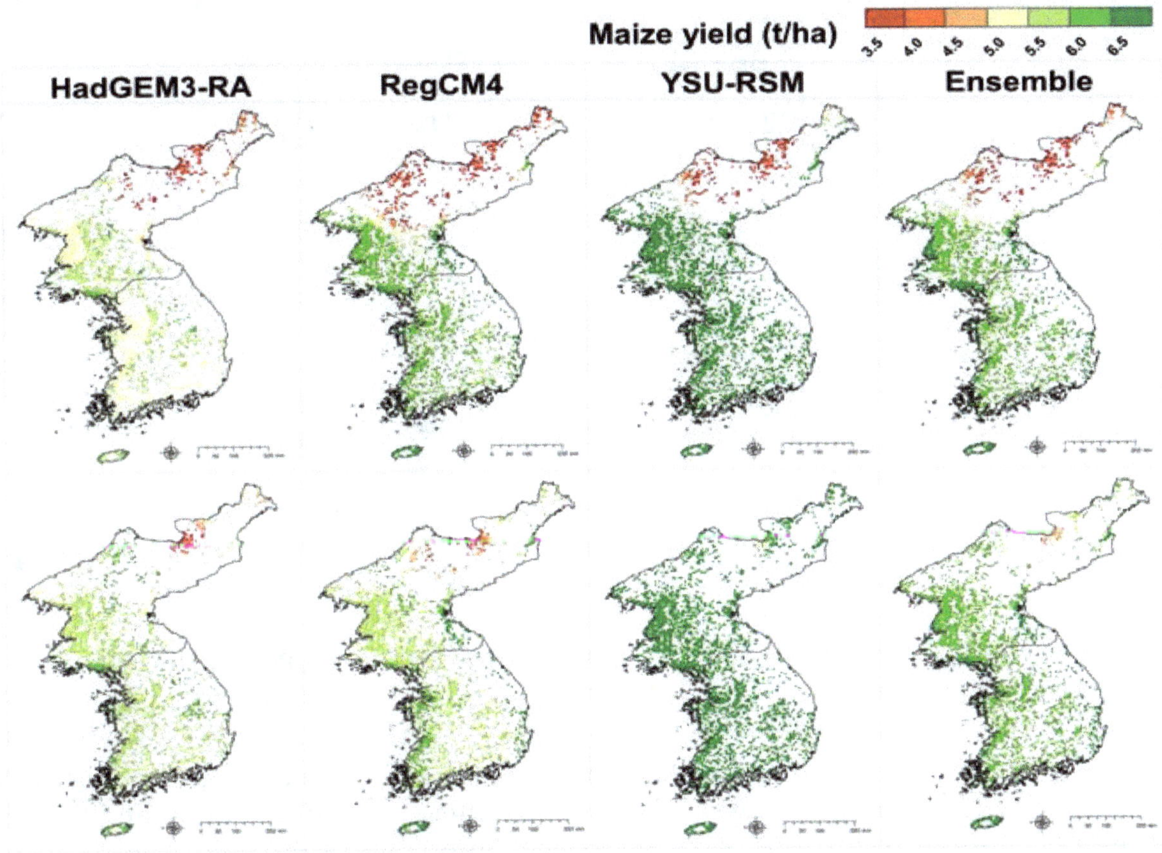

Figure 5. Spatial distribution of maize yield in the past and future for each climate model.

3.2. Estimation of PET Using Multiple Methods

Three RCMs and five PET methods were used to estimate a total of fifteen PETs in the Korean Peninsula. The PET estimation results were greater than the RCM results by over 1000 mm per year. The annual variations were minimal, and the differences between the RCM or PET method results were greater. Although all five methods were representative PET methods, the PET was largely overestimated and the BR was largely underestimated. The trends in the results for each past and future model (RCP 8.5) were similar, but the values were predicted to increase by 100 mm in the future (Figure 6). The ensemble values of the fifteen results confirmed that latitudinal differences were prevalent. In the past ensemble results, South Korea showed a PET level of 1000–1300 mm and North Korea 700–1100 mm. In the future, the PET levels for both South and North Korea are predicted to increase by an average of 100 mm (Figure 7).

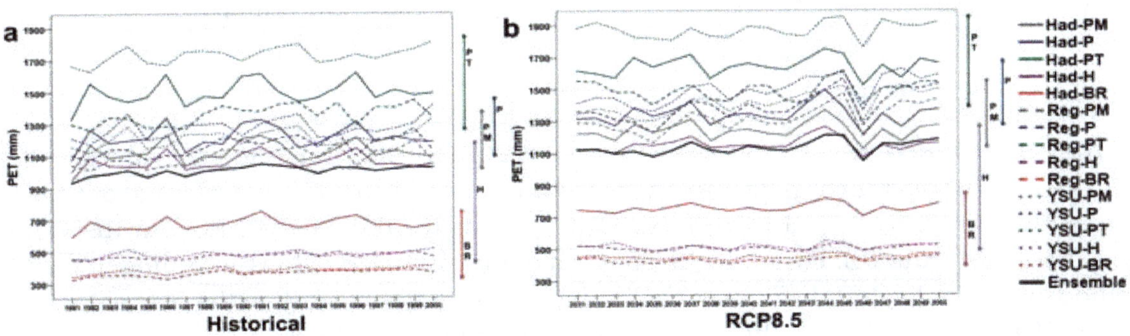

Figure 6. Estimated PET using multiple RCMs and PET methods for the past and future (**a**) Historical period (1981–2000); (**b**) RCP8.5 scenario (2031–2050).

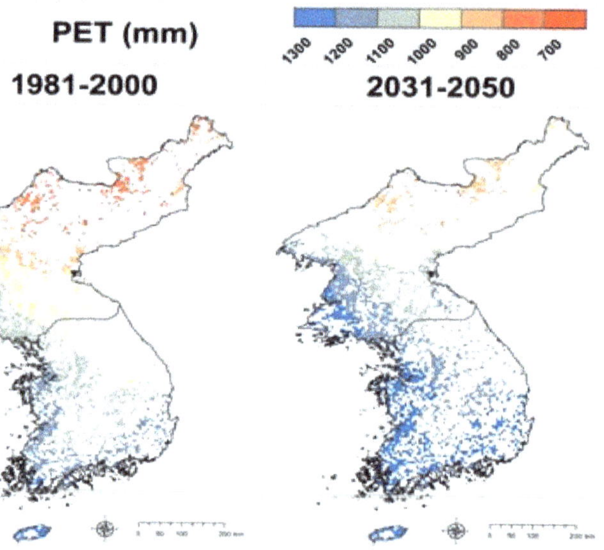

Figure 7. Spatial distribution of potential evapotranspiration for the past and future using the ensemble result.

The ensemble values were similar to the results of the existing PET studies in the Korean Peninsula or the satellite-based estimates when compared to the fifteen individual results, and the ensemble results for the future analysis period had a low level of uncertainty [41,54].

3.3. *Estimation of Consumptive Water Use by Growing Season Evapotranspiration*

The CWU was estimated using the growing season evapotranspiration (GET) data of each crop in the cropland. Similar to the PET estimation, three RCMs and five PET methods were applied, revealing fifteen results for the past and future periods. The GET estimation results differed largely from the RCM and PET method results, but the interval of the difference decreased as the total estimated amount decreased.

For the past data, the YSU-RSM-PT results (the most overestimated ones in PET) showed that the GET was 450 mm for both rice and maize, much higher than the other models. The RegCM4-BR results (the least underestimated PET results for rice) showed that the GET was less than 200 mm. The RegCM4-H results (least underestimated PET results for maize) showed that the GET was 165 mm on average. The average of the fifteen results was 307 mm for rice and 278 mm for maize (Figures 8 and 9). Initially, the water consumption of rice was much higher, but the results for the two crops showed a small difference in average values due to the basic latitudinal differences in GET and the relatively long growing period in North Korea.

For the future period, the estimated GET of rice was generally similar to past values, and showed a slight increase with rising temperatures (Figure 8). From 2031 to 2050, the average of the ensemble values was 312 mm. The spatial pattern was the same as for the past values, but the values were slightly decreased (Figure 10). For maize, the pattern of the estimated values according to the RCM and PET methods was similar, but the overall GET value was slightly lower than the past values (Figure 9). This lower GET is caused by the shorter growing period in North Korea following the temperature increase. The ensemble value of the fifteen GETs of maize was 271 mm. The spatial distribution results also confirmed that the GET decreased significantly in North Korea (Figure 10).

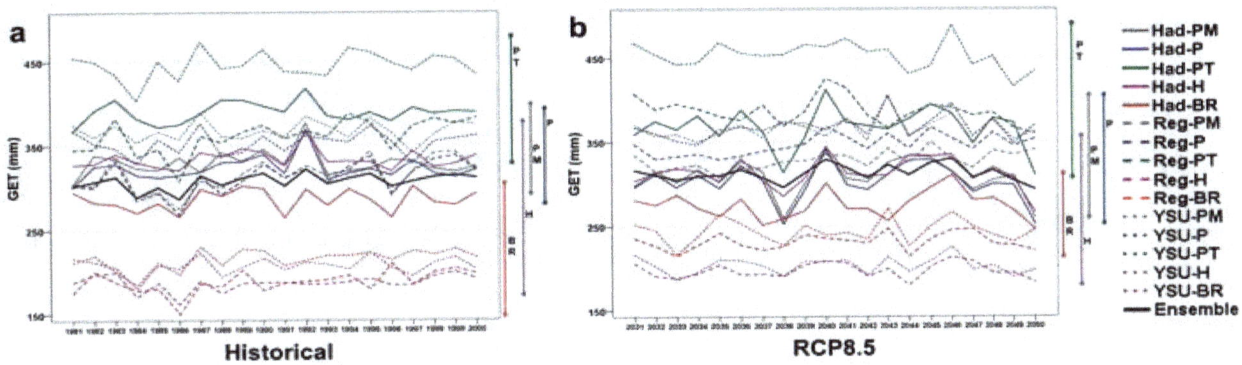

Figure 8. Estimated GET of rice using multiple RCMs and PET methods in the past and future (**a**) Historical period (1981–2000); (**b**) RCP8.5 scenario (2031–2050).

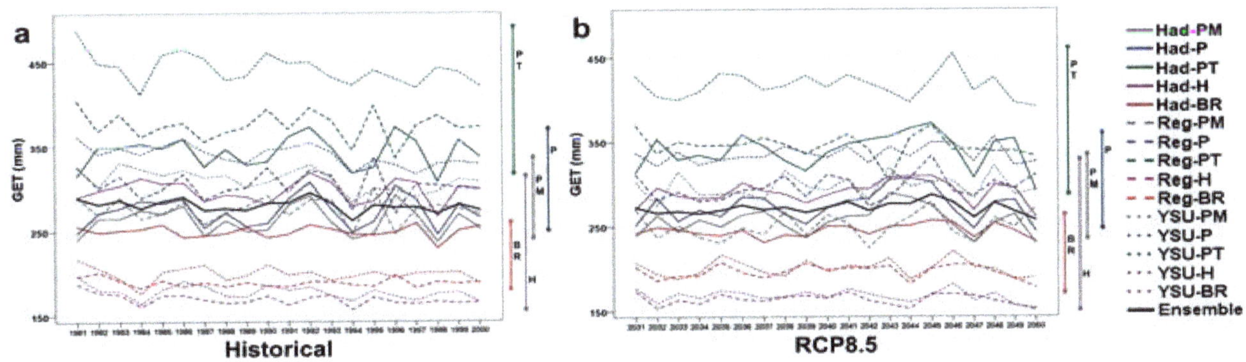

Figure 9. Estimated GET of maize using multiple RCMs and PET methods in the past and future (**a**) Historical period (1981–2000); (**b**) RCP8.5 scenario (2031–2050).

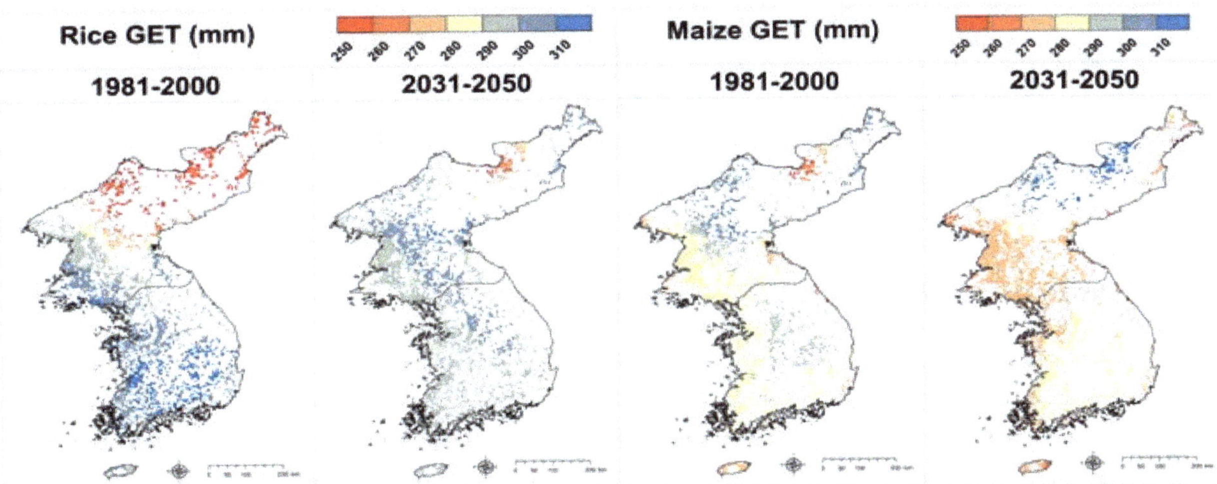

Figure 10. Spatial distribution of growing season evapotranspiration in the past and future using the ensemble result.

3.4. Virtual Water Content of Past and Future Using Multiple Data Sources

The crop yields and GET were used to calculate the VWC for each crop. Only the results from the three RCMs were estimated for the crop yield; therefore, the VWC was calculated using the five GETs and one crop yield, estimated by each RCM.

Reflected in the crop yield, the quantitative differences in the PET and the GET by the PET method significantly decreased. The VWC of rice ranged from 0.7 to 2.1 m^3 kg^{-1} and the average ensemble value was 1.18 m^3 kg^{-1}, consistent with the past data (Figure 11). In the future estimates, the range decreases to 0.4~1.1 m^3 kg^{-1} and the mean value of the ensemble is 0.76 m^3 kg^{-1}, significantly lower than the past data (Figure 11). These results indicate slight increases under the climate change scenario, but the rice yield is expected to increase greatly. The degree of change in the VWC calculated by the ratio of the two values is large. Spatially, there was a pattern of change similar to the rice yield result, with low VWC regions expanding northward (Figure 13).

The past values of the VWC of maize ranged between 0.3 and 1.0 m^3 kg^{-1}, and the ensemble average was 0.58 m^3 kg^{-1} (Figure 12). Under the climate change scenario, the range decreased to 0.2–0.7 m^3 kg^{-1} and the ensemble average to 0.48 m^3 kg^{-1}, slightly lower than the past values (Figure 12). Maize showed a smaller decrease in VWC than rice, and the maize yield is estimated to slightly increase while the GET slightly decreases. The degree of change in the VWC calculated by the ratio of the two values is relatively small. Spatial changes caused by climate change tend to be similar to the maize yield results, and the regions with lower VWCs are predicted to expand across the Korean Peninsula (Figure 13).

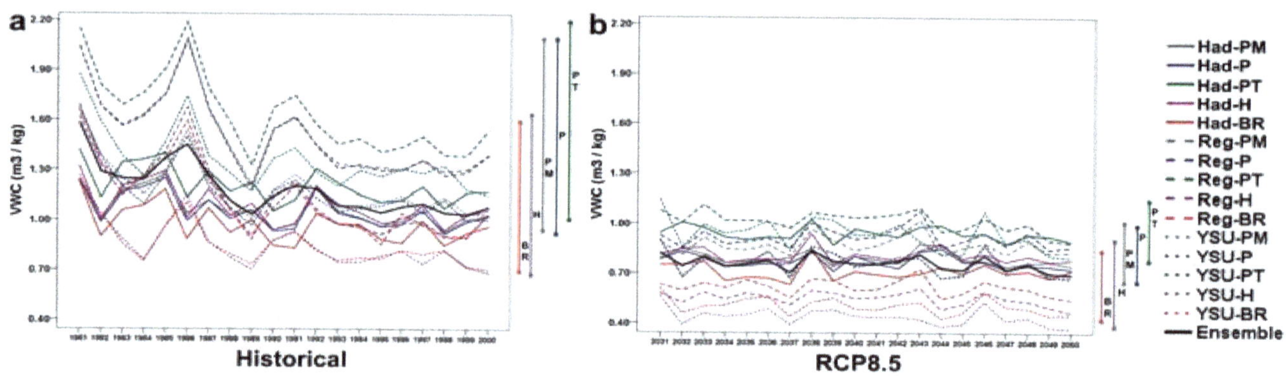

Figure 11. Estimated VWC of rice using multiple RCMs and PET methods in past and future (**a**) Historical period (1981–2000); (**b**) RCP8.5 scenario (2031–2050).

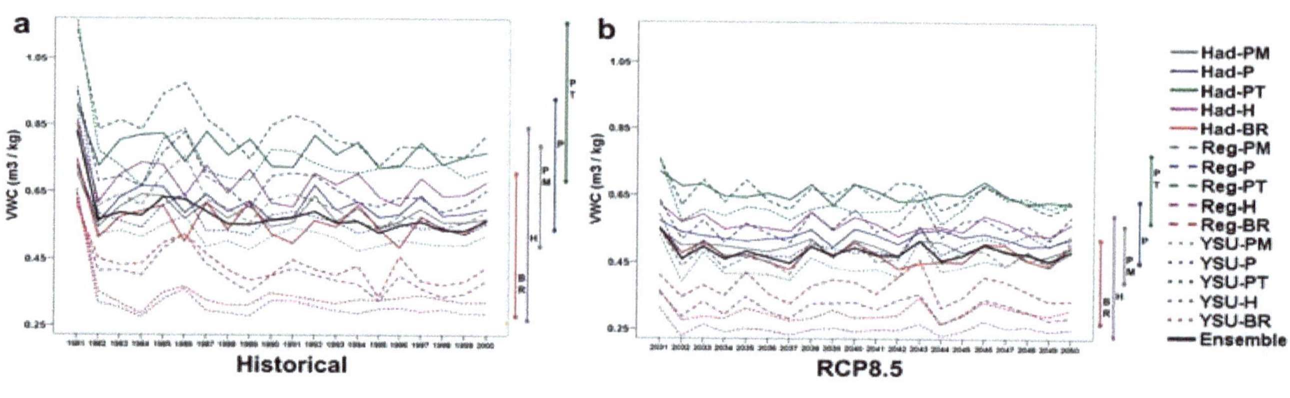

Figure 12. Estimated VWC of maize using multiple RCMs and PET methods in the past and the future (**a**) Historical period (1981–2000); (**b**) RCP8.5 scenario (2031–2050).

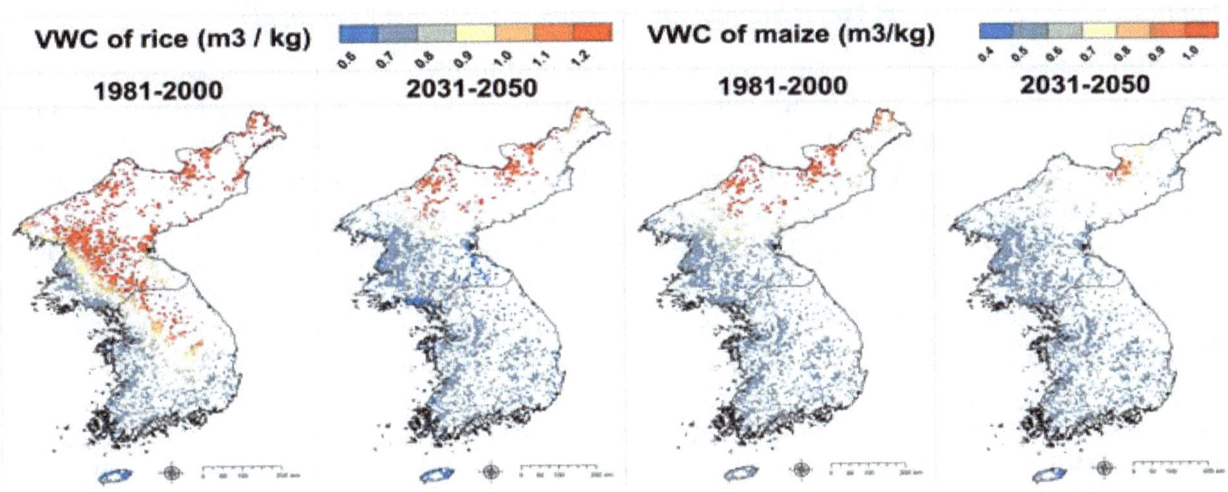

Figure 13. Spatial distribution of virtual water contents of crops in the past and future using the ensemble result.

4. Discussion

4.1. Assessing the Ensemble Result of Crop Yield, PET, GET, and VWC

The four outputs (crop yield, PET, GET, and VWC) estimated in this study were calculated using multiple methods and multiple data sources, so the ensemble results presented here have a lower level of uncertainty. This ensemble approach is also emphasized in the future climate change research outlook [19,21].

One of the uncertainties addressed in this study is the uncertainty in crop yields in the climate data. Since crop yields have a high dependence on climate, the accuracy of the climate data is very important, even if the same crop model is used [17,53]. In this study, three RCM data sets were used, derived from one global climate model (GCM). The climate values calculated by each RCM are different from the previous studies, reflected in the crop yield results of this study [10,17]. Ultimately, three crop yields from each crop and period could be used to estimate a more realistic crop productivity for the Korean Peninsula.

There are many ways to estimate the evapotranspiration required for calculating VWC and to lower the uncertainty in the results of those methods [13,21]. The data required for the five PET methods used in this study were different, and the estimated results also showed great differences. The results of the H and BR methods, estimated using only the temperature, tend to be underestimated, and the results of the PT method, estimated using temperature and solar radiation, tend to be overestimated. Hence, the methodological uncertainty is lowered by the analysis of the differences in the methodologies and the ensemble result of evapotranspiration using multiple methodologies.

In particular, we calculated fifteen VWCs for each crop, which have inherent uncertainties due to the climate models and the PET method. Presenting the average of fifteen VWCs as an ensemble value resulted in mitigating the differences in data and methods by up to 294% (0.33–0.97 m^3 kg^{-1} (in 1986)) (historical maize data, annual average). The maximum difference of 252% (0.87–2.19 m^3 kg^{-1} (in 1986)) in the historical rice data was alleviated. In the future projections, the total VWC decreased compared to the past values, but the effect of the ensemble results was similar between the past and future values. The maximum reduction in the RCP85-rice value was 262% (0.42–1.10 m^3 kg^{-1} (in 2043)) and a reduction of up to 245% (0.31–0.76 m^3 kg^{-1} (in 2031)) was observed in RCP85-maize results (Figures 10 and 11). In other words, the ensemble VWC reduced the difference by over 200% (according to data and methods) in all of the periods and crops. Although there are no VWC studies available for the Korean Peninsula, ensemble results are more similar than individual results when compared with the results of Zhao et al. [15], which calculated the VWC for China.

4.2. Implications for Agricultural Water Supply and Demand in the Korean Peninsula

Precipitation in the Korean Peninsula is concentrated in the summer monsoon season: 50–60% of annual precipitation occurs during this season. The water supply is not constant, as it is highly dependent on the river regime [10]. Most agricultural products in South Korea have been replaced by imported products due to changes at the economic level, but the self-sufficiency rate of food crops is so high that the consumption of water in agriculture is still at significant levels. In North Korea, the demand for water has increased because of the increase in food production owing to limited imports [24]. Overall, the Korean Peninsula, having a relatively high population density (South Korea: 519; North Korea: 208) and limited water supply, has a high demand for municipal, industrial, and agricultural water [55].

The average amount of water consumed by crops per unit production of major crops can be spatially determined by quantifying the agricultural water consumption using the results of the ensemble model approach employed in this study. This study also provided the future projections of the average agricultural water consumption, reflecting the impact of climate change, which has positive effects on productivity; as climate change accelerates, the amount of water consumed in the production of major crops will decrease. Climate change will have a negative impact on water supply as it affects precipitation patterns and amounts [10], but it can positively change water demand. However, this positive change in water demand due to climate change is the result of optimized adaptation modeling to water and nutrient stress. The consequences of climate change impacts without adaptation can vary significantly.

5. Conclusions

Multiple data sources and methods were employed in this study to estimate the past and future VWC of each crop in the Korean Peninsula. The EPIC crop model was used, and three RCMs and five PET methods were applied to reduce uncertainties in the data and methods. The rice and maize productivity varied significantly in the RCM results, confirming the increased potential production of both crops in the future. Positive changes in the northern part of the Korean Peninsula are noticeable, and maize is predicted to have high productivity in the entire peninsula in the future. The fifteen PETs and GETs from the RCM and PET methods were significantly different, and the water consumption of the crops was estimated by minimizing the errors by calculating the average values. The VWC was calculated for past and future crop yields and consumptive water use, with over 200% difference between them according to the RCM and PET methods. Computing the ensemble VWC for each period and crop by averaging fifteen VWCs could reduce the errors. The VWCs of the crops in the future projections were lower than those using the past data, which reflected a positive change in productivity and a decline in the length of the growth period. The past and future ensemble VWCs presented in this study provide quantitative data to shape the overall water demand for agriculture in the Korean Peninsula. We conclude that these results can be useful to improve agricultural sustainability, including food and water security, in the Korean Peninsula.

Acknowledgments: This study was supported by the Korean Ministry of Environment as part of the "Public Technology Development Project based on Environmental Policy" (Project Number: 2016000210001) and "Climate Change Correspondence Program" (Project Number: 2014001310008). We also thank Hanbin Kwak for technical support.

Author Contributions: Chul-Hee Lim designed this research, analyzed the result and wrote the paper; Seung Hee Kim and Yuyoung Choi participated in the data work. All authors, including Menas C. Kafatos and Woo-kyun Lee, gave comments and approved the final manuscript.

References

1. Sachs, J.D. From millennium development goals to sustainable development goals. *Lancet* **2012**, *379*, 2206–2211. [CrossRef]

2. IPCC. *Climate Change 2014: Impacts, Adaptation, and Vulnerability. Part A: Global and Sectoral Aspects. Contribution of Working Group II to the Fifth Assessment Report of the Intergovernmental Panel on Climate Change*; Cambridge University Press: Cambridge, UK, 2014.
3. Griggs, D.; Stafford-Smith, M.; Gaffney, O.; Rockström, J.; Öhman, M.C.; Shyamsundar, P.; Noble, I. Policy: Sustainable development goals for people and planet. *Nature* **2013**, *495*, 305–307. [CrossRef] [PubMed]
4. Wheeler, T.; Von Braun, J. Climate change impacts on global food security. *Science* **2013**, *341*, 508–513. [CrossRef] [PubMed]
5. Godfray, H.C.J.; Garnett, T. Food security and sustainable intensification. *Philos. Trans. R. Soc. B* **2004**, *369*, 20120273. [CrossRef] [PubMed]
6. Elliott, J.; Deryng, D.; Müller, C.; Frieler, K.; Konzmann, M.; Gerten, D.; Glotter, M.; Flörke, M.; Wada, Y.; Best, N.; et al. Constraints and potentials of future irrigation water availability on agricultural production under climate change. *Proc. Natl. Acad. Sci. USA* **2014**, *111*, 3239–3244. [CrossRef] [PubMed]
7. Calzadilla, A.; Rehdanz, K.; Betts, R.; Falloon, P.; Wiltshire, A.; Tol, R.S. Climate change impacts on global agriculture. *Clim. Chang.* **2013**, *120*, 357–374. [CrossRef]
8. Haddeland, I.; Heinke, J.; Biemans, H.; Eisner, S.; Flörke, M.; Hanasaki, N.; Stacke, T. Global water resources affected by human interventions and climate change. *Proc. Natl. Acad. Sci. USA* **2014**, *111*, 3251–3256. [CrossRef] [PubMed]
9. Huang, B.; Polanski, S.; Cubasch, U. Assessment of precipitation climatology in an ensemble of CORDEX-East Asia regional climate simulations. *Clim. Res.* **2015**, *64*, 141–158. [CrossRef]
10. Park, C.; Min, S.K.; Lee, D.; Cha, D.H.; Suh, M.S.; Kang, H.S.; Kwon, W.T. Evaluation of multiple regional climate models for summer climate extremes over East Asia. *Clim. Dyn.* **2016**, *46*, 2469–2486. [CrossRef]
11. Hoekstra, A.Y.; Hung, P.Q. Globalisation of water resources: International virtual water flows in relation to crop trade. *Glob. Environ. Chang.* **2005**, *15*, 45–56. [CrossRef]
12. Zhuo, L.; Mekonnen, M.M.; Hoekstra, A.Y. The effect of inter-annual variability of consumption, production, trade and climate on crop-related green and blue water footprints and inter-regional virtual water trade: A study for China (1978–2008). *Water Res.* **2016**, *94*, 73–85. [CrossRef] [PubMed]
13. Liu, J.; Zehnder, A.J.; Yang, H. Global consumptive water use for crop production: The importance of green water and virtual water. *Water Resour. Res.* **2009**, *45*. [CrossRef]
14. Liu, J.; Folberth, C.; Yang, H.; Röckström, J.; Abbaspour, K.; Zehnder, A.J. A global and spatially explicit assessment of climate change impacts on crop production and consumptive water use. *PLoS ONE* **2013**, *8*, e57750. [CrossRef] [PubMed]
15. Zhao, Q.; Liu, J.; Khabarov, N.; Obersteiner, M.; Westphal, M. Impacts of climate change on virtual water content of crops in China. *Ecol. Inform.* **2014**, *19*, 26–34. [CrossRef]
16. Zhuo, L.; Mekonnen, M.M.; Hoekstra, A.Y. Consumptive water footprint and virtual water trade scenarios for China—With a focus on crop production, consumption and trade. *Environ. Int.* **2016**, *94*, 211–223. [CrossRef] [PubMed]
17. Chun, J.A.; Li, S.; Wang, Q.; Lee, W.S.; Lee, E.J.; Horstmann, N.; Park, H.; Veasna, T.; Vanndy, L.; Pros, K.; et al. Assessing rice productivity and adaptation strategies for Southeast Asia under climate change through multi-scale crop modeling. *Agric. Syst.* **2016**, *143*, 14–21. [CrossRef]
18. Baek, H.-J.; Lee, J.; Lee, H.-S.; Hyun, Y.-K.; Cho, C.; Kwon, W.-T.; Marzin, C.; Gan, S.-Y.; Kim, M.-J.; Choi, D.-H. Climate change in the 21st century simulated by HadGEM2-AO under representative concentration pathways. *Asia-Pac. J. Atmos. Sci.* **2013**, *49*, 603–618. [CrossRef]
19. Rosenzweig, C.; Elliott, J.; Deryng, D.; Ruane, A.C.; Müller, C.; Arneth, A.; Boote, K.J.; Folberth, C.; Glotter, M.; Khabarov, N. Assessing agricultural risks of climate change in the 21st century in a global gridded crop model intercomparison. *Proc. Natl. Acad. Sci. USA* **2014**, *111*, 3268–3273. [CrossRef] [PubMed]
20. Wallach, D.; Rivington, M. A framework for assessing the uncertainty in crop model predictions. *FACCE MACSUR Rep.* **2014**, *3*, 1–5.
21. Liu, W.; Yang, H.; Folberth, C.; Wang, X.; Luo, Q.; Schulin, R. Global investigation of impacts of PET methods on simulating crop-water relations for maize. *Agric. For. Meteorol.* **2016**, *221*, 164–175. [CrossRef]
22. Choi, I.H.; Woo, J.C. Developmental process of forest policy direction in Korea and present status of forest desolation in North Korea. *J. For. Sci.* **2007**, *23*, 14.
23. Jeon, Y.; Kim, Y. Land reform, income redistribution, and agricultural production in Korea. *Econ. Dev. Cult. Chang.* **2000**, *48*, 253–268. [CrossRef]

24. Korea Rural Economic Institute (KREI). *KREI Quarterly Agriculture Trends in North Korea*; Korea Rural Economic Institute: Seoul, Korea, 2014.
25. Lee, J.W.; Hong, S.Y.; Chang, E.C.; Suh, M.S.; Kang, H.S. Assessment of future climate change over east Asia due to the RCP scenarios downscaled by GRIMs-RMP. *Clim. Dyn.* **2014**, *42*, 733–747. [CrossRef]
26. Giorgi, F.; Coppola, E.; Solmon, F.; Mariotti, L.; Sylla, M.B.; Bi, X.; Elguindi, N.; Diro, G.T.; Nari, V.; Giuliani, G.; et al. RegCM4: Model description and preliminary tests over multiple CORDEX domains. *Clim. Res.* **2012**, *52*, 7–29. [CrossRef]
27. Williams, J.R.; Jones, C.A.; Dyke, P.T. A modeling approach to determining the relationship between erosion and soil productivity. *Trans. ASABE* **1984**, *27*, 129–144. [CrossRef]
28. Williams, J.R.; Jones, C.A.; Kiniry, J.R.; Spanel, D.A. The EPIC crop growth model. *Trans. ASAE* **1989**, *32*, 497–511. [CrossRef]
29. Knisel, W.G. *GLEAMS: Groundwater Loading Effects of Agricultural Management Systems 2006*; Version 2.10 (No. 5); University of Georgia: Athens, GA, USA, 2006.
30. Izaurralde, R.; Williams, J.R.; McGill, W.B.; Rosenberg, N.J.; Jakas, M.Q. Simulating soil C dynamics with EPIC: Model description and testing against long-term data. *Ecol. Model.* **2006**, *192*, 362–384. [CrossRef]
31. Renard, K.G.; Foster, G.R.; Weesies, G.A.; Porter, J.P. Rusle: Revised Universal Soil Loss Equation. *J. Soil Water Conserv.* **1991**, *46*, 30–33.
32. Rinaldi, M. Application of EPIC model for irrigation scheduling of sunflower in Southern Italy. *Agric. Water Manag.* **2001**, *49*, 185–196. [CrossRef]
33. Williams, J.R. The EPIC model. In *Computer Models of Watershed Hydrology*; Singh, V.P., Ed.; Water Resources Publications: Littleton, CO, USA, 1995.
34. Xiong, W.; Balkovič, J.; van der Velde, M.; Zhang, X.; Izaurralde, R.C.; Skalský, R.; Lin, E.; Mueller, N.; Obersteiner, M. A calibration procedure to improve global rice yield simulations with EPIC. *Ecol. Model.* **2014**, *273*, 128–139. [CrossRef]
35. Lim, C.H.; Lee, W.K.; Song, Y.; Eom, K.C. Assessing the EPIC model for estimation of future crops yield in South Korea. *J. Clim. Chang. Res.* **2015**, *6*, 21–31. [CrossRef]
36. Lim, C.H.; Kim, M.; Lee, W.K.; Folberth, C. Spatially Explicit Assessment of Agricultural Water Equilibrium in the Korean Peninsula. *Agric. Water Manag.* **2017**, under review.
37. Balkovič, J.; van der Velde, M.; Schmid, E.; Skalský, R.; Khabarov, N.; Obersteiner, M.; Xiong, W. Pan-European crop modelling with EPIC: Implementation, up-scaling and regional crop yield validation. *Agric. Syst.* **2013**, *120*, 61–75. [CrossRef]
38. Lim, C.H.; Choi, Y.; Kim, M.; Jeon, S.W.; Lee, W.K. Impact of deforestation on agro-environmental variables in cropland, North Korea. *Sustainability* **2017**, under review.
39. Yoo, B.H.; Kim, K.S. Development of a gridded climate data tool for the Coordinated Regional climate Downscaling Experiment data. *Comput. Electron. Agric.* **2017**, *133*, 128–140. [CrossRef]
40. Nash, J.E.; Sutcliffe, J.V. River flow forecasting through conceptual models part I—A discussion of principles. *J. Hydrol.* **1970**, *10*, 282–290. [CrossRef]
41. Niu, X.; Easterling, W.; Hays, C.J.; Jacobs, A.; Mearns, L. Reliability and input-data induced uncertainty of the EPIC model to estimate climate change impact on sorghum yields in the US Great Plains. *Agric. Ecosyst. Environ.* **2009**, *129*, 268–276. [CrossRef]
42. Balkovič, J.; van der Velde, M.; Skalský, R.; Xiong, W.; Folberth, C.; Khabarov, N.; Smirnov, A.; Mueller, N.D.; Obersteiner, M. Global wheat production potentials and management flexibility under the representative concentration pathways. *Glob. Planet. Chang.* **2014**, *122*, 107–121. [CrossRef]
43. Monteith, J. Evaporation and environment. *Symp. Soc. Exp. Biol.* **1965**, *19*, 205–234. [PubMed]
44. Penman, H.L. Natural evaporation from open water, bare soil and grass. *Proc. R. Soc. Lond. Ser. A* **1948**, *193*, 120–145. [CrossRef]
45. Priestley, C.; Taylor, R. On the assessment of surface heat flux and evaporation using large-scale parameters. *Mon. Weather Rev.* **1972**, *100*, 81–92. [CrossRef]
46. Hargreaves, G.H.; Samani, Z.A. Reference crop evapotranspiration from temperature. *Appl. Eng. Agric.* **1985**, *1*, 96–99. [CrossRef]
47. Baier, W.; Robertson, G.W. Estimation of latent evaporation from simple weather observations. *Can. J. Plant Sci.* **1965**, *45*, 276–284. [CrossRef]

48. Liu, J.; Williams, J.R.; Zehnder, A.J.; Yang, H. GEPIC—Modelling wheat yield and crop water productivity with high resolution on a global scale. *Agric. Syst.* **2007**, *94*, 478–493. [CrossRef]
49. Food Agriculture Organization. *FAO Digital Soil Map of the World*; FAO: Rome, Italy, 1995.
50. Batjes, N.H. *ISRIC-WISE Derived Soil Properties on a 5 by 5 Arc-Minutes Global Grid*; ISRIC—World Soil Information: Wageningen, The Netherlands, 2006.
51. Chen, J.; Chen, J.; Liao, A.; Cao, X.; Chen, L.; Chen, X.; Zhang, W. Global land cover mapping at 30 m resolution: A POK-based operational approach. *ISPRS J. Photogramm. Remote Sens.* **2015**, *103*, 7–27. [CrossRef]
52. Kim, H.Y.; Ko, J.; Kang, S.; Tenhunen, J. Impacts of climate change on paddy rice yield in a temperate climate. *Glob. Chang. Biol.* **2013**, *19*, 548–562. [CrossRef] [PubMed]
53. Shin, Y.; Lee, E.J.; Im, E.S.; Jung, I.W. Spatially distinct response of rice yield to autonomous adaptation under the CMIP5 multi-model projections. *Asia-Pac. J. Atmos. Sci.* **2017**, *53*, 21–30. [CrossRef]
54. Liaqat, U.W.; Choi, M. Accuracy comparison of remotely sensed evapotranspiration products and their associated water stress footprints under different land cover types in Korean peninsula. *J. Clean. Prod.* **2017**, *155*, 93–104. [CrossRef]
55. Chung, E.S.; Won, K.; Kim, Y.; Lee, H. Water resource vulnerability characteristics by district's population size in a changing climate using subjective and objective weights. *Sustainability* **2014**, *6*, 6141–6157. [CrossRef]

14

New Breeding Techniques for Greenhouse Gas (GHG) Mitigation: Plants may Express Nitrous Oxide Reductase

Jordan J. Demone [1,†], Shen Wan [1,†], Maryam Nourimand [1], Asbjörn Erik Hansen [1], Qing-yao Shu [2] and Illimar Altosaar [1,*]

1 Department of Biochemistry, Microbiology, and Immunology, University of Ottawa, Ottawa, ON K1H 8M5, Canada; jdemone@uottawa.ca (J.J.D.); swa2@uottawa.ca (S.W.); mnourima@uottawa.ca (M.N.); aerik.hansen@gmail.com (A.E.H.)

2 National Key Laboratory of Rice Biology, Institute of Crop Sciences, Zhejiang University, Hangzhou 310058, China; qyshu@zju.edu.cn

* Correspondence: altosaar@uottawa.ca

† These authors contributed equally to this work.

Abstract: Nitrous oxide (N_2O) is a potent greenhouse gas (GHG). Although it comprises only 0.03% of total GHGs produced, N_2O makes a marked contribution to global warming. Much of the N_2O in the atmosphere issues from incomplete bacterial denitrification processes acting on high levels of nitrogen (N) in the soil due to fertilizer usage. Using less fertilizer is the obvious solution for denitrification mitigation, but there is a significant drawback (especially where not enough N is available for the crop via N deposition, irrigation water, mineral soil N, or mineralization of organic matter): some crops require high-N fertilizer to produce the yields necessary to help feed the world's increasing population. Alternatives for denitrification have considerable caveats. The long-standing promise of genetic modification for N fixation may be expanded now to enhance dissimilatory denitrification via genetic engineering. Biotechnology may solve what is thought to be a pivotal environmental challenge of the 21st century, reducing GHGs. Current approaches towards N_2O mitigation are examined here, revealing an innovative solution for producing staple crops that can 'crack' N_2O. The transfer of the bacterial nitrous oxide reductase gene (*nosZ*) into plants may herald the development of plants that express the nitrous oxide reductase enzyme (N_2OR). This tactic would parallel the precedents of using the molecular toolkit innately offered by the soil microflora to reduce the environmental footprint of agriculture.

Keywords: radiative warming; atmospheric phytoremediation; N_2O; nitrous oxide reductase; N_2OR; *nosZ*; fertilizer; crop breeding; transgenic; GHG

---

1. Introduction—Nitrous Oxide Continues to Bloom Unabated

Atmospheric nitrogen (N) deposition is a pressing matter for climate change scientists concerned with the increasing danger that nitrous oxide (N_2O), a noxious greenhouse gas (GHG), poses. Reactive nitrogen (Nr)—ammonia (NH_3), nitrogen oxides (NO_x), nitrates (NO_3^-), and N_2O—enters the biosphere from its original form of atmospheric N as at least three derivatives: gas, dry deposit, and precipitation (wet deposition) [1,2]. The sources of N_2O are largely anthropogenic [3]. Many crops must receive N-based fertilizer to reach yield targets, which is supplied by inorganic fertilizers and animal manure [4]. In an effort to boost the yield in crop staples like wheat, corn, and soybeans, farmers apply N fertilizers at rates and times that are not always properly synchronized with crop demand [5]. While crops thrive when fertilized, experimental analysis has demonstrated that up to

40% of fertilizer N can be lost via leaching [6,7]. Other routes of N loss include soil erosion, NH_3 volatilization and oxidation, and bacterial/fungal denitrification [8], although N losses through NH_3 volatilization are higher than those via N leaching [9]. Around 62% of total global N_2O issues from natural and agricultural soils, and the bulk of this production, mainly results from the processes of bacterial nitrification and denitrification [10].

Nr compounds enter the atmosphere through biological processes, but the invention of the Haber-Bosch process in 1908 was a critical moment for the sudden increase in Nr and GHG production globally [11]. This process of artificial N-fixation allowed for the large-scale reduction of N_2 to NH_3, producing massive amounts of synthetic N-based fertilizers that supported dramatic increases in high-yield farming [12]. This process now accounts for 80% of anthropogenic N-fixation (the remaining 20% resulting from combustion [13], with anthropogenic N-fixation in turn accounting for 60% of global N-fixation [14]). Haber-Bosch remains the industry standard synthetic N fertilizer today and as a result, has contributed to the ~2% increase in atmospheric levels of N_2O [15,16]. This effect is also magnified by the global emissions of N_2O produced by fossil fuel combustion [17] and the natural ability of legumes to fix N through symbiotic relationships with soil bacteria [18].

N_2O is the third most prevalent GHG, behind carbon dioxide (CO_2) and methane (CH_4) [19]. The concentration of this gas in the atmosphere has been steadily increasing since the early 1900s (Figure 1), and it is 265 times more radiative than CO_2 [19]. N_2O also has an atmospheric lifetime of 121 years; by comparison, CH_4 has an atmospheric lifetime of only 12 years, but CO_2 also has a long half-life and can take anywhere from 20–200 years to be absorbed by the ocean [19], compounding the 'greenhouse gas' effect. Since chlorofluorocarbons (CFCs) were banned in 1989, N_2O has become the leading cause of ozone layer depletion [20].

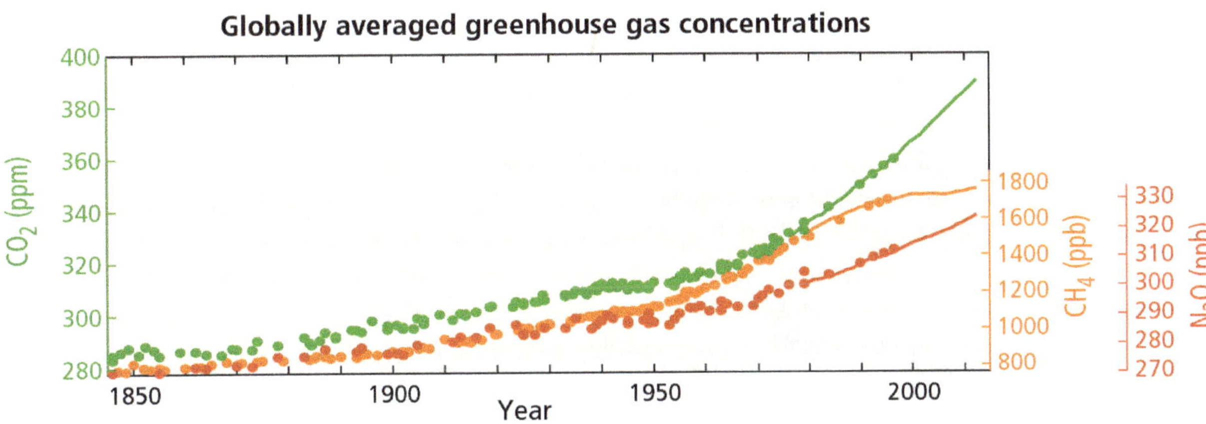

Figure 1. GHG levels since 1850. The green line represents the increase in CO_2 concentration since 1850; the orange line represents the increase in CH_4 concentration since 1850; lastly, the red line represents the increase in N_2O since 1850 [19].

N_2O emission results from the coupled oxidation and reduction of N performed by heterotrophic [21] (and some autotrophic) soil proteobacteria: (1) the nitrification pathway is catalyzed by autotrophs (*Nitrosomonas* spp. and other genera [22]) and also heterotrophs, and involves the oxidation of NH_3/ammonium(NH_4^+) to nitrite (NO_2^-) [23] and nitric oxide (NO) [24]), which is followed by the oxidation of NO_2^- to NO_3^- by *Nitrobacter* spp. [25]; and (2) the denitrification pathway, whereby NO_3^- is reduced to N_2O and ultimately inert N_2 gas [26]. As many as a third of soil bacterial species [27] lack the *nosZ* gene that reduces N_2O to inert N_2 [28], which leads to a sizeable amount of incomplete denitrification reactions and the subsequent buildup of N_2O since it is an obligate intermediate [29]. This N_2O diffuses out of the soil and into the atmosphere, contributing to the greenhouse effect, contaminating water, and leading to serious human health implications [30,31].

2. Combating GHGs: Current N_2O Mitigation Strategies and Limitations

Demands for crop-borne food must be met, and so researchers must address the hazards of N-based fertilizers [32]. There are multiple N_2O mitigation strategies either currently in commercial use or in development (summarized in Table 1).

Table 1. Summary of current N_2O mitigation strategies.

Strategy	Mechanism of Action	Pros	Cons
(1) Conservation tillage and crop rotation [33]	Tillage, rotation of N-fixing crops, cover cropping [33]	Prevent NH_3 volatilization and eventual N_2O emissions [34,35]	Unreliable N_2O mitigation [36,37]. Yield reduction [38]. Not effective at scrubbing N_2O from the air
(2) Best management practices (BMPs) [39]	Correct source, placement, time, and rate of fertilization [40]. Proper irrigation (fertigation) [41]	Proven to reduce N_2O emissions [41] and other N losses [42]	Technical constraints [43]
(3) EENFs [44]	Multiple types: stable, short-release (SRFs), and constant-release (CRFs); rely on enrichment of chemical inhibitors or coated N-compounds that are released into the soil over a period of time [45]; urease inhibitors (UIs) [46]	Proven to reduce N_2O emissions [47,48]	Inconsistent yields from year to year [48]. More expensive than standard N fertilizers [49]. Long lifetime of N-compounds in soil can lead to NH_3 volatilization [50,51]. Not effective at scrubbing N_2O from the air
(4) Synthetic N_2O mitigators	SNIs suppress activity of nitrifying bacteria in the soil [52]. SDIs operate by unknown mechanism [44,53,54]	SNIs and SDIs reduce N_2O emissions [52,54]	Effectiveness depends on environmental conditions, prefer low temperature and sandy soils [55]. Not effective at scrubbing N_2O from the air
(5) Biological N_2O mitigators	BNIs suppress activity of nitrifying bacteria in the soil by releasing compounds that inhibit NH_3-oxidizing pathways [56]. BDIs inhibit nitrate reductase to inhibit N_2O production [57]	BNIs demonstrated to reduce N_2O emission [56]; BDIs inhibit denitrification and can conceivably mitigate N_2O emissions [57]	BNI-exuding plants must be grown in rotation with other crops [58]. Little work done on BDI-exuding plants [57]. Not effective at scrubbing N_2O from the air
(6) Microbial bioremediation	Proper water table management to facilitate growth of rhizobia [59]; inoculation of plant roots with genetically modified N_2O-cracking rhizobia [60,61]	Enables plants to degrade contaminants in the soil; N_2O-cracking rhizobia demonstrated to reduce N_2O emissions [60,61]	Most effective on crops that naturally cultivate a rhizosphere of N_2O-reducing [62] microorganisms, i.e., soybean [63]. Not effective at scrubbing N_2O from the air
(7) Rhizosecretion	Transformation of amenable crops to express recombinant bacterial proteins that reduce N_2O [64]	Plants that secrete N_2O-cracking enzyme could target N_2O in soil [64]	Plant transformation is a time-consuming process [65]. Bacterial proteins may not function efficiently in heterologous hosts [66]. Not effective at scrubbing N_2O from the air
(8) Atmospheric phytoremediation	Transformation of amenable crops with genes expressing recombinant bacterial proteins that reduce N_2O [67]	Arm crops and other plant species to mop up N_2O in the atmosphere [67], including N_2O emitted by other non-agricultural sources	Plant transformation is a time-consuming process [65]. Bacterial genes may not function in a heterologous system [66]. Not yet experimentally validated via gas analysis

BDI, biological denitrification inhibitor; BNI, biological nitrification inhibitor; EENFs, enhanced efficiency nitrogen fertilizers; SDI, synthetic denitrification inhibitor; SNI, synthetic nitrification inhibitor.

(1) **Conservation tillage and crop rotation**. Mechanical incorporation (tillage) of N-based fertilizer into the soil may also be effective [68], but this is affected by many other parameters, such as the method of N application (i.e., broadcast vs surface urea ammonium nitrate). These techniques also result in a reduced yield [38]. Conservation tillage increases N_2O emissions compared with no-till and conventional tillage techniques using broadcast application, while tillage in general does not reduce N_2O emissions produced from surface urea ammonium nitrate-treated fields [69]. Other studies have shown that conservation tillage *reduces* N_2O emissions [70], underscoring the lack of reliability of this N management technique [36,37]. Crop rotation with N-acquisitive plant species can also reduce N_2O emissions following the application of high N-fertilizer treatment [33]; cover cropping can also control N_2O emissions, but the results are often variable and in some cases can increase N_2O emissions [71];

(2) **Best management practices (BMP) [39]**. Such nitrogen use efficiency techniques are myriad and involve simple steps such as proper fertilizer placement, timing of fertilizer application, the right type of N-compound, and so on. Others involve the proper incorporation of N-compounds into the soil so that they may be taken up by the plant more effectively and will be less likely to volatilize [72]. Fertigation, a technique involving careful irrigation of fields following the application of N fertilizer, is effective at mitigating N_2O emissions [41]. Such knowledge-based N management practices have been shown to be effective at both increasing crop yield and reducing immediate N_2O emissions [73], but some approaches may also increase N_2O production in the long term [55]. Their effectiveness also depends heavily on proper practices put in place by the farmers themselves, which requires proper training [43];

(3) **Fertilizer management using enhanced efficiency nitrogen fertilizers (EENFs)**. These fertilizer cocktails are concocted in such a way that they prevent the volatilization of NH_3 and inhibit nitrification/denitrification [46]. EENFs generally fall into one of three categories: (a) stabilized fertilizers, which contain nitrification and/or urease inhibitors; (b) slow-release fertilizers (SRFs), whereby the N source in the fertilizer is released over time from encapsulated granules, although the release rates can be variable; and (c) controlled-release fertilizers (CRFs), where the release rate is constant [45]. Urease inhibitors (UIs) are also a common EENF component. N-(*n*-butyl) thiophosphoric triamide (NBPT), phenylphosphorodiamidate (PPD), and hydroquinone are used worldwide and act by inhibiting the bacterial hydrolysis of urea into NH_3 in fertilizer [46,74,75]. UIs are typically used in conjunction with nitrification inhibitor (NIs) for maximum effectiveness [76,77], but NBPT alone can reduce N_2O emissions from N-treated soil [78]. There is controversy regarding the effectiveness of EENFs; while reductions in N_2O emissions from the soil have been recorded [47,48], recent studies have shown that crop yields are only marginally higher when EENFs are used in place of standard N fertilizers [79]. Those studies that demonstrated reduced N_2O emissions also reported inconsistent results from year to year [50]. Questionable effectiveness notwithstanding, EENFs are more expensive than conventional N-containing fertilizers and require special handling and storage [49,80], which are all features that make these fertilizers less attractive to farmers;

(4) **Synthetic N_2O mitigators**. Synthetic nitrification inhibitors (SNIs) and UIs are both used in EENFs and can be applied to crops in conjunction with standard N fertilizer. NIs inhibit the activity of *Nitrosomonas* to block the nitrification of N in fertilizer (the oxidation of NH_3 to hydroxylamine via ammonia monooxygenase (AMO)) [23,52]. The efficacy of the inhibitors is also dependent on environmental conditions, as they are unstable; 3,4-dimethylpyrazole phosphate (DMPP), for example, exhibited reduced activity in hot, dry conditions [81]. The use of these inhibitors can also lead to less than desirable results: DMPP and 3-methylpyrazole 1,2,4-triazole (3MP + TZ) have been shown to increase N_2O emissions in vegetable crop systems, as the inhibitors promote the buildup of N in the fraction of the soil most available to bacteria during the breakdown of vegetative matter. Synthetic denitrification inhibitors (SDIs) suppress denitrification via unknown mechanisms [82], although some are known to inhibit the activity of fungal copper reductase [83].

SDIs nitrapyrin [84], toluidine [54], and acetylene [44] all effectively mitigate N_2O emission, albeit with toxic side-effects [55], and they do not technically inhibit nitric oxide reductase;

(5) **Biological N_2O mitigators**. This category is comprised of compounds produced by plants that inhibit enzymes in either the bacterial nitrification or denitrification pathway. The exploitation of such inhibiting root exudates is another intriguing approach towards N_2O mitigation [82]. Biological nitrification inhibitors (BNIs) are compounds that block the activity of NO_2^- producing enzymes. The roots of the tropical grass *Brachiaria humidicola* exude brachialactone, a compound that can mitigate N_2O emission from soil [85]. Attempts at developing BNI-producing cultivated wheat by crossing *Triticum aestivum* with BNI-producer *Leymus racemosus*, a wild wheat, have imparted some BNI activity, but also made the lines susceptible to rust infection [86]. The use of BNIs as an effective N_2O mitigator is also severely limited by the fact that the enactor of nitrification is a plant itself and cannot be applied to growing crops, although growing *B. humidicola* in rotation with maize saw a four-fold increase in yield [87]. Biological denitrification inhibitors (BDIs) are a relatively new discovery. Currently, the only example of such an inhibitor is the procyanidin produced by the invasive Fallopia spp. (Asian knotweed). This compound has been demonstrated to be an allosteric inhibitor of *Pseudomonas brassicacearum* nitrate reductase and while it does reduce denitrification in the soil, it has not yet been proven to mitigate N_2O levels [57];

(6) **Microbial bioremediation [88]**. The success of N fertilizer management techniques and proper irrigation is largely due to the creation of a microsphere conducive to denitrifying bacteria flourishing [89]. Proper water table management techniques can promote the growth of N_2O-cracking bacteria in the soil and reduce N_2O emissions from the managed soil regions [59]. Another type of microbial bioremediation takes advantage of the ability of certain bacterial species to inhabit the root nodules of leguminous crops. Field peas [62], broad beans [90], and soybean [63] house bacteria (or rhizobia) that fix N and, unfortunately, also produce N_2O gas. While maintaining the rhizosphere, N_2O emissions can be mitigated by inoculating the roots of leguminous plants with rhizobia modified to express higher levels of a bacterial N_2O-cracking enzyme [60]. Genetically engineered strains of *Bradyrhizobium japonicum* have been used to inoculate the roots of soybean and reduced N_2O emissions [61]. Needless to say, this method is far more effective on crops that naturally cultivate a rhizosphere of N_2O-reducing microorganisms. It is also another technique that cannot target atmospheric N_2O;

(7) **Rhizosecretion**. This is a biotechnology-based approach, involving the transformation of amenable crop plants with genes expressing recombinant bacterial proteins that reduce N_2O by secreting N_2O-cracking enzymes [64,91]. Plants can be engineered to express proteins under the control of promoters that induce hairy root formation in plants. This rooting response results from the presence of the *rolABCD* genes from *Agrobacterium rhizogenes*, the bacterium that induces hairy root disease [92]. The rhizosecretion expression system harnesses the ability of *A. rhizogenes* to both target gene expression to the roots and to increase root biomass, subsequently increasing the amount of recombinant protein secreted into the soil [91]. Tobacco plants expressing a bacterial N_2O-cracking enzyme tagged for secretion under the control of the *A. rhizogenes rolD* promoter have been successful in demonstrating reducing activity [64,93]. Gas analysis was not performed to confirm that these plants mitigated N_2O emission. Ultimately, this approach arrives at a similar problem as other 'rhizoremediative' techniques: the N_2O-reducing ability of such a transgenic plant would be limited to the rhizosphere. This system would not have access to the bulk of N_2O gas, much of which comes from other sources;

(8) **Atmospheric phytoremediation using genetically engineered plants**. The potential of transgenic plants for environmental phytoremediation is well-documented: several fungal and bacterial oxidoreductases have been functionally expressed in plants as phytoremediation strategies including pentaerythritol tetranitrate reductase [94], mercuric reductase [95], and arsenate reductase [96]. This type of plant-based decontamination strategy provides advantages,

such as stable cultivation and control of the remediant organism and atmospheric exposure of the gas-cracking enzyme [97].

Atmospheric phytoremediation may ameliorate problems created by the other N_2O mitigation strategies described. The concept here is to develop crops with the ability to "crack" N_2O in both the soil and the atmosphere by incorporating the bacterial *nosZ* gene into their genomes. This gene encodes the nitrous oxide reductase enzyme (N_2OR), an oxidoreductase that catalyzes the removal of N_2O from the atmosphere, a process performed naturally by both denitrifying and non-denitrifying bacteria in the soil [98]. While conventional N_2O mitigation strategies aim to control N_2O production at earlier stages in the nitrification/denitrification pathway, this approach will target the atmospheric sum of N_2O emitted by all sources (Figure 2).

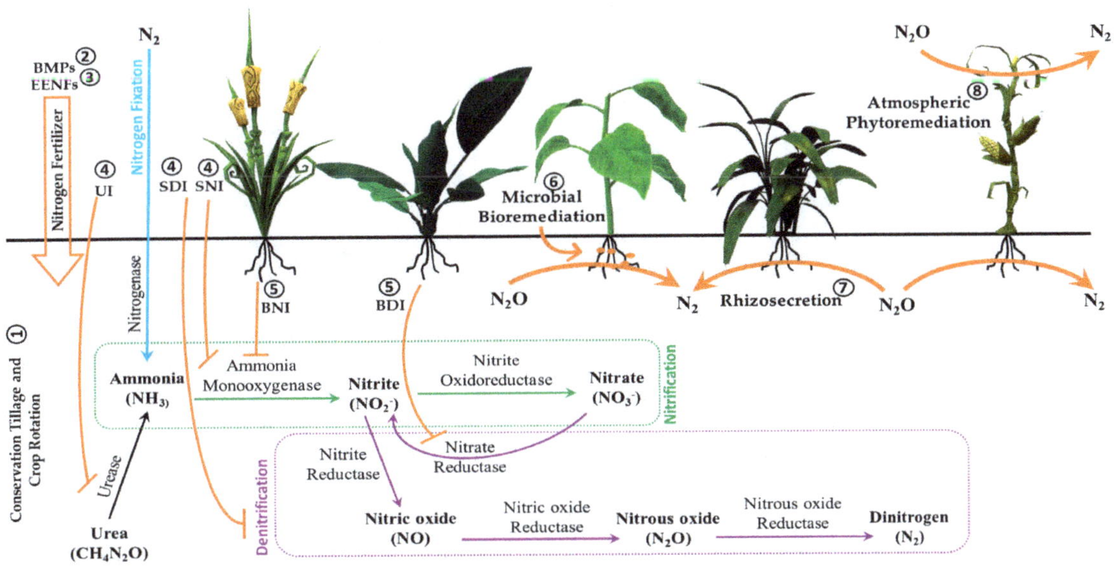

Figure 2. Nitrification-denitrification pathway and overview of current N_2O mitigation strategies. Orange arrows and lines show eight N_2O mitigation strategies described in Table 1. Green arrows show nitrification and purple arrows represent denitrification reactions. BDI, biological denitrification inhibitor; BMPs, best management practices; BNI, biological nitrification inhibitor; EENFs, enhanced efficiency nitrogen fertilizers; SDI, synthetic denitrification inhibitor; SNI, synthetic nitrification inhibitor; UI, urease inhibitor. **O** Encircled numbers refer to Table 1 strategies.

3. Nitrous Oxide Reductase—An Orphaned Soil Protein?

The *nosZ* gene can be categorized as either 'clade I' or 'clade II' based on sequence and *nos* operon organization, including the lack of an accessory *nosR* gene in the clade II members [99]. Clade II *nosZ* genes are also known as 'atypical' *nos* genes since they are found in non-denitrifying bacterial species. The N_2OR enzyme that the clade II gene encodes catalyzes the same reaction performed by the clade I-encoded enzyme, but has a higher affinity for N_2O [100], an important factor to consider when conceptualizing the development of an *nosZ*-expressing plant.

N_2OR is a multi-copper protein encoded by the *nosZ* gene (which is accompanied by an operon cluster of additional genes (*nosRDFYL*) [101]) and is the only enzyme that can catalyze the conversion of N_2O into N_2. The first active N_2OR was characterized from the soil bacterium *Pseudomonas stutzeri* and similar enzyme structures were resolved in bacterial species *Marinobacter hydrocarbonoclasticus* (formerly *Pseudomonas nautica*) (Figure 3), *Achromobacter cyclocastes*, and *Paracoccus denitrificans*. N_2OR is a head-to-tail homodimer and each monomer contains two domains: an electron transferring domain (binuclear Cu_A centre) and a catalytic domain (tetranuclear Cu_Z centre) [102]. There is some variability between the species regarding Cu_Z bridging and cupric coordination in the catalytic centre, suggesting that N_2OR substrate binding is species-specific. Regardless, the catalytic mechanism of N_2O reduction in N_2OR is still unclear [103].

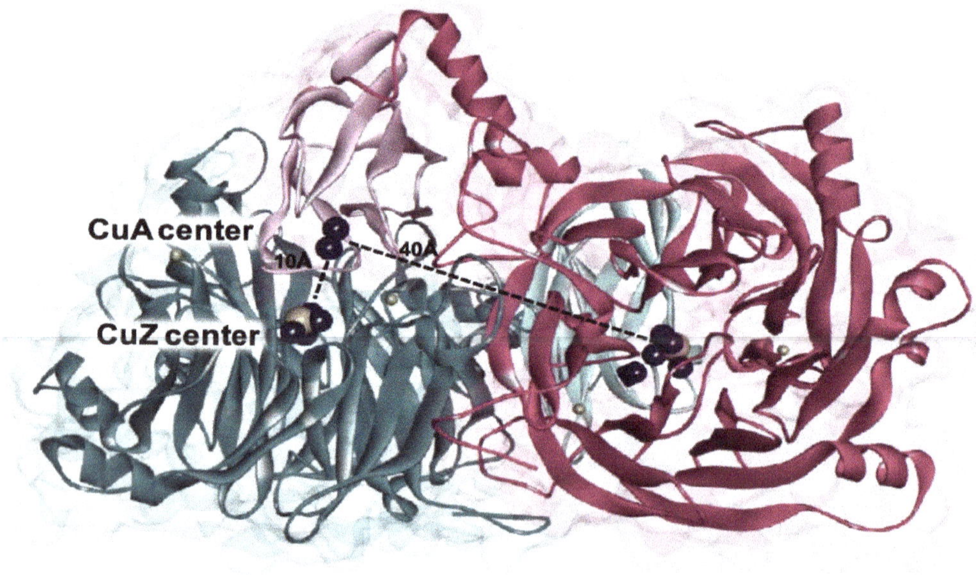

Figure 3. Structure of *Marinobacter hydrocarbonoclasticus* nitrous oxide reductase (N_2OR) homodimer. N_2OR is organized as a head-to-tail homodimer. Monomers are coloured differently so that they can be distinguished. In both monomers, the N-terminal domain is dark-coloured. The N-terminal domain forms a seven-bladed β-propeller fold that coordinates the catalytic tetranuclear active site Cu_Z through seven histidine residues at its hub. The C-terminal domain forms a cupredoxin fold and binds the dinuclear mixed-valent Cu_A centre [104].

The proven ability of N_2OR to "crack" the N_2O molecule raises the question of why the protein has not yet been incorporated into a commercially available transgenic cropping choice for environmentally motivated producers and small-plot farmers. Work has been done on this gene and its potential role in plant biotechnology since it was originally isolated in 1998 from the anaerobic soil bacterium *A. cyclocastes* [105,106], but it has yet to be converted into a commercially valuable tool. In this sense, N_2OR may be considered an "orphaned" protein, neglected among a veritable molecular toolkit of genes in the soil microflora [107,108]. Such forays into integrating soil and air sciences are demonstrative of the possibilities of what the soil microbiome offers biotechnologists [27]; it has already been discussed regarding the N-management possibilities offered by the microbiome and the current practice of 'bioprospecting' is also revealing a plethora of beneficial bacterial products, which is only accelerating thanks to whole-system approaches involving computational analyses [109].

Web of Science reports that between 1900 and 1991, there are no records binned under the combined topics "nitrous oxide reductase" and "microb*". The scientific literature blossomed from its first occurrence of 1992 to the present day, witnessing at least 175 publications dealing with the science of this important enzyme in our total environment. The scientific community waited until 1996 to start discussing denitrification in a plant context, according to these same search terms. With the search terms "nitrous oxide reductase" and "plant", the scientific record shows that soil microbiologists have taken a growing interest in the movement of N into the atmosphere (Figure 4). It is encouraging to note that in the same time period, the linkage between N_2OR and climate began its nascent phase.

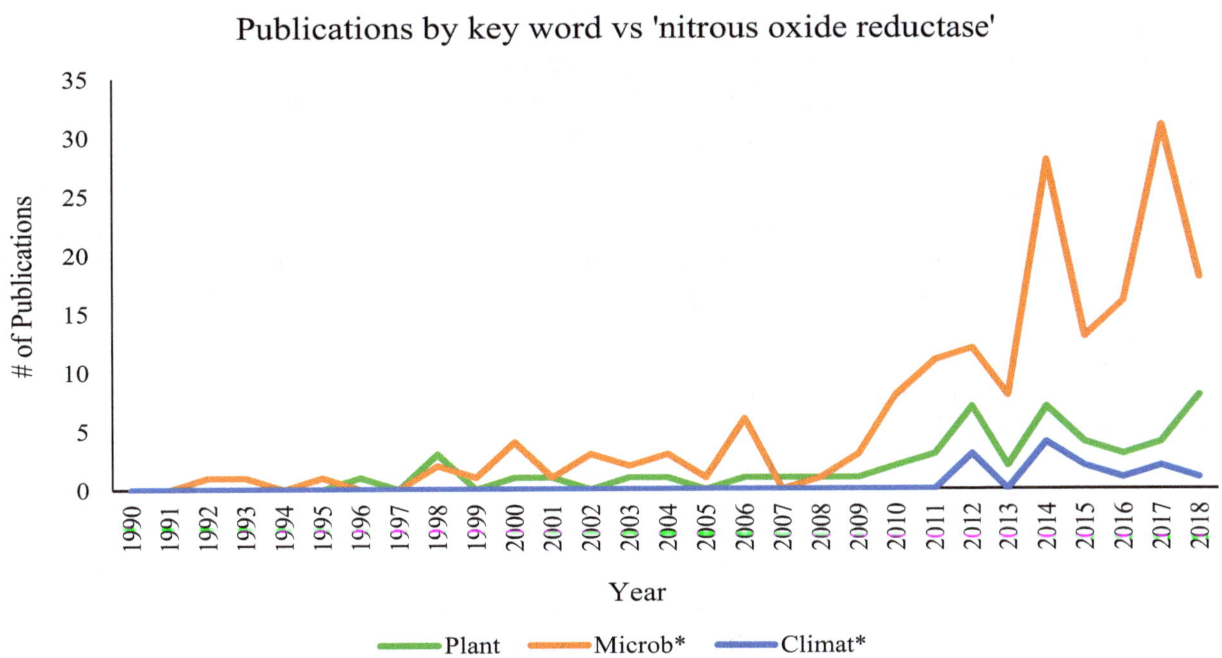

Figure 4. Nitrous oxide reductase-related publications released since 1990 on Web of Science (Clarivate Analytics). Publications by key word vs "nitrous oxide reductase" from 1990 to 2018. The orange line indicates "nitrous oxide reductase" + "microb*"; green: "nitrous oxide reductase" + "plant"; blue: "nitrous oxide reductase" + "climat*".

4. Catch Me If You Can: Can Plants Catalytically Convert N_2O *in planta*?

Rather than a 'cat and robin redbreast' conundrum, we are confronted with an opportunity to deploy protein engineering to ensure that more N_2OR molecules are attracted to the substrate binding site of the copper enzyme. Protein engineering offers ways to sidestep the challenges of expressing a complex bacterial protein in a plant [110]. There are potential issues with a recombinant metalloprotein like N_2OR, such as whether the ABC transporter can assemble within a plant cell, or the plant can incorporate copper into the electron transferring and catalytic domains [111,112]. It is possible to re-engineer N_2OR and produce a functional product [66], so there is precedent for designing an artificial metalloenzyme through rational protein design. This approach may be key to engineering a plant-compatible N_2OR protein.

A principle challenge associated with imparting N_2OR functionality to plants is that transforming the *nosZ* sequence alone may not be effective [113]; in *P. stutzeri*, the transcription of *nosZ* was dependent on the *nosDFY* genes being expressed, as they encode components of a putative ABC transporter system for the biogenesis of the Cu_Z centre [114]. Therefore, catalytically active N_2OR may not be produced when only *nosZ* is expressed in a heterologous host [28]. Nevertheless, a model N_2O-expressing plant has been engineered [64,93]. The clade I *nosZ* gene from soil bacterium *Pseudomonas stutzeri* was successfully expressed in a heterologous system—in this case, the tobacco plant (*Nicotiana tabacum*). In those proof-of-concept experiments the *nosZ*-expressing tobacco plants reduced 826 µg N_2O/min/gram of leaf tissue [115]. Assuming the tobacco yield to be 0.50 tonne/ha [116], the calculated N_2O-cracking ability of the *nosZ*-expressing tobacco could be as high as 600 kg of N_2O/ha/day [115], or 60 tonnes/ha/year (100 day growing season). This value surpasses the calculated N_2O flux of 0.05–1.98 kg N_2O/ha/year [117]. In other words, if every tobacco plant in the world produced N_2OR, this industrial crop (6.6 million tonnes were produced worldwide in 2016 [118]) could conceivably crack 785 Tg of N_2O (1 Tg = 1 million metric tonnes) during an average growing season of 100 days, far surpassing the estimated ~30 Tg of N_2O emitted per year [119]. Such catalytic capacity would give the 'Stop Smoking' campaigns a whole new flavour.

Although these transgenic plants produced a functional N_2OR enzyme, no gas analysis was performed to quantifiably ensure that these plants could reduce N_2O to N_2 using a recombinant N_2OR. In the future, it is imperative that such analyses be performed to properly judge the efficacy of such a gene-stacking trait system for atmospheric phytoremediation.

An associated issue rests with *P. stutzeri* being an anaerobic species that produces enzymes that function optimally in a low-oxygen environment. While expressing *nosZ* in plants to reduce N_2O appears to be an elegant solution, the N_2OR enzyme was not evolutionarily engineered to be functional in the presence of oxygen. Most soil bacteria that produce N_2OR do so in an anaerobic environment [102].

In the past five years, studies have identified several prokaryotic species that may express an oxygen-compatible N_2OR. Aerobic N_2O reducers may be undertaking an important role in mitigating the amounts of N_2O emitted to the atmosphere in events of oxic-to-anoxic transitions, but these systems have not yet been validated in plants. Here, we discuss two candidates for an oxygen-compatible *nosZ* expression system: clade II-*nosZ* member *Gemmatimonas aurantiaca gen nov.*, spp. *nov.* strain T-27, a polyphosphate-accumulating soil aerobe that is strongly represented in many oxygen-rich soil samples [120]; and *Azospira oryzae*, another clade II N-fixing bacterium originally isolated from the roots of rice (*Oryza sativa*) [121]. N_2O reduction by the *G. aurantiaca* strain T-27 was observed in both the absence and presence of oxygen [120]. The inability of this organism to consume N_2O in the complete absence of oxygen and the observed oxygen-induced activation of *nosZ* expression compels one to consider *in planta* overexpression, whereby the diurnal fluctuation of photosynthetic oxygen production may offer an egress for N_2O accumulation. The *A. oryzae* strains I09 and I13 also show more rapid N_2OR recovery rates and tolerance against oxygen inhibition than *P. stutzeri* [121] and so may be appropriate candidates for crop plant transformation and N_2OR expression.

If the ideal *nosZ* sequence were to be identified and transformed into commercially important crop plants, the benefits would be numerous and profound: seed-borne GHG technology foresees the transgenic cassette passed on from generation to generation, meaning that constant application of the beneficial catalyst would not be required (as with NI application and rhizoremediation); the expression of *nosZ* in the aerial tissues of the plants allows the reducing enzyme to confront N_2O much more easily than when the enzyme is expressed in the soil.

5. Novel Breeding Task: "Gas Cracking" Plants

The challenge of expressing heterologous bacterial proteins in plants necessitates codon optimization due to differences in GC content and codon bias with eukaryotes [122]. Altering the codon bias (or applying 'directed evolution' [123]) of a bacterial gene to be expressed in plants has been highly successful: *P. stutzeri nosZ* in tobacco [115], 5-enolpyruvylshikimate-3-phosphate (EPSP) synthase from *Agrobacterium tumefaciens* in Roundup Ready crops [124], and *Bacillus thuringiensis Cry* genes in maize [125] and rice [126]. Indeed, the global advance promulgating engineered crops is pillared on today's artificial intelligence-guided plant codon optimization rules offered by both large and small boutique DNA houses. However, there has been success expressing native bacterial sequences in plants, i.e., in the case of cotton expressing the native sequence of the *P. stutzeri* gene *ptxd* (*PHOSPHONATE DEHYDROGENASE*) [127,128]. One can dare to fathom how a universally-functional *nosZ* expression system could conceivably redirect some aspects of GHG mitigation research. Such a plant transformation cassette could theoretically be applied to any plant—wheat, rice, soybean, peat moss [129]—recruiting these species for the purpose of denitrification mitigation.

Even with an effective *nosZ* expression system, there are additional challenges in developing *nosZ*-expressing plant lines. There are relatively few powerful monocot-optimized expression systems available [130] (although *Bt* corn, LibertyLink wheat, and Roundup Ready wheat can attest to the effectiveness of the 35S promoter system in monocots), and there is difficulty in transforming monocots [65]. With the advent of new plant transformation technologies like the soil bacterium

Ochrobactrum haywardense [131] and the *BABYBOOM/WUSCHEL2* system [132], the production of genetically modified crops with stacked or pyramided GHG genes may be expedited in the near future.

6. Conclusions—Challenges to the Future Success of *nosZ*

We must address what may be the greatest challenge of all for the modern molecular plant breeder: convincing the general public that transgenic crops may be beneficial for all the plant-planet's denizens, as modified crops that enter the food stream may appear unpopular in some boroughs. Regardless, there is a clear, urgent need to control soil N_2O losses due to the detrimental effects of this potent GHG in the atmosphere. Climate-smart crops should be given a crack at directly addressing this issue and tackling climate change. Such GHG-reducing plant lines, endowed with the ability to catalytically "crack" N_2O in the air, could be vital in the battle to shift public perception towards the acceptance of "GMOs" in agricultural research.

Involvement of N_2O in climate change and global warming has been the subject of increasing investigations due to its potential heat-trapping properties [3]. N_2O emission from soil is primarily the result of an incomplete enzymatic reaction which is mediated by the bacterial enzyme, N_2OR [98]. Therefore, in the late 1990s [105,106], the development of N_2OR-positive transgenic plants was proposed as an environmental phytoremediation strategy with promise to remove N_2O from soil and the atmosphere (Figure 2). However, producing a foreign protein in a plant cell is often a serious challenge. For example, different codon usage [133] and cellular properties between eukaryotic and prokaryotic cells are considered as unknown aspects of this strategy. At least two key questions need to be addressed in future studies to probe the probability for success of this green gene de-toxic tactic for accelerating the destruction of nitrous oxide via canopy catalysis: (1) Which candidate is the best source-organism to donate *nosZ* sequence for plant transformation? Activity of bacterial N_2OR is associated with the anaerobic conditions in soil [101], whereas the plant cell is mostly an aerobic environment. Photosynthesis and respiration cause different levels of oxygen content in plant cells in a diurnal cycle which is not consistent with the enzymatic activity of N_2OR in anaerobic soil bacteria. Therefore, selecting obligate or facultative aerobic bacteria containing active N_2OR enzymes as 'the source code' would be pivotal; (2) Which plant cell compartment is the best destination for targeting N_2OR accumulation? The native enzyme N_2OR in bacteria is directed to the periplasm, where Cu chaperones provide enough Cu for the assembly of metal centres [134]. The absence of periplasmic space in plant cells reinforces the notion that subcellular localization of N_2OR may influence its enzymatic activity *in planta*. Moreover, the important role of Cu in the functional assembly of N_2OR posits whether the transformation of bacterial *nosDFY*, along with *nosZ*, is essential for a functional enzyme. Urgent exploration of how the cellular pool of metal nutrients and proteins (pseudo chaperones) in eukaryotic cells may suffice to activate N_2OR *in planta* may compel the use of such climate-smart plants.

Author Contributions: J.J.D. constructed the initial paper draft; Q.-y.S., J.J.D., S.W., A.E.H., M.N., and I.A. contributed to study design and data analysis; J.J.D., S.W., and I.A. wrote the paper with input from A.E.H., and M.N.; M.N. constructed Figure 2; Q.-y.S. conceptualized the original approach to amplifying soil denitrification *in planta*; I.A. coordinated and supervised the project; all authors contributed to and approved the final draft.

Acknowledgments: We thank three anonymous reviewers for taking the time to critique and strengthen this manuscript. We are grateful to Quinn Ingram for technical expertise with plant growth.

References

1. Dentener, F.; Drevet, J.; Lamarque, J.F. Nitrogen and sulfur deposition on regional and global scales: A multimodel evaluation. *Glob. Biogeochem. Cycles* **2006**, *20*, 1–21. [CrossRef]

2. Vet, R.; Artz, R.S.; Carou, S. A global assessment of precipitation chemistry and deposition of sulfur, nitrogen, sea salt, base cations, organic acids, acidity and pH, and phosphorus. *Atmos. Environ.* **2014**, *93*, 3–100. [CrossRef]
3. Forster, P.; Ramaswamy, V. *Changes in Atmospheric Constituents and in Radiative Forcing*; Solomon, S.Q.D., Manning, M., Chen, Z., Marquis, M., Averyt, K.B., Tignor, M., Miller, H.L., Eds.; Cambridge University Press: Cambridge, UK, 2007; pp. 129–234.
4. Reay, D.S. Fertilizer 'solution' could turn local problem global. *Nature* **2004**, *427*, 485. [CrossRef] [PubMed]
5. Shapiro, C.A.; Ferguson, R.B.; Hergert, G.W.; Wortmann, C.S.; Walters, D.T. *Fertilizer Suggestions for Corn*; EC117; Nebraska Extension: Lincoln, NE, USA, 2008; pp. 1–6.
6. Froment, M.A.; Chalmers, A.G.; Collins, C.; Grylls, J.P. Rotational set-aside; influence of vegetation and management for one-year plant covers on soil mineral nitrogen during and after set-aside at five sites in England. *J. Agric. Sci.* **1999**. [CrossRef]
7. Quinton, J.N.; Govers, G.; Van Oost, K.; Bardgett, R.D. The impact of agricultural soil erosion on biogeochemical cycling. *Nat. Geosci.* **2010**, *3*, 311–314. [CrossRef]
8. Zhu, Z.L.; Cai, G.X.; Simpson, J.R.; Zhang, S.L.; Chen, D.L.; Jackson, A.V.; Freney, J.R. Processes of nitrogen loss from fertilizers applied to flooded rice fields on a calcareous soil in north-central China. *Fertil. Res.* **1988**, *18*, 101–115. [CrossRef]
9. Zhou, M.; Zhu, B.; Brüggemann, N.; Dannenmann, M.; Wang, Y.; Butterbach-Bahl, K. Sustaining crop productivity while reducing environmental nitrogen losses in the subtropical wheat-maize cropping systems: A comprehensive case study of nitrogen cycling and balance. *Agric. Ecosyst. Environ.* **2016**, *231*, 1–14. [CrossRef]
10. Zumft, W.G. Cell biology and molecular basis of denitrification. *Microbiol. Mol. Biol. Rev.* **1997**, *61*, 533–616. [PubMed]
11. Gu, B.; Chang, J.; Min, Y.; Ge, Y.; Zhu, Q.; Galloway, J.N.; Peng, C. The role of industrial nitrogen in the global nitrogen biogeochemical cycle. *Sci. Rep.* **2013**, *3*, 2579. [CrossRef] [PubMed]
12. Erisman, J.W.; Sutton, M.A.; Galloway, J.; Klimont, Z.; Winiwarter, W. How a century of ammonia synthesis changed the world. *Nat. Geosci.* **2008**, *1*, 636–639. [CrossRef]
13. Fowler, D.; Pyle, J.A.; Raven, J.A.; Sutton, M.A. The global nitrogen cycle in the twenty-first century: Introduction. *Philos. Trans. R. Soc. Lond. B Biol. Sci.* **2013**, *368*, 20130164. [CrossRef] [PubMed]
14. Kanakidou, M.; Myriokefalitakis, S.; Daskalakis, N.; Fanourgakis, G.; Nenes, A.; Baker, A.R.; Tsigaridis, K.; Mihalopoulos, N. Past, present, and future atmospheric nitrogen deposition. *J. Atmos. Sci.* **2016**, *73*, 2039–2047. [CrossRef]
15. Richardson, D.; Felgate, H.; Watmough, N.; Thomson, A.; Baggs, E. Mitigating release of the potent greenhouse gas N_2O from the nitrogen cycle—Could enzymic regulation hold the key? *Trends Biotechnol.* **2009**, *27*, 388–397. [CrossRef] [PubMed]
16. Burney, J.A.; Davis, S.J.; Lobell, D.B. Greenhouse gas mitigation by agricultural intensification. *Proc. Natl. Acad. Sci. USA* **2010**, *107*, 12052–12057. [CrossRef] [PubMed]
17. Dignon, J.; Hameed, S. Global emissions of nitrogen and sulfur-oxides from 1860 to 1980. *Japca* **1989**, *39*, 180–186. [CrossRef]
18. Larsson, L.; Ferm, M.; Kasimir-Klemedtsson, A.; Klemedtsson, L. Ammonia and nitrous oxide emissions from grass and alfalfa mulches. *Nutr. Cycl. Agroecosyst.* **1998**, *51*, 41–46. [CrossRef]
19. Intergovernmental Panel on Climate Change. *Climate Change 2014: Mitigation of Climate Change*; Working Group III Contribution to the IPCC Fifth Assessment Report. 9781107654815; Cambridge University Press: Cambridge, UK, 2015.
20. Ravishankara, A.R.; Daniel, J.S.; Portmann, R.W. Nitrous oxide N_2O: The dominant ozone-depleting substance emitted in the 21st century. *Science* **2009**, *326*, 123–125. [CrossRef] [PubMed]
21. Chen, Z.; Ding, W.; Xu, Y.; Müller, C.; Rütting, T.; Yu, H.; Fan, J.; Zhang, J.; Zhu, T. Importance of heterotrophic nitrification and dissimilatory nitrate reduction to ammonium in a cropland soil: Evidences from a 15N tracing study to literature synthesis. *Soil Biol. Biochem.* **2015**, *91*, 65–75. [CrossRef]
22. Ruser, R.; Schulz, R. The effect of nitrification inhibitors on the nitrous oxide (N_2O) release from agricultural soils—A review. *J. Plant Nutr. Soil Sci.* **2015**, *178*, 171–188. [CrossRef]
23. Lees, H. Hydroxylamine as an intermediate in nitrification. *Nature* **1952**, *169*, 156–157. [CrossRef]

24. Caranto, J.D.; Lancaster, K.M. Nitric oxide is an obligate bacterial nitrification intermediate produced by hydroxylamine oxidoreductase. *Proc. Natl. Acad. Sci. USA* **2017**, *114*, 8217–8222. [CrossRef] [PubMed]
25. Cébron, A.; Garnier, J. *Nitrobacter* and *Nitrospira* genera as representatives of nitrite-oxidizing bacteria: Detection, quantification and growth along the lower Seine River (France). *Water Res.* **2005**, *39*, 4979–4992. [CrossRef] [PubMed]
26. Shaw, L.J.; Nicol, G.W.; Smith, Z.; Fear, J.; Prosser, J.I.; Baggs, E.M. *Nitrosospira* spp. can produce nitrous oxide via a nitrifier denitrification pathway. *Environ. Microbiol.* **2006**, *8*, 214–222. [CrossRef] [PubMed]
27. Kole, M.M.; Page, W.J.; Altosaar, I. Distribution of *Azotobacter* in Eastern Canadian soils and in association with plant rhizospheres. *Can. J. Microbiol.* **1988**, *34*, 815–817. [CrossRef]
28. Philippot, L.; Andert, J.; Jones, C.M.; Bru, D.; Hallin, S. Importance of denitrifiers lacking the genes encoding the nitrous oxide reductase for N_2O emissions from soil. *Glob. Chang. Biol.* **2011**, *17*, 1497–1504. [CrossRef]
29. Zumft, W.G.; Kroneck, P.M.H. Respiratory transformation of nitrous oxide (N_2O) to dinitrogen by *Bacteria* and *Archaea*. *Adv. Microb. Physiol.* **2007**, *52*, 107–227. [PubMed]
30. Brodsky, J.B.; Cohen, E.N. Adverse effects of nitrous oxide. *Med. Toxicol.* **1986**, *1*, 362–374. [CrossRef] [PubMed]
31. Fluegge, K. Does environmental exposure to the greenhouse gas, N_2O, contribute to etiological factors in neurodevelopmental disorders? A mini-review of the evidence. *Environ. Toxicol. Pharmacol.* **2016**, *47*, 6–18. [CrossRef] [PubMed]
32. Edgerton, M.D. Increasing crop productivity to meet global needs for feed, food, and fuel. *Plant Physiol.* **2009**, *149*, 7–13. [CrossRef] [PubMed]
33. Abalos, D.; van Groenigen, J.W.; De Deyn, G.B. What plant functional traits can reduce nitrous oxide emissions from intensively managed grasslands? *Glob. Chang. Biol.* **2018**, *24*, e248–e258. [CrossRef] [PubMed]
34. Terman, G.L. Volatilization losses of nitrogen as ammonia from surface-applied fertilizers, organic amendments, and crop residues. *Adv. Agron.* **1980**, *31*, 189–223.
35. Holcomb, J.C.; Sullivan, D.M.; Horneck, D.A.; Clough, G.H. Effect of irrigation rate on ammonia volatilization. *Soil Sci. Soc. Am. J.* **2011**, *75*, 2341–2347. [CrossRef]
36. Van Kessel, C.; Venterea, R.; Six, J.; Adviento-Borbe, M.A.; Linquist, B.; van Groenigen, K.J. Climate, duration, and N placement determine N_2O emissions in reduced tillage systems: A meta-analysis. *Glob. Chang. Biol.* **2013**, *19*, 33–44. [CrossRef] [PubMed]
37. Mei, K.; Wang, Z.; Huang, H.; Zhang, C.; Shang, X.; Dahlgren, R.A.; Zhang, M.; Xia, F. Stimulation of N_2O emission by conservation tillage management in agricultural lands: A meta-analysis. *Soil Till. Res.* **2018**, *182*, 86–93. [CrossRef]
38. Pittelkow, C.M.; Liang, X.; Linquist, B.A.; van Groenigen, K.J.; Lee, J.; Lundy, M.E.; van Gestel, N.; Six, J.; Venterea, R.T.; van Kessel, C. Productivity limits and potentials of the principles of conservation agriculture. *Nature* **2015**, *517*, 365–368. [CrossRef] [PubMed]
39. Bruulsema, T.; Lemunyon, J.; Herz, B. Know your fertilizer rights. *Crops Sci.* **2009**, *42*, 13–18.
40. Bruulsema, T.W.; Ketterings, Q. *Best Management Practices for Fertilizers on Northeastern Dairy Farms*; Fertilize BMP Item 30-3220; Ref. 08052; International Plant Nutrition Institute: Norcross, GA, USA, 2008.
41. Guardia, G.; Cangani, M.T.; Andreu, G.; Sanz-Cobena, A.; García-Marco, S.; Álvarez, J.M.; Recio-Huetos, J.; Vallejo, A. Effect of inhibitors and fertigation strategies on GHG emissions, NO fluxes and yield in irrigated maize. *Field Crops Res.* **2017**, *204*, 135–145. [CrossRef]
42. Cayuela, M.L.; Aguilera, E.; Sanz-Cobena, A.; Adams, D.C.; Abalos, D.; Barton, L.; Ryals, R.; Silver, W.L.; Alfaro, M.A.; Pappa, V.A.; et al. Direct nitrous oxide emissions in Mediterranean climate cropping systems: Emission factors based on a meta-analysis of available measurement data. *Agric. Ecosyst. Environ.* **2017**, *238*, 25–35. [CrossRef]
43. Sanz-Cobena, A.; Lassaletta, L.; Aguilera, E.; Del Prado, A.; Garnier, J.; Billen, G.; Iglesias, A.; Sanchez, B.; Guardia, G.; Abalos, D.; et al. Strategies for greenhouse gas emissions mitigation in Mediterranean agriculture: A review. *Agric. Ecosyst. Environ.* **2017**, *238*, 5–24. [CrossRef]
44. Hynes, R.; Knowles, R. Production of nitrous oxide by *Nitrosomonas europaea*: Effects of acetylene, pH, and oxygen. *Can. J. Microbiol.* **2011**, *30*, 1397–1404. [CrossRef]
45. Halvorson, A.D.; Snyder, C.S.; Blaylock, A.D.; Del Grosso, S.J. Enhanced-efficiency nitrogen fertilizers: Potential role in nitrous oxide emission mitigation. *Agron. J.* **2014**, *106*, 715–722. [CrossRef]

46. Thapa, R.; Chatterjee, A.; Awale, R.; McGranahan, D.A.; Daigh, A. Effect of enhanced efficiency fertilizers on nitrous oxide emissions and crop yields: A meta-analysis. *Soil Sci. Soc. Am. J.* **2016**, *80*, 1121–1134. [CrossRef]
47. Shoji, S.; Delgado, J.; Mosier, A.; Miura, Y. Use of controlled release fertilizers and nitrification inhibitors to increase nitrogen use efficiency and to conserve air and water quality. *Commun. Soil Sci. Plant Anal.* **2001**, *32*, 1051–1070. [CrossRef]
48. McTaggart, I.P.; Tsuruta, H. The influence of controlled release fertilisers and the form of applied fertiliser nitrogen on nitrous oxide emissions from an andosol. *Nutr. Cycl. Agroecosyst.* **2003**, *67*, 47–54. [CrossRef]
49. Sanz-Cobena, A.; Abalos, D.; Meijide, A.; Sanchez-Martin, L.; Vallejo, A. Soil moisture determines the effectiveness of two urease inhibitors to decrease N_2O emission. *Mitig. Adapt. Strateg. Glob. Chang.* **2016**, *21*, 1131–1144. [CrossRef]
50. Kim, M.; Zhang, Z.; Okano, H.; Yan, D.; Groisman, A.; Hwa, T. Need-based activation of ammonium uptake in *Escherichia coli*. *Mol. Syst. Biol.* **2012**, *8*, 616. [CrossRef] [PubMed]
51. Soares, J.; Cantarella, H.; Menegale, M. Ammonia volatilization losses from surface-applied urea with urease and nitrification inhibitors. *Soil Biol. Biochem.* **2012**, *52*, 82–89. [CrossRef]
52. McCarty, G.W. Modes of action of nitrification inhibitors. *Biol. Fertil. Soils* **1999**, *29*, 1–9. [CrossRef]
53. Liu, C.; Wang, K.; Zheng, X. Effects of nitrification inhibitors (DCD and DMPP) on nitrous oxide emission, crop yield and nitrogen uptake in a wheat-maize cropping system. *Biogeosciences* **2013**, *10*, 2427–2437. [CrossRef]
54. Bollag, J.-M.; Kurek, E.J. Nitrite and nitrous oxide accumulation during denitrification in the presence of pesticide derivatives. *Appl. Environ. Microbiol.* **1980**, *39*, 845–849. [PubMed]
55. Lam, S.K.; Suter, H.; Mosier, A.R.; Chen, D. Using nitrification inhibitors to mitigate agricultural N_2O emission: A double-edged sword? *Glob. Chang. Biol.* **2016**, *23*, 485–489. [CrossRef] [PubMed]
56. Subbarao, G.V.; Nakahara, K.; Hurtado, M.P.; Ono, H.; Moreta, D.E.; Salcedo, A.F.; Yoshihashi, A.T.; Ishikawa, T.; Ishitani, M.; Ohnishi-Kameyama, M.; et al. Evidence for biological nitrification inhibition in *Brachiaria* pastures. *Proc. Natl. Acad. Sci. USA* **2009**, *106*, 17302. [CrossRef] [PubMed]
57. Bardon, C.; Poly, F.; Piola, F.; Pancton, M.; Comte, G.; Meiffren, G.; Haichar Fel, Z. Mechanism of biological denitrification inhibition: Procyanidins induce an allosteric transition of the membrane-bound nitrate reductase through membrane alteration. *FEMS Microbiol. Ecol.* **2016**, *92*, fiw034. [CrossRef] [PubMed]
58. Subbarao, G.V.; Sahrawat, K.L.; Nakahara, K.; Rao, I.M.; Ishitani, M.; Hash, C.T.; Kishii, M.; Bonnett, D.G.; Berry, W.L.; Lata, J.C. A paradigm shift towards low-nitrifying production systems: The role of biological nitrification inhibition (BNI). *Ann. Bot.* **2013**, *112*, 297–316. [CrossRef] [PubMed]
59. Jacinthe, P.-A.; Dick, W.A.; Brown, L.C. Bioremediation of nitrate-contaminated shallow soils and waters via water table management techniques: Evolution and release of nitrous oxide. *Soil Biol. Biochem.* **2000**, *32*, 371–382. [CrossRef]
60. Itakura, M.; Tabata, K.; Eda, S.; Mitsui, H.; Murakami, K.; Yasuda, J.; Minamisawa, K. Generation of *Bradyrhizobium japonicum* mutants with Increased N_2O reductase activity by selection after introduction of a mutated *dnaQ* gene. *Appl. Environ. Microbiol.* **2008**, *74*, 7258–7264. [CrossRef] [PubMed]
61. Itakura, M.; Uchida, Y.; Akiyama, H.; Hoshino, Y.T.; Shimomura, Y.; Morimoto, S.; Tago, K.; Wang, Y.; Hayakawa, C.; Uetake, Y.; et al. Mitigation of nitrous oxide emissions from soils by *Bradyrhizobium japonicum* inoculation. *Nat. Clim. Chang.* **2012**, *3*, 208–212. [CrossRef]
62. Houlden, A.; Timms-Wilson, T.M.; Day, M.J.; Bailey, M.J. Influence of plant developmental stage on microbial community structure and activity in the rhizosphere of three field crops. *FEMS Microbiol. Ecol.* **2008**, *65*, 193–201. [CrossRef] [PubMed]
63. Xu, Y.; Wang, G.; Jin, J.; Liu, J.; Zhang, Q.; Liu, X. Bacterial communities in soybean rhizosphere in response to soil type, soybean genotype, and their growth stage. *Soil Biol. Biochem.* **2009**, *41*, 919–925. [CrossRef]
64. Wan, S.; Johnson, A.M.; Altosaar, I. Expression of nitrous oxide reductase from *Pseudomonas stutzeri* in transgenic tobacco roots using the root-specific *rolD* promoter from *Agrobacterium rhizogenes*. *Ecol. Evol.* **2012**, *2*, 286–297. [CrossRef] [PubMed]
65. Sood, P.; Bhattacharya, A.; Sood, A. Problems and possibilities of monocot transformation. *Biol. Plant* **2011**, *55*, 1–15. [CrossRef]
66. Matsubara, T.; Frunzke, K.; Zumft, W.G. Modulation by copper of the products of nitrite respiration in *Pseudomonas perfectomarinus*. *Biol. Fertil. Soils* **1982**, *149*, 816–823.

67. Wan, S.; Ward, T.L.; Altosaar, I. Strategy and tactics of disarming GHG at the source: N_2O reductase crops. *Trends Biotechnol.* **2012**, *30*, 410–415. [CrossRef] [PubMed]
68. Rochette, P.; Angers, D.A.; Chantigny, M.H.; Gasser, M.-O.; MacDonald, J.D.; Pelster, D.E.; Bertrand, N. NH_3 volatilization, soil concentration and soil pH following subsurface banding of urea at increasing rates. *Can. J. Soil Sci.* **2013**, *93*, 261–268. [CrossRef]
69. Venterea, R.T.; Burger, M.; Spokas, K.A. Nitrogen oxide and methane emissions under varying tillage and fertilizer management. *J. Environ. Qual.* **2005**, *34*, 1467–1477. [CrossRef] [PubMed]
70. Chen, G.; Kolb, L.; Cavigelli, M.A.; Weil, R.R.; Hooks, C.R.R. Can conservation tillage reduce N_2O emissions on cropland transitioning to organic vegetable production? *Sci. Total Environ.* **2018**, *618*, 927–940. [CrossRef] [PubMed]
71. Mitchell, D.C.; Castellano, M.J.; Sawyer, J.E.; Pantoja, J. Cover crop effects on nitrous oxide emissions: Role of mineralizable carbon. *Soil Sci. Soc. Am. J.* **2013**, *77*, 1765. [CrossRef]
72. Snyder, C.S.; Bruulsema, T.W.; Jensen, T.L.; Fixen, P.E. Review of greenhouse gas emissions from crop production systems and fertilizer management effects. *Agric. Ecosyst. Environ.* **2009**, *133*, 247–266. [CrossRef]
73. Abalos, D.; Jeffery, S.; Sanz-Cobena, A.; Guardia, G.; Vallejo, A. Meta-analysis of the effect of urease and nitrification inhibitors on crop productivity and nitrogen use efficiency. *Agric. Ecosyst. Environ.* **2014**, *189*, 136–144. [CrossRef]
74. Cantarella, H.; Otto, R.; Soares, J.R.; Silva, A.G.d.B. Agronomic efficiency of NBPT as a urease inhibitor: A review. *J. Adv. Res.* **2018**, *13*, 19–27. [CrossRef] [PubMed]
75. Bremner, J.M. Recent research on problems in the use of urea as a nitrogen fertilizer. *Fertil. Res.* **1995**, *42*, 321–329. [CrossRef]
76. Zaman, M.; Zaman, S.; Nguyen, M.L.; Smith, T.J.; Nawaz, S. The effect of urease and nitrification inhibitors on ammonia and nitrous oxide emissions from simulated urine patches in pastoral system: A two-year study. *Sci. Total Environ.* **2013**, *465*, 97–106. [CrossRef] [PubMed]
77. Ding, W.X.; Yu, H.Y.; Cai, Z.C. Impact of urease and nitrification inhibitors on nitrous oxide emissions from fluvo-aquic soil in the North China Plain. *Biol. Fertil. Soils* **2011**, *47*, 91–99. [CrossRef]
78. Singh, S.; Bakshi, B.R. Accounting for the biogeochemical cycle of nitrogen in input-output life cycle assessment. *Environ. Sci. Technol.* **2013**, *47*, 9388–9396. [CrossRef] [PubMed]
79. Rose, T.J.; Quin, P.; Morris, S.G.; Kearney, L.J.; Kimber, S.; Rose, M.T.; Van Zwieten, L. No evidence for higher agronomic N use efficiency or lower nitrous oxide emissions from enhanced efficiency fertilisers in aerobic subtropical rice. *Field Crops Res.* **2018**, *225*, 47–54. [CrossRef]
80. Chaves, B.; Opoku, A.; De Neve, S.; Boeckx, P.; Van Cleemput, O.; Hofman, G. Influence of DCD and DMPP on soil N dynamics after incorporation of vegetable crop residues. *Biol. Fertil. Soils* **2006**, *43*, 62–68. [CrossRef]
81. Yang, M.; Fang, Y.; Sun, D.; Shi, Y. Efficiency of two nitrification inhibitors (dicyandiamide and 3,4-dimethypyrazole phosphate) on soil nitrogen transformations and plant productivity: A meta-analysis. *Sci. Rep.* **2016**, *6*, 1–10.
82. Coskun, D.; Britto, D.T.; Shi, W.; Kronzucker, H.J. How plant root exudates shape the nitrogen cycle. *Trends Plant Sci.* **2017**, *22*, 661–673. [CrossRef] [PubMed]
83. Matsuoka, M.; Kumar, A.; Muddassar, M.; Matsuyama, A.; Yoshida, M.; Zhang, K.Y. Discovery of fungal denitrification inhibitors by targeting copper nitrite reductase from *Fusarium oxysporum*. *J. Chem. Inf. Model.* **2017**, *57*, 203–213. [CrossRef] [PubMed]
84. Liu, T.; Liang, Y.; Chu, G. Nitrapyrin addition mitigates nitrous oxide emissions and raises nitrogen use efficiency in plastic-film-mulched drip-fertigated cotton field. *PLoS ONE* **2017**, *12*, e0176305. [CrossRef] [PubMed]
85. Byrnes, R.C.; Nùñez, J.; Arenas, L.; Rao, I.; Trujillo, C.; Alvarez, C.; Arango, J.; Rasche, F.; Chirinda, N. Biological nitrification inhibition by *Brachiaria* grasses mitigates soil nitrous oxide emissions from bovine urine patches. *Soil Biol. Biochem.* **2017**, *107*, 156–163. [CrossRef]
86. Subbarao, G.V.; Arango, J.; Masahiro, K.; Hooper, A.M.; Yoshihashi, T.; Ando, Y.; Nakahara, K.; Deshpande, S.; Ortiz-Monasterio, I.; Ishitani, M.; et al. Genetic mitigation strategies to tackle agricultural GHG emissions: The case for biological nitrification inhibition technology. *Plant Sci. Int. J. Exp. Plant Biol.* **2017**, *262*, 165–168. [CrossRef] [PubMed]

87. Karwat, H.; Moreta, D.; Arango, J.; Núñez, J.; Rao, I.; Rincón, Á.; Rasche, F.; Cadisch, G. Residual effect of BNI by *Brachiaria humidicola* pasture on nitrogen recovery and grain yield of subsequent maize. *Plant Soil.* **2017**, *420*, 389–406. [CrossRef]
88. Kuiper, I.; Lagendijk, E.L.; Bloemberg, G.V.; Lugtenberg, B.J.J. Rhizoremediation: A beneficial plant-microbe interaction. *Mol. Plant Microbe Interact.* **2004**, *17*, 6–15. [CrossRef] [PubMed]
89. Qian, J.H.; Doran, J.W.; Weier, K.L.; Mosier, A.R.; Peterson, T.A.; Power, J.F. Soil denitrification and nitrous oxide losses under corn irrigated with high-nitrate groundwater. *J. Environ. Qual.* **1997**, *26*, 348–360. [CrossRef]
90. Köpke, U.; Nemecek, T. Ecological services of faba bean. *Field Crops Res.* **2010**, *115*, 217–233. [CrossRef]
91. Gaume, A.; Komarnytsky, S.; Borisjuk, N.; Raskin, I. Rhizosecretion of recombinant proteins from plant hairy roots. *Plant Cell Rep.* **2003**, *21*, 1188–1193. [CrossRef] [PubMed]
92. Gelvin, S.B. Crown gall disease and hairy root disease. *Plant Physiol.* **1990**, *92*, 281–285. [CrossRef] [PubMed]
93. Wan, S.; Mottiar, Y.; Johnson, A.M.; Goto, K.; Altosaar, I. Expression of the *nos* operon proteins from *Pseudomonas stutzeri* in transgenic plants to assemble nitrous oxide reductase. *Transgenic Res.* **2012**, *21*, 593–603. [CrossRef] [PubMed]
94. French, C.E.; Rosser, S.J.; Davies, G.J.; Nicklin, S.; Bruce, N.C. Biodegradation of explosives by transgenic plants expressing pentaerythritol tetranitrate reductase. *Nat. Biotechnol.* **1999**, *17*, 491–494. [CrossRef] [PubMed]
95. Rugh, C.L.; Wilde, H.D.; Stack, N.M.; Thompson, D.M.; Summers, A.O.; Meagher, R.B. Mercuric ion reduction and resistance in transgenic *Arabidopsis thaliana* plants expressing a modified bacterial *merA* gene. *Proc. Natl. Acad. Sci. USA* **1996**, *93*, 3182–3187. [CrossRef] [PubMed]
96. Nahar, N.; Rahman, A.; Nawani, N.N.; Ghosh, S.; Mandal, A. Phytoremediation of arsenic from the contaminated soil using transgenic tobacco plants expressing *ACR2* gene of *Arabidopsis thaliana. J. Plant Physiol.* **2017**, *218*, 121–126. [CrossRef] [PubMed]
97. Heaton, A.C.P.; Rugh, C.L.; Wang, N.J.; Meagher, R.B. Phytoremediation of mercury- and methylmercury-polluted soils using genetically engineered plants. *Soil Sediment Contam.* **1998**, *7*, 497–509. [CrossRef]
98. Maeda, K.; Hanajima, D.; Toyoda, S.; Yoshida, N.; Morioka, R.; Osada, T. Microbiology of nitrogen cycle in animal manure compost. *Microb. Biotechnol.* **2011**. [CrossRef] [PubMed]
99. Sanford, R.A.; Wagner, D.D.; Wu, Q.; Chee-Sanford, J.C.; Thomas, S.H.; Cruz-Garcia, C.; Rodriguez, G.; Massol-Deya, A.; Krishnani, K.K.; Ritalahti, K.M.; et al. Unexpected nondenitrifier nitrous oxide reductase gene diversity and abundance in soils. *Proc. Natl. Acad. Sci. USA* **2012**, *109*, 19709–19714. [CrossRef] [PubMed]
100. Yoon, S.; Nissen, S.; Park, D.; Sanford, R.A.; Löffler, F.E. Nitrous oxide reduction kinetics distinguish bacteria harboring clade I *NosZ* from those harboring clade II *NosZ*. *Appl. Environ. Microbiol.* **2016**, *82*, 3793–3800. [CrossRef] [PubMed]
101. Cuypers, H.; Berghofer, J.; Zumft, W.G. Multiple *nosZ* promoters and anaerobic expression of *nos* genes necessary for *Pseudomonas stutzeri* nitrous oxide reductase and assembly of its copper centers. *Biochim. Biophys. Acta* **1995**, *1264*, 183–190. [CrossRef]
102. Rosenzweig, A.C. Nitrous oxide reductase from Cu_A to Cu_Z. *Nat. Struct. Biol.* **2000**, *7*, 169–171. [CrossRef] [PubMed]
103. Wang, Y.; Wang, Z.; Duo, Y.; Wang, X.; Chen, J.; Chen, J. Gene cloning, expression, and reducing property enhancement of nitrous oxide reductase from *Alcaligenes denitrificans* strain TB. *Environ. Pollut.* **2018**, *239*, 43–52. [CrossRef] [PubMed]
104. Carreira, C.; Pauleta, S.R.; Moura, I. The catalytic cycle of nitrous oxide reductase—The enzyme that catalyzes the last step of denitrification. *J. Inorg. Biochem.* **2017**, *177*, 423–434. [CrossRef] [PubMed]
105. Inatomi, K.I. Analysis of the nitrous oxide reduction genes, *nosZDFYL*, of *Achromobacter cycloclastes. DNA Res.* **1998**, *5*, 365–371. [CrossRef] [PubMed]
106. Inatomi, K.-I. Gene Encoding Nitrous Oxide Reductase and the Nitrous Oxide Reductase. Patent Application JP H1175842A, 29 August 1997.
107. Cheng, J.; Romantsov, T.; Engel, K.; Doxey, A.C.; Rose, D.R.; Neufeld, J.D.; Charles, T.C. Functional metagenomics reveals novel β-galactosidases not predictable from gene sequences. *PLoS ONE* **2017**, *12*, e0172545. [CrossRef] [PubMed]

108. Hill, P.; Heberlig, G.W.; Boddy, C.N. Sampling terrestrial environments for bacterial polyketides. *Molecules* **2017**, *22*, E707. [CrossRef] [PubMed]
109. Kolter, R.; Chimileski, S. The end of microbiology. *Environ. Microbiol.* **2018**. [CrossRef] [PubMed]
110. Lin, Y.W. Rational design of metalloenzymes: From single to multiple active sites. *Coord. Chem. Rev.* **2017**, *336*, 1–27. [CrossRef]
111. Zumft, W.G. Biogenesis of the bacterial respiratory Cu_A, Cu-S enzyme nitrous oxide reductase. *J. Mol. Microbiol. Biotechnol.* **2005**, *10*, 154–166. [CrossRef] [PubMed]
112. Dell'Acqua, S.; Pauleta, S.R.; Moura, I.; Moura, J.J.G. The tetranuclear copper active site of nitrous oxide reductase: The Cu_Z center. *J. Biol. Inorg. Chem.* **2011**, *16*, 183–194. [CrossRef] [PubMed]
113. Kusnadi, A.R.; Nikolov, Z.L.; Howard, J.A. Production of recombinant proteins in transgenic plants: Practical considerations. *Biotechnol. Bioeng.* **1997**, *56*, 473–484. [CrossRef]
114. Honisch, U.; Zumft, W.G. Operon structure and regulation of the *nos* gene region of *Pseudomonas stutzeri*, encoding an ABC-Type ATPase for maturation of nitrous oxide reductase. *J. Bacteriol.* **2003**, *185*, 1895–1902. [CrossRef] [PubMed]
115. Wan, S.; Greenham, T.; Goto, K.; Mottiar, Y.; Johnson, A.M.; Staebler, J.M.; Zaidi, M.A.; Shu, Q.; Altosaar, I. A novel nitrous oxide mitigation strategy: Expressing nitrous oxide reductase from *Pseudomonas stutzeri* in transgenic plants. *Can. J. Plant Sci.* **2014**, *94*, 1013–1023. [CrossRef]
116. Andri, K.B.S.; Santosa, P.; Arifin, Z. An empirical study of supply chain and intensification program on Madura tobacco industry in East Java. *Int. J. Agric. Res.* **2011**, *6*, 58–66. [CrossRef]
117. Sgouridis, F.; Ullah, S. Soil greenhouse gas fluxes, environmental controls, and the partitioning of N_2O sources in UK natural and seminatural land use types. *J. Geophys. Res. Biogeosci.* **2017**, *122*, 2617–2633. [CrossRef]
118. FAO. FAOSTAT. 2016; Available online: fao.org.
119. Intergovernmental Panel on Climate Change. *Climate Change 2014: Synthesis Report*; Contribution of Working Groups I, II and III to the Fifth Assessment Report of the Intergovernmental Panel on Climate Change; Pachauri, R.K., Meyer, L.A., Eds.; Cambridge University Press: Cambridge, UK, 2014.
120. Park, D.; Kim, H.; Yoon, S. Nitrous oxide reduction by an obligate aerobic bacterium, *Gemmatimonas aurantiaca* strain T-27. *Appl. Environ. Microbiol.* **2017**, *80*, e00502–e00517. [CrossRef] [PubMed]
121. Suenaga, T.; Aoyagi, T.; Hosomi, M.; Hori, T.; Terada, A. Draft genome sequence of *Azospira* sp. strain I13, a nitrous oxide-reducing bacterium harboring clade II type *nosZ*. *Genome Annouc.* **2018**, *6*, e00414–e00418. [CrossRef] [PubMed]
122. Plotkin, J.B.; Kudla, G. Synonymous but not the same: The causes and consequences of codon bias. *Nat. Rev. Genet.* **2011**, *12*, 32–42. [CrossRef] [PubMed]
123. Hartley, B.S. Evolution of enzyme structure. *Proc. R. Soc. Lond. B Biol. Sci.* **1979**, *205*, 443–452. [CrossRef] [PubMed]
124. Padgette, S.R.; Kolacz, K.H.; Delannay, X.; Re, D.B.; LaVallee, B.J.; Tinius, C.N.; Rhodes, W.K.; Otero, Y.I.; Barry, G.F.; Eichholtz, D.A.; et al. Development, identification, and characterization of a glyphosate-tolerant soybean line. *Crop Sci.* **1995**, *35*, 1451–1461. [CrossRef]
125. Sardana, R.; Dukiandjiev, S.; Giband, M.; Cheng, X.; Cowan, K.; Sauder, C.; Altosaar, I. Construction and rapid testing of synthetic and modified toxin gene sequences *Cry*IA (*b&c*) by expression in maize endosperm culture. *Plant Cell Rep.* **1996**, *15*, 677–681. [PubMed]
126. Cheng, X.; Sardana, R.; Kaplan, H.; Altosaar, I. *Agrobacterium*-transformed rice plants expressing synthetic *cryIA*(*b*) and *cryIA*(*c*) genes are highly toxic to striped stem borer and yellow stem borer. *Proc. Natl. Acad. Sci. USA* **1998**, *95*, 2767–2772. [CrossRef] [PubMed]
127. Pandeya, D.; Campbell, L.A.M.; Nunes, E.; Lopez-Arredondo, D.L.; Janga, M.R.; Herrera-Estrella, L.; Rathore, K.S. *ptxD* gene in combination with phosphite serves as a highly effective selection system to generate transgenic cotton (*Gossypium hirsutum* L.). *Plant Mol. Biol.* **2017**, 566–567. [CrossRef] [PubMed]
128. Pandeya, D.; López-Arredondo, D.L.; Janga, M.R.; Campbell, L.M.; Estrella-Hernández, P.; Bagavathiannan, M.V.; Herrera-Estrella, L.; Rathore, K.S. Selective fertilization with phosphite allows unhindered growth of cotton plants expressing the *ptxD* gene while suppressing weeds. *Proc. Natl. Acad. Sci. USA* **2018**, *115*, E6946–E6955. [CrossRef] [PubMed]
129. Liimatainen, M.; Voigt, C.; Martikainen, P.J.; Hytönen, J.; Regina, K.; Óskarsson, H.; Maljanen, M. Factors controlling nitrous oxide emissions from managed northern peat soils with low carbon to nitrogen ratio.

Soil Biol. Biochem. **2018**, *122*, 186–195. [CrossRef]
130. Park, S.-H.; Yi, N.; Kim, Y.S.; Jeong, M.-H.; Bang, S.-W.; Choi, Y.D.; Kim, J.-K. Analysis of five novel putative constitutive gene promoters in transgenic rice plants. *J. Exp. Bot.* **2010**, *61*, 2459–2467. [CrossRef] [PubMed]
131. Anand, A.; Bass, S.H.; Bertain, S.M.; Cho, H.-J.; Kinney, A.J. Ochrobactrum-Mediated Transformation of Plants. US Patent App. WO/2017/040343, 26 August 2016.
132. Hofmann, N.R. A breakthrough in monocot transformation methods. *Plant Cell* **2016**, *28*, 1989. [CrossRef] [PubMed]
133. Webster, G.R.; Teh, A.Y.H.; Ma, J.K.C. Synthetic gene design—The rationale for codon optimization and implications for molecular pharming in plants. *Biotechnol. Bioeng.* **2017**, *114*, 492–502. [CrossRef] [PubMed]
134. Wunsch, P.; Herb, M.; Wieland, H.; Schiek, U.M.; Zumft, W.G. Requirements for Cu-A and Cu-S center assembly of nitrous oxide reductase deduced from complete periplasmic enzyme maturation in the nondenitrifier *Pseudomonas putida*. *J. Bacteriol.* **2003**, *185*, 887–896. [CrossRef] [PubMed]

Permissions

All chapters in this book were first published in MDPI; hereby published with permission under the Creative Commons Attribution License or equivalent. Every chapter published in this book has been scrutinized by our experts. Their significance has been extensively debated. The topics covered herein carry significant findings which will fuel the growth of the discipline. They may even be implemented as practical applications or may be referred to as a beginning point for another development.

The contributors of this book come from diverse backgrounds, making this book a truly international effort. This book will bring forth new frontiers with its revolutionizing research information and detailed analysis of the nascent developments around the world.

We would like to thank all the contributing authors for lending their expertise to make the book truly unique. They have played a crucial role in the development of this book. Without their invaluable contributions this book wouldn't have been possible. They have made vital efforts to compile up to date information on the varied aspects of this subject to make this book a valuable addition to the collection of many professionals and students.

This book was conceptualized with the vision of imparting up-to-date information and advanced data in this field. To ensure the same, a matchless editorial board was set up. Every individual on the board went through rigorous rounds of assessment to prove their worth. After which they invested a large part of their time researching and compiling the most relevant data for our readers.

The editorial board has been involved in producing this book since its inception. They have spent rigorous hours researching and exploring the diverse topics which have resulted in the successful publishing of this book. They have passed on their knowledge of decades through this book. To expedite this challenging task, the publisher supported the team at every step. A small team of assistant editors was also appointed to further simplify the editing procedure and attain best results for the readers.

Apart from the editorial board, the designing team has also invested a significant amount of their time in understanding the subject and creating the most relevant covers. They scrutinized every image to scout for the most suitable representation of the subject and create an appropriate cover for the book.

The publishing team has been an ardent support to the editorial, designing and production team. Their endless efforts to recruit the best for this project, has resulted in the accomplishment of this book. They are a veteran in the field of academics and their pool of knowledge is as vast as their experience in printing. Their expertise and guidance has proved useful at every step. Their uncompromising quality standards have made this book an exceptional effort. Their encouragement from time to time has been an inspiration for everyone.

The publisher and the editorial board hope that this book will prove to be a valuable piece of knowledge for researchers, students, practitioners and scholars across the globe.

List of Contributors

Abdul-Aziz Ibn Musah, Jianguo Du, Thomas Bilaliib Udimal and Mohammed Abubakari Sadick
School of Management, Jiangsu University, 301 Xuefu Road, Zhenjiang 212013, Jiangsu, China

Heitor Luís Costa Coutinho, Ana Paula Dias Turetta and Joyce Maria Guimarães Monteiro
Embrapa Solos, Rua Jardim Botânico 1024, Rio de Janeiro-RJ 22460-000, Brazil

Selma Simões de Castro
Laboratory of Geomorphology, Pedology and Physical Geography—LABOGEF, IESA, UFG, Campus Samambaia, Goiânia, GO 74.690-900, Brazil

José Paulo Pietrafesa
Education Faculty of UFG—Program of Pos Graduation in Education (PPGE). Rua 235, s/n-Setor Universitário, Goiânia, GO 74.605-050, Brazil

Temitope S. Egbebiyi, Olivier Crespo and Chris Lennard
Climate System Analysis Group (CSAG), Department of Environmental and Geographical Science, University of Cape Town, Private Bag X3, Rondebosch, Cape Town 7701, South Africa

Dan Wang, Guangyong Li and Mingkun Cai
College of Water Resources & Civil Engineering, China Agriculture University, Beijing 100083, China

Yan Mo
College of Water Resources & Civil Engineering, China Agriculture University, Beijing 100083, China
Department of Irrigation and Drainage, China Institute of Water Resources and Hydropower Research, Beijing 100091, China

Xinyang Bian
Scientific Research Department, Water-Saving and Equipment Limited Company of Kingland Muhe, Chifeng 024000, China

Noppol Arunrat, Sukanya Sereenonchai and Nathsuda Pumijumnong
Faculty of Environment and Resource Studies, Mahidol University, Nakhon Pathom 73170, Thailand

Wenling Gao and Xinmin Bian
College of Resources and Environmental Sciences, Nanjing Agricultural University, 1st Weigang Road, Xuanwu District, Nanjing 210095, China

Tafesse Matewos
Geography & Environmental Studies & PhD Fellow in Development Studies at Institute of Policy & Development Research (IPDR), Hawassa University, Ethiopia. Po Box 05 Hawassa, Ethiopia

Pamela Berry, Fuad Yassin and Karl-Erich Lindenschmidt
School of Environment and Sustainability, University of Saskatchewan, 11 Innovation Boulevard, Saskatoon, SK S7N 3H5, Canada

Kenneth Belcher
Department of Agricultural and Resource Economics, University of Saskatchewan, Saskatoon, SK S7N 5A8, Canada

Billy Tusker Haworth
Humanitarian and Conflict Response Institute, University of Manchester, Ellen Wilkinson Building, Oxford Road, Manchester M13 9PL, UK

Eloise Biggs, John Duncan and Bryan Boruff
UWA School of Agriculture and Environment, University of Western Australia, Perth, WA 6009, Australia

Nathan Wales
School of Geography, Earth Science & Environment, University of the South Pacific, Suva, Fiji

Eleanor Bruce
School of Geosciences, University of Sydney, Sydney, NSW 2006, Australia

Bangyou Zheng
CSIRO Agriculture and Food, Queensland Bioscience Precinct, 306 Carmody Road, St. Lucia, QLD 4067, Australia

Scott Chapman
CSIRO Agriculture and Food, Queensland Bioscience Precinct, 306 Carmody Road, St. Lucia, QLD 4067, Australia
School of Agriculture and Food Sciences, The University of Queensland, Building 8117A NRSM, Gatton, QLD 4343, Australia

Karine Chenu
The University of Queensland, Queensland Alliance for Agriculture and Food Innovation (QAAFI), 203 Tor Street, Toowoomba, QLD 4350, Australia

Hugo Herrera
Geography Department—System Dynamics Group, Bergen University, Fosswinckelsgate 6, 5007 Bergen, Norway
Department DEMS, University of Palermo, 90100 Palermo, Italy

Lillian Kay Petersen
Los Alamos High School, Los Alamos, NM 87544, USA

Yuyoung Choi and Woo-Kyun Lee
Department of Environmental Science and Ecological Engineering, Korea University, Seoul 02841, Korea

Chul-Hee Lim
Department of Environmental Science and Ecological Engineering, Korea University, Seoul 02841, Korea
Center of Excellence in Earth Systems Modeling and Observations, Chapman University, Orange, CA 92866, USA

Seung Hee Kim and Menas C. Kafatos
Center of Excellence in Earth Systems Modeling and Observations, Chapman University, Orange, CA 92866, USA

Jordan J. Demone, Shen Wan, Maryam Nourimand, Asbjörn Erik Hansen and Illimar Altosaar
Department of Biochemistry, Microbiology, and Immunology, University of Ottawa, Ottawa, ON K1H 8M5, Canada

Qing-yao Shu
National Key Laboratory of Rice Biology, Institute of Crop Sciences, Zhejiang University, Hangzhou 310058, China

Index

A

Active Accumulated Temperature, 58, 63-64, 69-70

Agricultural Inputs, 76-77, 107, 119-120

Agriculture Production, 1-3, 9-10, 15-19, 121

Agriculture Sector, 20, 120, 146-148, 213

Alleles, 164, 167

Atmospheric Phytoremediation, 230, 234-235, 238

B

Baseline Scenario, 29, 168, 170-171, 175

Biomass, 18, 38, 59, 89, 105, 123-125, 138-140, 142, 216, 234

C

Chilling Injury, 58-59, 65, 68-71, 73

Climate Adaptation, 140, 145-146, 152, 155-156, 158, 163-164

Climate Change, 1-4, 6, 17, 20-25, 33, 41, 43-45, 52-57, 71, 74-77, 80, 87-89, 91, 102-103, 106-111, 114-124, 126-127, 129-131, 133, 138, 140-141, 143-152, 154, 156, 158-160, 162-163, 176, 179-181, 185, 191-194, 196-198, 204, 209-214, 218-219, 224-230, 239-240, 245

Cocoa Yields, 17-18, 20

Crop Growth, 17, 40-42, 44-45, 49-50, 52-53, 59, 69, 92, 97-98, 105, 128-129, 140, 145, 209-210, 216, 220, 228

Crop Model, 164-165, 213, 225-227

Crop Prices, 128, 130, 204

Crop Production, 3, 6, 16-19, 22, 24, 41, 43, 55-57, 59, 71, 78, 86, 91-92, 96-98, 101, 128-129, 136, 143, 159, 209-213, 217, 227, 243

Crop Suitability, 40-45, 47-53, 56

Crop-climate Departure, 40-41, 43, 47, 52-53

Cropping Systems, 52, 72, 79, 87, 89, 91-95, 102, 104-105, 138, 164, 240-241

Crops Rotation, 91-92, 95

E

Ecocrop, 40, 42-45, 47-49, 53, 56

Ensemble Results, 134, 213, 219, 221-222, 225

Environmental Gains, 123, 139-140

Extreme Value Theory, 1-3, 9, 19, 21

Extreme Weather, 1-3, 10, 15-16, 18, 20, 23, 41, 107, 121, 151, 158, 163, 196-197, 204, 209, 212

F

Fertilization, 24, 93, 103-104, 196-197, 199-200, 204-210, 212, 220, 245

Fertilizer, 56, 61-62, 69, 74-75, 78, 80, 84, 87-88, 91-93, 96-105, 111, 115-119, 128, 167, 218, 230-231, 233-234, 240-241, 243

Food Security, 1-3, 21, 40-41, 52-53, 55, 57, 59, 91-92, 95, 112, 116, 119-120, 122, 148, 150, 152-153, 159-160, 162, 179-182, 185, 187, 189, 193-194, 197, 213, 227

Future Crop Yields, 196-197, 199, 209, 226

G

Genotypes, 164, 167-168, 171-172

Geographic Information, 145-147, 149, 153, 155, 157-160, 162-163

Global Warming, 4, 41, 44, 52, 55, 58, 76, 87, 91-92, 103, 107, 122, 196, 210, 230, 239

Greenhouse Gas Emissions, 25, 41, 74, 76, 88, 103, 107, 196, 209, 241, 243

Gross Income, 128, 130, 135-137, 140

Gross Margin, 123, 134-136, 138, 140, 164, 167-168, 172-174

Growing Seasons, 18, 20, 91-92, 95, 102, 112, 115, 136

H

Harvesting, 17-18, 26, 31-32, 76, 94, 125, 138, 141-142, 212

Herbicides, 76-77

Humid Regions, 92, 95, 97-99

I

Incentive Measures, 74, 85-87

Insecticides, 76-77

Irrigation Application, 129, 134-138

Irrigation Water, 6, 83, 128-129, 134-138, 141, 218, 227, 230

L

Life Cycle Assessment, 74, 76, 87, 243

Livelihoods, 1, 49, 52, 56, 107-108, 115, 145-146, 148, 150-152, 154, 157, 163

M

Maize Growth, 58-59, 61-65, 69-71, 211

Maize Production, 22, 58, 92-95, 97-98, 102, 104, 180-181, 185, 187-189, 211, 216, 218-219

Management Practices, 91-94, 102, 125, 164-165, 168, 171-172, 175, 233, 235, 241

Mitigation Techniques, 74-76, 79-80, 82-87

Mulched Drip Irrigation, 58-61, 63, 65, 67, 69-70, 72

N

Nitrous Oxide Reductase, 230, 235-237, 241-242, 244-246

P

Phenology, 56, 71-72, 160, 164-165, 167-168, 171-172

Plant Available Water, 164-165, 167-168, 171, 173-175

Planting Date, 58-59, 62, 64-65, 67-72

Post-harvest Burning, 36, 76

Problem Structuring Process, 179-180, 184

Production Indexes, 1-3, 9, 16-18

R

Reservoir, 123, 125-129, 134, 136-141

Response Ratios, 93, 96-97, 100, 104
Retention Systems, 123-126, 138-141
Return Level, 1, 8, 11-15, 19
Rice Field, 74, 83, 88-89

S
Simulation Periods, 134-135, 138
Smallholder Farmers, 88, 106-108, 110, 116, 119, 121-122, 145-148, 152, 155-160, 162-163
Soil Management, 91-94
Soil Water, 17, 62, 72, 90, 100, 128, 141, 164, 167-168, 171, 175, 228
Southern Oscillation Index, 164-165, 174, 176-177
Sowing Date, 70-72, 165, 170-171, 177
Soybeans, 17, 196-198, 201, 204, 207-209, 230
Summer Rainfall, 125, 138, 168
Sustainable Development, 26, 34, 37, 92, 101, 106-107, 126, 140, 142, 153, 163, 210, 213, 226-227

T
Tillage, 78, 91, 93, 100-103, 105, 124, 233, 241, 243

V
Virtual Water Content, 213-215, 217, 223, 227

W
Water Availability, 32, 129-130, 136-138, 227
Water Management, 20, 72, 74, 78, 86-87, 123-126, 138, 140-141, 152
Water Requirements, 129-130
Watershed, 38, 123, 125-128, 138-139, 141-142, 144, 151, 153, 162, 228
Weather Extremes, 1, 9-10, 18-20, 159
Wheat Crops, 135, 165, 167-168
Wheatbelt, 164-166, 168-170, 174, 176-177
Winter Precipitation, 132, 137

Printed in the USA
CPSIA information can be obtained
at www.ICGtesting.com
LVHW080037211023
761596LV00010BA/1796